DYNAMICS IN
METAZOAN EVOLUTION

DYNAMICS IN METAZOAN EVOLUTION

The origin of the coelom and segments

BY

R. B. CLARK

PROFESSOR OF ZOOLOGY
UNIVERSITY OF NEWCASTLE UPON TYNE

CLARENDON PRESS · OXFORD

Oxford University Press, Ely House, London W.1
GLASGOW NEW YORK TORONTO MELBOURNE WELLINGTON
CAPE TOWN SALISBURY IBADAN NAIROBI LUSAKA ADDIS ABABA
BOMBAY CALCUTTA MADRAS KARACHI LAHORE DACCA
KUALA LUMPUR HONG KONG TOKYO

© *Oxford University Press*, 1964

FIRST PUBLISHED 1964
REPRINTED LITHOGRAPHICALLY IN GREAT BRITAIN
FROM CORRECTED SHEETS OF THE FIRST EDITION
BY WILLIAM CLOWES AND SONS, LIMITED
LONDON AND BECCLES
1967

PREFACE

THERE has been a great revival of interest in problems of metazoan phylogeny during the last decade and, indeed, speculation about the course of evolution of the major groups of animals flourishes now as it never has since the turn of the century. These speculations are inevitably inconclusive and unsatisfying in some degree and we can probably never hope to arrive at a final and complete solution of these problems, but this is no reason for ignoring them and attempts must constantly be made to reconcile current theory with the mounting body of information, drawn from a wide variety of disciplines, that is at our disposal.

This book is not intended as a direct contribution to the debate on metazoan phylogeny and so I have not attempted to construct phylogenetic trees or to discuss in detail the interrelationships of existing animals. Instead, I have considered some of the principles of comparative morphology which must be taken into account when phylogenies are proposed, but which have hitherto escaped serious discussion in this context. In particular, as a student of the annelids, I have been exercised by two outstanding problems: the nature and origin of the coelom and of metameric segmentation. I have therefore been more concerned with the gross structural organization of animals than with the minutiae which are very properly the preoccupation of systematists who have been the chief contributors to phylogenetic studies in recent years.

Movement and the relationship between the organism and the physical world appear to have been important determinants of structure and for this reason I have directed my attention particularly to the dynamic and mechanical aspects of animals of different grades of structural organization. This forms the greater part of the book and constitutes an attempt to state some of the principles of animal morphology, especially those relating to the secondary body cavity and to metamerism. It is, of course, impossible to discuss these matters *in vacuo* and I have necessarily also considered how the conclusions from this study can be applied to some of the problems of metazoan phylogeny. This, the final chapter, is the most tentative and seems the most likely to require modification in the future as more information becomes available. Indeed several important works have appeared since the completion of the manuscript. Consequently, I have not presented an exhaustive discussion in this section of the book, nor have I offered detailed conclusions about the course of evolution of the Metazoa, which must ultimately depend upon the familiar classical approach through embryology and comparative anatomy; that is beyond the scope of this book.

R. B. CLARK

Bristol,
December, 1963.

ACKNOWLEDGEMENTS

It is inevitable that in writing this book I have considered matters with which I have little or no first-hand experience, and I owe much to those who have so willingly discussed problems with me, have devoted time to reading parts or all of the book in draft form, and have given me the benefit of their comments. They include Mr. Q. Bone, Dr. C. Burdon-Jones, Dr. H. D. Crofton, Professor J. E. Harris, F.R.S., Professor L. A. Harvey, Dr. H. E. Hinton, F.R.S., Professor G. P. Wells, F.R.S. and Dr. H. P. Whiting. I should, perhaps, have taken their advice more often than I did and, although I have frequently modified my views as a result of conversations with them, I doubt if they will find themselves in agreement with everything I have written. But whatever I may owe to these and other friends, the errors that remain are my own.

I am grateful to the following for permission to reproduce or copy illustrations from their publications. Akademische Verlagsgesellschaft (Figs. 12, 17, 27); Cambridge University Press (Fig. 44); Carnegie Institution of Washington (Fig. 93); Company of Biologists (Figs. 1, 11, 15, 16, 18–22, 34, 36–38, 43, 45, 54, 59, 66, 68–73, 78, 87b, 89, 94–97, 99, 101–104, 111); Connecticut Academy of Arts and Science (Fig. 23); Danish Science Press (Fig. 123); Editions Desoer (Fig. 4); VEB Gustav Fischer Verlag (Figs. 46, 47, 49, 62, 63, 108); Walter de Gruyter and Company (Fig. 29); Koninklijke Nederlandse Akademie van Wetenschappen (Fig. 110); Linnaean Society of London (Fig. 64); Linnaean Society of New South Wales (Fig. 48); Council of the Marine Biological Association (Fig. 88); Masson et Cie (Figs. 80, 81); McGraw-Hill Book Company (Figs. 50, 61, 122); National Academy of Science, Washington (Fig. 31); New York Zoological Society (Figs. 87a, c); Oliver and Boyd (Figs. 90–92); Pergamon Press (Fig. 41); Royal Society (Figs. 82, 83, 98, 100); Royal Society of Edinburgh (Fig. 79); Springer Verlag (Figs. 28, 33, 56); Taylor and Francis (Fig. 53); VEB Georg Thiele (Figs. 39, 40, 51); Universidad de São Paulo (Fig. 61); Wistar Institute (Fig. 57); Zoological Society of London (Figs. 76, 77, 84, 85, 106, 107, 109); Zoologiske Bidrag från Uppsala (Figs. 7–9, 65); XXI Session of the International Geological Congress, Norden, 1960 (Figs. 5, 13).

I also owe a debt of gratitude to all those authors whose names are cited in the captions to the figures for permitting me to make use of their published illustrations, and in particular, to Mr. J. B. Cowey, Professor Sir James Gray, F.R.S., Professor J. E. Harris, F.R.S. and Dr. K. H. Mann who have lent me original drawings for reproduction. Mr. R. A. Hammond kindly prepared Fig. 86 from his own unpublished observations and I am grateful to Mr. R. L. Blackman and Mr. A. Gardarsson for investigating the musculature and locomotion of *Sabella* and *Aphrodite*, respectively, for me and on whose work Figs. 74 and 75c are based.

I cannot adequately express my indebtedness to my wife, Dr. Mary E. Clark. Not only has she discussed many of the problems that emerged during the writing of the book, but she has also taken time from her own work to translate literature, prepare figures and help with proof-reading and the preparation of indices. I doubt if the book would ever have been completed in its present form without her help.

Finally, I am grateful to Mrs. M. L. Frost for typing the manuscript with such patience and care, and to Mr. K. Wood for all his advice and assistance in photographic matters.

CONTENTS

                                                           PAGE

1. THE PROBLEM 1
   - The coelom 2
   - Metamerism 17
   - The present position 27

2. THE ACOELOMATE CONDITION 31
   - The hydrostatic skeleton 31
   - Limitations of change of shape set by the basement membrane . . 35
   - Locomotion of turbellarians and nemerteans 42
   - Gastropod locomotion 53
   - Pedal locomotion in anemones 64
   - Locomotion of some bryozoans 65
   - The parenchyma of turbellarians and nemerteans 66
   - Conclusions 68

3. THE COELOMATE CONDITION 71
   - Constancy of volume 71
   - The cuticle 75
   - Nematode locomotion 78
   - The functioning of the body-wall musculature in coelomate worms . 83
   - The burrowing habit 84
   - Locomotion in the burrow and over the substratum 93
   - Peristalsis of the body wall of worms 98
   - The locomotion of some echinoderms 99
   - Other mechanical functions of the coelom 102
   - Hydrostatic pressure of body fluids 107
   - Conclusions 115

4. THE SEPTATE CONDITION 118
   - Locomotion and the musculature of earthworms 118
   - Locomotion of errant polychaetes 124
   - The musculature of errant polychaetes 130
   - The septa of polychaetes 134
   - Structure of the primitive septum 140
   - Reduction and loss of septa 141
   - Retention of septa for specialized purposes 145
   - Oligomerous animals 150
   - The pseudocoelomate phyla 159
   - Conclusions 161

5. SWIMMING 165
   - The incidence of undulatory swimming 165
   - The mechanics of swimming 174
   - The behaviour of the musculature 186
   - The co-ordination of muscular activity 196
   - The segmentation of vertebrates and invertebrates 198
   - The evolution of chordate segmentation 200
   - Conclusions 202

6. THE PHYLOGENY OF THE METAZOA 205
- The origin of the Bilateria 206
- The Protobilateria 212
- The evolution of the coelom 214
- The pseudocoelomate phyla 219
- The oligomerous phyla 222
- The evolution of metamerism 228
- Advance and regression 235
- Conclusions 257

BIBLIOGRAPHY 262

AUTHOR INDEX 289

SUBJECT INDEX 293

1
THE PROBLEM

Two events of major importance in the history of the Metazoa were the evolution of a secondary body cavity—a coelom, pseudocoel or haemocoel—and the evolution of metameric segmentation. These events appear to have represented essential stages in the development of more complicated structural organization. Nearly all metazoan animals are coelomate or pseudocoelomate and the most highly organized and successful groups, the arthropods and the chordates, are segmented. It is not surprising, therefore, than an understanding of the nature of the coelom and of metamerism should be regarded as fundamental to an understanding of animal morphology, and not even the most elementary textbook of zoology fails to insist upon the dogma that has grown up around these two morphological concepts. One begins to suspect that our understanding of them is not as great as it might be, when it is recalled that metamerism and the secondary body cavity are now rarely discussed except in elementary terms, despite the fact that there is no generally accepted theory of the origin of either. What is worse, many of the concepts that are accepted, are without logical basis or reasoned analysis. The comment that it is difficult to find a term in the whole of zoology that is so vaguely defined but, at the same time, so universally employed as the coelom (Sarvaas, 1933), applies as much now as when it was made and it might be applied equally to the concept of metamerism.

The final decades of the last century were a time of intense speculation about the origin of the coelom and of metamerism. The debate was confused and many different hypotheses were advanced, often without reference to each other, often, indeed, without any points of mutual contact at all, and eventually interest shifted from this inconclusive discussion, leaving the problems still unresolved. As the subject received almost no serious discussion after the early years of the century, until, during the last few years, it has been revived almost accidentally in discussions of metazoan origins, the problems remain unresolved and these fundamental concepts of animal morphology are still only vaguely understood.

To give but three examples of the type of confusion that still exists:

(1) The cestodes are almost universally regarded as lacking 'true' segments, though for no better reason than that the proglottids are proliferated immediately behind the scolex while the segments of annelids are formed caudally at the anterior face of the pygidium. The difficulty lies in framing a morphological definition of segmentation that excludes the cestode type

of organization. This difficulty has been frankly acknowledged by Hyman (1951a) who concludes, reasonably, that cestodes must be regarded as metamerically segmented animals. Most zoologists would probably still prefer to distinguish between the type of segmentation displayed by cestodes and that of annelids, even though they have no reasoned or convincing grounds for doing so.

(2) There is no disagreement, however, that both annulates and chordates are metamerically segmented. It is still widely supposed that metamerism in the two great phyla is homologous, though this implies that both were evolved from a common segmented ancestor, which, in turn, entails acceptance of the proposition that chordates evolved from inverted annelids, a hypothesis that was proposed by Dohrn (1875) and Semper (1876) but which never gained much support and was discarded many years ago. With the acceptance of the view that metamerism was acquired during the emergence of the chordates from an unsegmented protochordate stock, it has become increasingly appreciated that metamerism must have originated independently in annelids and chordates (and possibly a third time in the cestodes, depending upon how metamerism is defined). This changed attitude, of course, demands a revaluation of the older theories since metamerism and generally also the coelom are presumed in them to be homologous throughout the Metazoa. Even the new position is equivocal, for it is not clear whether metamerism is analogous in annulates and chordates, having evolved independently in the two groups but in response to similar evolutionary pressures, or if metamerism in the two phyla is only accidentally and superficially similar, and has evolved independently and in different ways to serve different functions in these two branches of the Metazoa.

(3) The evolution of the pseudocoel has never been satisfactorily explained. Insofar as it represents a secondary body cavity between the gut and body wall, it is comparable to a coelom in many of its functions, but its fundamentally different morphological nature is constantly emphasized. How or why the pseudocoel evolved has never been considered in theories concerned with the origin of the coelom, though clearly it represents a parallel development which can scarcely be ignored.

Difficulties of this sort have arisen chiefly because re-examination of the fundamental concepts of morphology has not kept pace with advances in the rest of zoology, and generalizations which were consistent with zoological knowledge and thought at the turn of the century have still to do service today. Our first task must therefore be to consider in which respects the older generalizations about the coelom and metamerism need revision.

THE COELOM

While all theories of the origin and evolution of the coelom differ from one another in detail, the majority can be regarded as variants of the gonocoel or enterocoel theories, and these two have won by far the greatest support from zoologists. The two remaining theories, the nephrocoel and schizocoel theories, have attracted much less attention and have therefore not generally been subjected to very searching or critical analysis.

The gonocoel theory

One of the most widely held theories of the origin of the coelom is based essentially on the very common association of the gonads with the coelomic

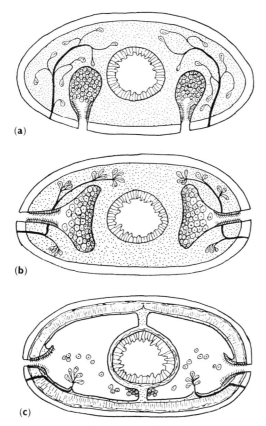

FIG. 1. The gonocoel theory: enlargement and cavitation of the gonads to form the coelomic pouches. (*a*) The platyhelminth condition, (*b*) the nemertean condition, (*c*) the annelid condition. (After Goodrich, 1946.)

epithelium. It is supposed that the paired coelomic pouches such as appear in the ontogeny of annelids and chordates, are derived phylogenetically from paired gonadial sacs (Fig. 1), and hence, that any cavity within the gonads or the genital ducts is coelomic in nature.

The theory was first enunciated by Bergh (1885) but it is based on Hatschek's

(1877, 1878) view that the mesodermal teloblasts can be regarded as equivalent to germinal primordia. This thesis was developed in detail by Meyer (1890, 1901). He accepted the clear distinction that the Hertwigs (1882) had made between primary mesoderm (mesenchyme, ectomesoblast) and the mesodermal bands, and he concluded that the separation of the mesodermal cells from the gut wall during embryological development recapitulated a migration of primitive gonadial cells into the interior of the animal, which had entailed the change from shedding the gametes through the mouth to releasing them through the body wall. The coelomate condition was evolved when, after the shedding of the gametes, the peripheral layer of the gonad remained and formed a hollow sac, and the muscle fibres in the wall of the gonad became the body-wall muscles of the animal. As a subsequent evolutionary process, there was a progressive sterilization of the germinal tissue, total in some parts of the body, partial in others, separating off follicular cells, the musculature, connective-tissue, and other somatic tissues, which eventually replaced the primary mesoderm altogether.

This interpretation of the evolution of the coelom can of course be attacked on the grounds that there is no embryological evidence for the appearance of a mature gonad before the development of the coelomic cavities, rather, the reverse is true; that it is incredible that the gonad should remain turgid and even enlarge, instead of regressing after the gametes are shed; that it is difficult to see how gonadial muscles could eventually become the body-wall muscles; and that it is disconcerting to find that in the early coelomates, the coelom and body-wall muscles should not have been formed until after the animals had reproduced, by which time most simple animals die. However, none of these objections is conclusive and in any case acceptance of the gonocoel theory does not depend upon a faithful adherence to Meyer's phylogeny. Lang (1903) suggested that the gonad might become dilated at the outset, before reproduction, as a nutritive adaptation. Or, as Goodrich (1946) pointed out, if the germinal epithelium grows faster than the gametes enlarge, a cavity inevitably appears within the gonad and this cavity—the coelom—might initially serve as a space within which the gametes mature, but later acquire additional functions associated with nutrition and excretion. Beauchamp (1911), considering the chief function of the coelom was that it freed the intestine from movements of the body wall, suggested that the enlargement of the gonadial cavities was caused simply by mechanical forces, in much the same way as serous pockets are formed in connective-tissue by friction and mechanical irritation.

Meyer (1890) originally supposed that the coelomates were derived phylogenetically from animals comparable to a turbellarian with a single pair of gonads which extended the whole length of the body. Cavitation of these resulted in the evolution of an unsegmented coelomate worm, and segmentation was seen as a secondary development conferring greater suppleness on the body of the animal as an adaptation to swimming. But in 1901, he abandoned this position in favour of one much closer to that of Bergh (1885) and Lang (1903) who saw the origin of paired meristic coelomic pouches of segmented animals in the pseudometameric arrangement of the gonads in some turbellarians and nemerteans (Fig. 2). Chiefly as a result of

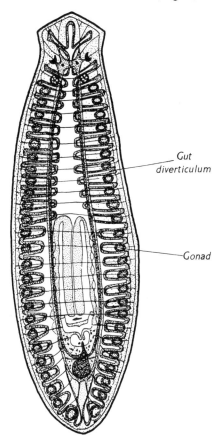

FIG. 2. The pseudometameric arrangement of gut diverticula and gonads in the triclad *Gunda segmentata*. (After Lang, 1881.)

Lang's (1903) persuasive writing, this version of the gonocoel theory has had a distinguished succession of supporters, although it implies that non-segmented coelomates have been secondarily derived from segmented ones. The alternative, that the coelom of segmented and unsegmented animals has been independently evolved, though in the same manner in each, has been suggested as a possible escape from this dilemma (e.g. Goodrich, 1946), though never with much enthusiasm. Meyer's (1890) original view that a segmented coelom evolved from bilateral but unsegmented gonocoels has received little support except from Hoffmann (1929–30) and also from Lam (1920) who described an archiannelid, *Protannelis meyeri*, with an unsegmented coelom which he regarded as primitive. Since Lam's description was based on a single, incomplete specimen, this evidence which might otherwise be crucial, is difficult to evaluate.

While it is probably impossible to refute the gonocoel theory conclusively, there are several obstacles to its whole-hearted acceptance.

(1) Most proponents of the theory have linked the origin of metamerism with the evolution of the coelom. This presents some difficulty in accounting for the evolution of unsegmented coelomates; either they must have been secondarily derived from segmented animals, or the coelom is polyphyletic in origin. It has been almost universally accepted that the evolution of the coelom was monophyletic and the former alternative is therefore implied in the theory. The sipunculids, which are coelomate but show no signs of segmentation during embryonic development or in adult structure (Dawydoff, 1928, 1959; Tétry, 1959), can be accommodated within the theory only with the greatest difficulty.

(2) The formation of endomesoderm by the inward migration of gonadial cells, and the formation of the coelom by cavitation of the gonads after the release of the gametes, are regarded as separate phylogenetic processes. There is no embryological support for this view and it encounters some difficulty when the enterocoelous method of coelom formation is considered. In terms of this theory, enterocoely is, of course, a secondary derivation from schizocoely, but it provides unwelcome evidence of coelom formation being linked with mesoderm formation and long preceding the appearance of the gonads.

(3) If the gonocoel theory is correct, the mesodermal teloblasts might be expected always to be associated with the gonads and the coelom and with nothing else, but this is not so. In most annelids, teloblasts derived from the cell 4d certainly give rise to such structures, except in the few cases in which 4d also contributes to endodermal tissues (though this is presumably permissible since the gonads are held to be derived, phylogenetically, from endoderm), but in platyhelminths, which are supposed to be primitive, the comparable cells give rise not only to gonadial structures and to part of the intestine, but also the mesenchyme. Indeed, the whole question of the distinction between primary and secondary mesoderm (Hertwig and Hertwig, 1882; Salensky, 1907), which appears to be crucial to the theory, is difficult so sustain (Braem, 1895; Sarvaas, 1933).

(4) It is by no means certain that the germ cells invariably originate in the genital follicle (i.e. the coelomic epithelium) in coelomates. Heymons (1891) claimed that the primary germ cells of *Blatta* were differentiated before the appearance of the coelomic pouches and, at a later stage, were intercalated between the coelomic vesicles in which the new germ cells were proliferated, and which later migrated through the peritoneum to join the

primary germ cells. Heymons' views on the embryonic germ layers of insects, on which his interpretation of primary and secondary germ cells depends, are no longer accepted, but there can be no question about the dual nature of the gonad: a mesodermal envelope and an enclosed germinal tissue. The latter may be, and generally is, associated with the former, but this is not invariably the case. Studies of the effect of selective damage to developing eggs by irradiation, have made it clear that the formation of a gonad is not dependent upon the presence of a functional germinal epithelium (Geigy, 1931). While none of this evidence conclusively refutes the gonocoel theory, there must obviously now be greater doubt that the association between germ cells and the coelom is essential or an invariably primitive feature.

The enterocoel theory

The chief rival of the gonocoel theory is the view that the enterocoelous method of formation of mesoderm and the secondary body cavity, such as occurs during the embryonic development of echinoderms, hemichordates, brachiopods and some other animals, is primitive and indicates the phylogenetic origin of the coelom. The originators of this theory were less concerned with phylogenetic speculation than with the embryological origin of the coelom, and it was only later that the evolutionary consequences of the theory were considered. Indeed, to begin with, the concept of a coelom had not been clearly formulated and much of the early discussion of the subject now appears rather confused. An additional feature of the enterocoel theory is that, to a much greater extent than in discussions of the gonocoel theory, the origin and development of a coelom in phylogeny has become closely associated with the evolution of metamerism.

Initially, ideas about the constitution of the coelom were vague. Haeckel (1872), who first introduced the term, included any cavity of the body that was not connected with the gut within the definition of coelom, and Lankester (1874) suggested that the excretory system of platyhelminths (at that time regarded as a 'water vascular system') was a rudimentary coelom. This view is clearly incompatible with the enterocoel theory and was never considered very seriously in the subsequent phylogenetic speculation, when the nature of the coelom was more completely understood.

By 1874, it had become clear that in some animals, the coelom was derived from a diverticulum of the archenteron before the primary and secondary body cavities were clearly delimited (Lankester, 1874, 1875a, b). This observation was given some phyletic significance and it was also appreciated that essentially the same manner of mesoderm and coelom formation might be manifested by a mass of cells separating from the primitive alimentary canal and later, by splitting, develop a coelomic cavity by schizocoely. 'In this way it is conceivable that the schizocoelous condition might develop from the enterocoelous and gradually lose all trace of its ancestral origin further than is afforded by the derivation of some mesoblastic cells from the hypoblast' (Lankester, 1875b). A similar view was also adopted by the Hertwig brothers (1882) when they developed their concept of primary (mesenchyme) and secondary mesoderm.

In 1877, Lankester suggested that the coelom originated phylogenetically as a pair of diverticula of the gut which became separated from it and then fused to form a single, spacious coelom. The evolution of metamerism clearly post-dates the appearance of the coelom in this theory, and Lankester, at that time, appears to have regarded the acquisition of segments as a piecemeal process. The segmental vertebrate kidney was derived by subdivision from a single pair of excretory organs of an unsegmented chordate ancestor, and other organs were similarly divided independently of one another. Later supporters of the enterocoel theory, however, have considered the evolution of the coelom and of metamerism as indissolubly linked. In 1881, Lang, as a result of his investigation of the triclad *Gunda* (= *Procerodes*) concluded that the coelomic pouches were derived by the separation of the regularly repeated diverticula of the digestive system from the main part of the gut of pseudometameric platyhelminths (Fig. 2). The development of coelomoducts and nephridia placed the coelom in communication with the exterior, and the body-wall musculature became segmentally arranged in relation to the underlying coelomic pouches. Later he abandoned this thesis in favour of the gonocoel theory (Lang, 1903), but continued to maintain that metameric coelomates evolved from pseudometameric acoelomates, with the implication that unsegmented coelomates were derived from segmented ancestors.

Sedgwick (1884) opened a fresh line of speculation when he suggested that the origin of the coelom must be sought, not in the acoelomate worms, but in the coelenterates. A similar view appears to have been proposed earlier by Leuckhart (1848), but Sedgwick's paper was the chief stimulus to the subsequent speculation. The gastric pockets of anthozoans, it was argued, became separated from the main gastric cavity to form the coelomic pouches (Fig. 3). The elongated mouth of the anemone indicates the future anteroposterior axis of the coelomate, the circum-oral nerve ring becomes the central nervous system, and the cinclides the coelomoducts. Hubrecht (1904) and Lameere (1932a) elaborated upon Sedgwick's theory by suggesting that the chordates and annulates have evolved independently from coelenterates in response to different adaptations (Fig. 4). According to Hubrecht, the annulates, with a ventral nervous system, evolved from pelagic coelenterates, the chordates from sessile coelenterates. Lameere's view was the exact converse, that the chordates evolved from a pelagic stock in which the marginal tentacles of the coelenterate became the lateral fin rays and the labial tentacles interdigitated to form the unpaired dorsal fin, while the annelids evolved from coelenterates that had adopted the habit of creeping and feeding,

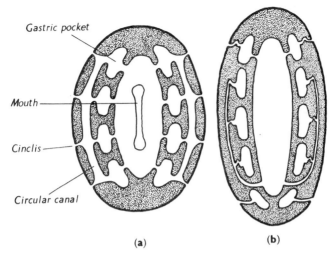

FIG. 3. The enterocoel theory: transformation of the gastric pockets of an anthozoan (*a*), into coelomic pouches (*b*). (After Sedgwick, 1884.)

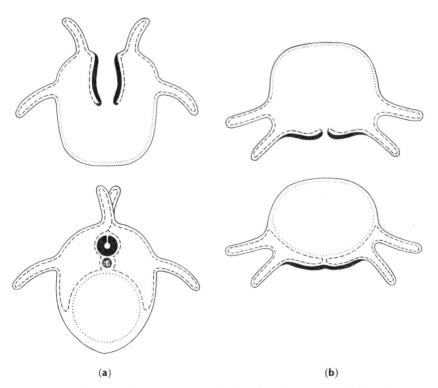

FIG. 4. Development of (*a*) animals with a hollow dorsal nerve cord and (*b*) animals with a solid ventral nerve cord from a cerianthid-like ancestor. (After Lameere, 1932b.)

inverted, on the sea bed, the coelenterate tentacles in this case becoming the parapodia.

Lemche (1959, 1960) has adopted a very similar point of view (Fig. 5), save that he accepts more recent interpretations of chordate evolution and regards them as of much later origin than the invertebrate coelomates. He homologizes the ring of outer tentacles of the Cnidaria with the ctenidia of molluscs (primitively disposed around the foot as in *Neopilina*), the parapodia of polychaetes, the gills of trilobites and the limbs of arthropods.

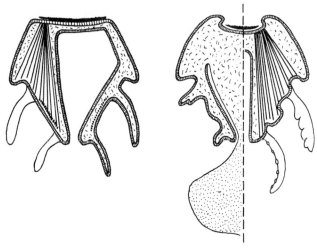

FIG. 5. Comparison between (*a*) a coelenterate (left-hand side, septal region; right-hand side, interseptal region) and (*b*) a cephalopod (left-hand side, young embryo; right-hand side, relevant structures in the adult). (After Lemche, 1960.)

The aboral shell gland of molluscs corresponds with the aboral pedal disc of cnidarians, which secretes the theca. The early and middle Cambrian monoplacophoran molluscs of the order Stenothecoidea, in which the muscle scars radiate in such a manner as to resemble cnidarian septa, are regarded as near the stem group from which both the Cnidaria and the Spiralia (molluscs, annelids, etc.) were derived.

Van Benenden (1891) appreciated that the mesenteries of most anthozoans multiply by intercalation and it is only in the cerianthids that they are laid down in an orderly succession that can in any way be compared with the antero-posterior succession of segments in annelids and chordates. Thus, the cerianthids must be regarded as the ancestors of all coelomates and presumably of all other anthozoans. This is a reversal of the most widely accepted view of the systematic position of the Cerianthidea (Hyman, 1940) and militates strongly against Sedgwick's version of the enterocoel theory instead of lending it the support that van Benenden supposed. It is possible to avoid this difficulty, however, by supposing as Lemche (1960) does, that the 'segments' of the short-bodied Monoplacophora are primary segments and the only ones to be formed in this way. Secondary segments, which are proliferated at the pygidium in annelids, follow a different pattern of growth. This introduces Iwanoff's (1928) concept of primary and secondary segments,

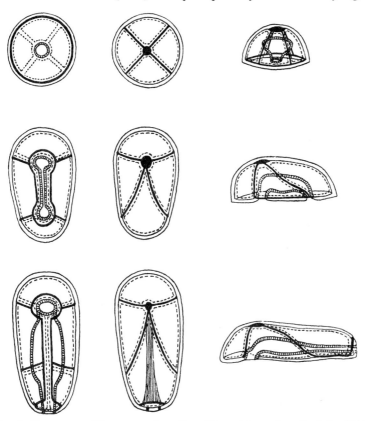

FIG. 6. The enterocoel theory: transformation of the gastric pouches of a medusoid into coelomic pouches in ventral (oral), dorsal (aboral) and lateral views. (After Masterman, 1897.)

and can be discussed more conveniently when we consider the origin of metamerism.

It is a corollary of Sedgwick's theory that both acoelomate worms and unsegmented coelomates have evolved from segmented coelomates (Lameere, 1932b), although Sedgwick himself suggested that unsegmented coelomates evolved as a result of a fusion of the coelomic pouches at an early stage before other organ systems became involved in the segmental organization. Lankester (1900) adopted a position much closer to Sedgwick's than that he had taken previously (Lankester, 1877), and in doing so he achieved a remarkable synthesis which avoided the problem of the origin of unsegmented coelomates and also bridged the gap between the rival gonocoel and enterocoel theories. He considered that the coelomic pouch rudiments that became separated from the coelenterate gastric cavity were gonocoels. In consequence, the turbellarians and nemerteans were regarded as coelomate animals, the coelom being represented by the gonads and gonoducts. In effect, Lankester accepted the gonocoel theory with the proviso that the gonads of 'acoelomate' animals were evolved from those of coelenterates by a separation of the gastric pockets from the main lumen of the coelenteron.

Masterman (1897, 1898) also derived the coelom from the gastric pouches of coelenterates, though in this case from those of medusoids. Unlike Sedgwick, he considered the ancestor of coelomates to have been radially symmetrical with four gastric pouches (Fig. 6). The separation of the gastric pouches from the main gastric cavity was accompanied by a change from vertical to horizontal movement, with a consequent development of bilateral symmetry. One of the coelomic pouches was now anterior and this tended to be reduced while the two lateral pouches undertook the nutritive and reproductive functions, previously shared by all four. The terminal coelomic pouch became divided into two by the posterior growth of an intestine. Such an organism now possessed a single anterior 'protomere', and paired 'meso-' and 'metameres', and this type of organization (oligomerous) can be found in hemichordates, echinoderms, phoronids and brachiopods. The subsequent evolution of a polymerous organization by subdivision of the metamere, often with the reduction or obliteration of the coelomic cavities of the proto- and mesomeres, has occurred independently in annelids and chordates.

The derivation of the triploblastic Metazoa from pelagic coelenterates had previously been considered by Balfour (1881), Hubrecht (1887) and others, and a similar theory to Masterman's was also developed by Söderström (1924, 1925a, b), but this view never received the attention devoted to Sedgwick's theory until, in recent years, it has been energetically revived and extended by Remane (1950, 1952, 1954, 1958), Ulrich (1950), Marcus (1958) and others. In a somewhat modified version it also appears in Jägersten's (1955) bilaterogastraea theory (Figs. 7, 8 and 9). In all these current theories, the early metazoans in which the coelomic cavities first appeared, are visualized as corresponding approximately with Haeckel's gastraea. In the versions of Remane, Ulrich and Marcus, as in Masterman's theory, four gastric pockets formed the rudiments of the coelomic pouches. In Jägersten's view, six pockets were formed, giving paired anterior and posterior pouches from the outset.

The tendency has always been for supporters of the enterocoel theory to associate inextricably the evolution of the coelom with that of metameric segmentation. This poses problems which relate particularly to the evolution of segments, but insofar as it can be restricted to the origin of the coelom,

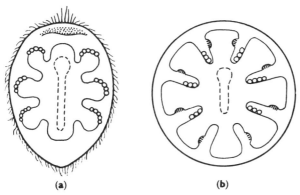

FIG. 7. The bilaterogastraea theory: (a) the bilaterogastraea with three pairs of gastric pockets, each with a gonad; (b) the organization of *Octocorallia*. (From Jägersten, 1955.)

the enterocoel theory can be criticized on several grounds. Lang's (1881) and Sedgwick's (1884) versions of the theory, as well as, to some extent, the recent modification of Sedgwick's theory by Lemche (1960), all involve

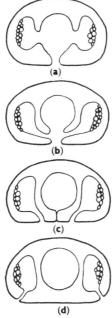

FIG. 8. The bilaterogastraea theory: the evolution of coelomic compartments from the gastric pockets of the bilaterogastraea. (From Jägersten, 1955.)

detailed and rather improbable comparison between the structure of modern turbellarians or anthozoans, and segmented animals. Lankester's (1874) original version of the enterocoel theory at least had the merit of regarding the origin of the coelom as an embryological modification. Lang (1903) himself abandoned the turbellarian theory and, as we have already seen, despite the powerful support given it by van Beneden (1891), Hubrecht (1904), MacBride (1914) and Naef (1926), Sedgwick's theory is untenable because the only anthozoans that fulfil the requirements of an ancestor of metamerically segmented animals are, in fact, highly specialized and are unlikely to have been the ancestors of anything (Hyman, 1940). This difficulty is avoided if it is assumed that the coelomic pouches originated from the gastric pockets of a medusoid (Masterman, 1897), or better, from the gastric pockets of a common ancestor of coelomates and cnidarians (Remane, 1950; Ulrich, 1950; Marcus, 1958), but even this modification of the theory, by

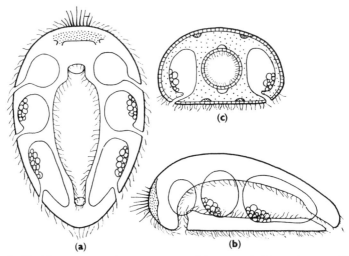

FIG. 9. The bilaterogastraea theory: structure of the most primitive coelomates. (From Jägersten, 1955.)

passing directly from cnidarians or protocnidarians to segmented coelomates, poses the serious problem of the origin of acoelomates and unsegmented coelomates. It is in the phylogenetic consequences of the enterocoel theory, as it is generally stated, that most of its opponents find the chief obstacle to its acceptance.

All versions of the enterocoel theory that envisage the origin of the coelom in the separation and sealing-off of gastric pockets in animals with the structure of a cnidarian, are faced with the problem of accounting for the selective advantage of such a change. The formation of pouches or the development of mesenteries in coelenterates is generally supposed to have been an adaptation to increasing the absorptive surface of the digestive system and to facilitate the distribution of nutrients to the tissues as the animals increased in size. These advantages are nullified if the gastric pouches are constricted from the main digestive cavity. Lameere's (1932a) answer to this problem, that the closing of the gastric pouches accompanied the evolution of extracellular digestion, falls to the ground since most of the lower coelomates continue to rely to a great extent on intracellular digestion (Yonge, 1937; Morton, 1960) rather than the reverse, as Lameere suggested.

The nephrocoel theory

Before the nature of the coelom was completely understood, Lankester (1874) had suggested that the nephridia of the Platyhelminthes were precursors of the coelom. This interpretation of the nephridia was not enlarged upon and was very quickly superseded by other theories of coelomic origins. The nephrocoel theory was later revived by Ziegler (1898, 1912), who believed that the coelomic pouches represent the expanded inner ends of platyhelminth nephridia. Presumably this implies an independent origin of the coelom in segmented and non-segmented coelomates as in the gonocoel theory, but the chief obstacle to this theory is the existence of protonephridia in coelomates. These were discovered by Goodrich (1897a) and up to the time when Ziegler wrote his first paper, flame cells and protonephridia were thought to be confined to acoelomates. The fact that echinoderms, though coelomate, possess no nephridia is an additional difficulty (Hyman, 1951a), if a secondary one.

A variant of the nephrocoel theory was proposed by Faussek (1899, 1911) and Snodgrass (1938) who regarded the coelom as an embryological adaptation, a cavity in which excretory products accumulated until they were voided to the exterior at a later stage of development when the coelom acquired ducts to the exterior. This interpretation, transposed to the phyletic level, suggests that the coelom appeared originally as a schizocoel into which the excreta drained and with which the nephridia became associated. The association of the gonads and genital ducts with the coelomic cavities was a subsequent development. It is of course likely that the coelom, however it was evolved, would serve an accessory excretory role, but there is little to suggest that this was ever its primary function. It is impossible to controvert this version of the nephrocoel theory because there is virtually no evidence to examine and weigh, but in view of the strong evidence that exists for other theories, there seems little reason to reject them in favour of this one.

The schizocoel theory

The schizocoel theory may be regarded as the converse of the enterocoel theory, in that schizocoely rather than enterocoely is regarded as the most primitive method of coelom formation in ontogeny, and so may be taken to be indicative of the manner in which the coelom evolved. But for most supporters of this theory, the difference between it and alternative views is more profound. It is fundamental to the theory that the sharp distinction between primary mesenchyme and secondary mesoderm, made by the Hertwigs (1882) and Salensky (1907), is rejected. The structural characteristics of the middle layer are held to be related to its function and not to its origin (Braem, 1895), and the embryological origins of the coelom and mesodermal structures may vary widely and have a mixed mesenchymal and mesodermal constitution (Sarvaas, 1933). In view of this, both the gonocoel and enterocoel theories are rejected and, instead, the coelom is regarded as being derived in phylogeny from the enlargement of the blastocoelic spaces. The association of the gonads with the coelom is a secondary phenomenon. No distinction is made between, for example, the haemocoel of molluscs and the coelom of annelids and, indeed, in a sense, the intercellular spaces in the parenchyma of turbellarians might also be regarded as constituting a rudimentary coelom.

Apart from Faussek (1899, 1911) and Snodgrass (1938), who regarded the schizocoel as having an excretory function, supporters of this theory have considered the coelom as having an essentially mechanical function, and its association with excretion or reproduction as purely secondary. Thiele (1902, 1910) and Sarvaas (1933) suggested that the coelom was a hydraulic organ employed for support or for locomotion. Beauchamp (1911) also regarded the coelom as serving a mechanical function in that it freed the gut from movements of the body wall, but he considered that the coelom was derived from a gonadial cavity rather than as a schizocoel.

The emphasis which these authors place upon the function of the coelom is an unusual feature of the schizocoel theory. Sarvaas is emphatic that the coelom is an organ with a specific function and is therefore inclined to include any cavity between the body wall and the gut that acts as a hydraulic organ of support or serves in locomotion within the definition of the coelom. She attaches little importance to the embryological origin or the morphological nature of the cavity, although she supposes it to be derived phylogenetically from the blastocoel. Cavities that may appear in mesodermal structures are not regarded as coelomic unless they satisfy the functional requirements of the definition or can be shown to be derived from cavities that did so. Thus, the molluscan pericardium, which is normally regarded as coelomic, is here considered to be no more than a dilation in the genital ducts, while the haemocoel is accepted as the true molluscan coelom.

Although this interpretation of the function of the coelom as a mechanical organ has gained limited support (Beauchamp, 1959), the fact that, as it is usually stated, the schizocoel theory pays scant attention to the morphological and embryological nature of the coelom, appears to have been a major objection to it. In fact, since it has so little common ground with the gonocoel and enterocoel theories and denies so many of the principles upon which they are founded, the schizocoel theory has been very largely ignored in subsequent writings.

METAMERISM

Several of the theories of the origin of the coelom that we have already discussed connect this event with the evolution of metamerism. This poses the problem of the origin of unsegmented coelomates and, in some cases, of acoelomate animals also. For if the coelom evolved at the same time that the metameric organization of the body was acquired, it follows that unsegmented animals which possess a coelom must be secondarily reduced to that condition. This problem is avoided if the origin of segments is divorced from the evolution of the coelom, and this is the view taken in some theories, but it is faintly surprising that the two theories concerning the origin of segments currently most in favour, the pseudometamerism and cyclomerism theories, are not numbered among them.

The pseudometamerism theory

Probably the most widely held view today is that the tendency towards a serial repetition of organs in some long turbellarians and nemerteans (pseudometamerism) foreshadows metameric segmentation (Figs. 2, 10 and 11). The evolution of segments, by a completion of the process already begun in these acoelomates, is generally firmly linked with the evolution of the coelom. Bergh (1885), Meyer (1901), Lang (1903) and Goodrich (1946) considered the coelom to have been derived from a pseudometameric series of gonads, Lang (1881) from a similar series of gastric pouches, and Ziegler (1898, 1912) from a series of nephridia. To most of these authors, metamerism was therefore an accidental consequence of the arrangement in acoelomate animals of the structures from which the coelomic pouches were deemed to have been derived. It was observed that the nervous system (Fig. 10), excretory organs (Fig. 11), gonads and gut diverticula (Fig. 2) of both turbellarians and nemerteans often show an apparently metameric organization, but that the body-wall musculature never does so. The final stage in the evolution of segmentation is therefore the subdivision of the musculature and this is generally considered to follow upon the acquisition of metameric coelomic pouches. It is argued that these turgid sacs, particularly when they are full of gametes, would have impeded locomotion by serpentine swimming movements, and that bending of the body would therefore have tended to take place at the regions between the coelomic compartments and so lead to segmentation of the musculature.

Metameric segmentation has been regarded as an adaptation to serpentine swimming in pseudometameric animals by a number of influential authors in recent years, including Snodgrass (1938), Goodrich (1946) and Hyman (1951a), and for this reason the pseudometameric theory has come to be widely accepted. Its main defect is that all ribbon-like animals swim in precisely the same manner whether they are segmented or not, and neither their swimming ability nor the frequency of occurrence of the swimming habit in any group appears to bear any relation to metamerism.

The origin of the pseudometameric arrangement of organs in turbellarians and nemerteans, which, according to this theory, represents a sort of piecemeal

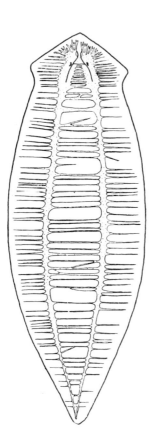

FIG. 10. Pseudometamerism of the nervous system of the triclad *Dugesia gonocephala*. (From Ude, 1908.)

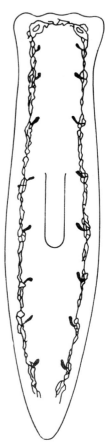

FIG. 11. Pseudometameric arrangement of the nephridial pores, and the longitudinal excretory ducts of *Dendrocoelum lacteum*. (From Goodrich, 1946, after Wilhelmi, 1906.)

segmentation, has not usually been discussed. The only reason that has been advanced for pseudometamerism, is that it is an adaptation to the likelihood of injury or breakage in very long worms (Hubrecht, 1887). The turbellarians are noted for their regenerative ability and, Hubrecht supposed, regeneration is facilitated if each fragment of the animal contains a set of all the essential organs from which the regeneration of new parts can start.

By an extension of this interpretation of the significance of pseudometamerism, it is possible to see the origins of complete, or metameric segmentation.

This is a rather naïve view of the regenerative process; new organs are not regenerated directly from fragments of old ones but from a blastema composed of undifferentiated or dedifferentiated cells. A simpler explanation of pseudometamerism is possible. In any large acoelomate, it is necessary to provide multiple excretory organs, particularly if the animals lack, or have a relatively ineffective circulatory system. If the worm is long and narrow, it is to be expected that the excretory organs will be more or less equidistantly spaced along the body. The same sort of argument can be applied to the serial repetition of the gonads, and the fact that they sometimes bear some spatial relationship to the gut diverticula may be for no better reason than that if gut diverticula exist, there is nowhere for the gonads to develop except between them. There is clearly no logical connexion between this interpretation of pseudometamerism and metameric segmentation.

The cyclomerism theory

The evolution of coelomic pouches from the gastric pouches of actinians postulated by Sedgwick (1884) implies the development of metameric segmentation from the fundamentally radial organization of the actinians, although it has been argued by supporters of this theory that actinians are, in fact, bilaterally symmetrical also. The acceptance of the cyclomerism theory therefore depends upon an acceptance of the enterocoel theory and, in this version of it, upon an unconventional view of primitive and advanced features in the Cnidaria.

In the version of the enterocoel-cyclomerism theory elaborated by Remane (1950, 1952), this difficulty does not arise because the four gastric pockets of a cnidarian ancestral form do not correspond with the coelomic compartments of metamerically segmented animals (Fig. 12). The unpaired anterior pocket and the two lateral gastric pockets become the proto- and mesocoels. These survive in oligomerous animals (e.g. hemichordates, echinoderms and pogonophorans) although even in these, the protocoel may be reduced or disappear (phoronids, ectoprocts and brachiopods). But in most animals, neither proto- nor mesocoels are represented in the adult and furthermore, since they give rise to totally different structures in the three divisions of the body of oligomerous animals, they cannot be regarded in the same sense as the segmental coelomic compartments of metameric animals. The mesodermal rudiments corresponding to the four pockets are represented in the Spiralia (most clearly in molluscs and annelids) by the fourth quartet of micromeres. Of these, 4a, 4b and 4c, corresponding to the anterior and middle archimetameres, become absorbed into the endoderm; only 4d, corresponding with the posterior archimetamere, produces mesoderm in which the coelomic pouches are formed. The paired mesodermal blocks, derived from the bilateral division and proliferation of 4d, then subdivide to form a small number of primary segments. The subsequent growth of additional, secondary segments occurs when a pygidial budding zone has been formed and the new segments are laid down in the familiar anterior-posterior sequence. In unsegmented coelomates such as molluscs and sipunculids, the posterior archimetamere does not undergo subdivision, and in turbellarians and nemerteans, it fails to form a coelomic cavity.

The notion that an initial group of segments is formed in annulates by the

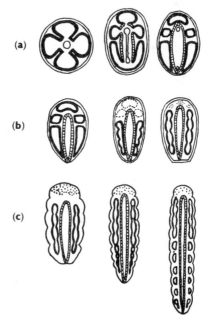

FIG. 12. The cyclomerism theory: (a) transformation of gastric pockets of a medusoid into the coelomic pouches of an oligomerous coelomate, (b) reduction of the proto- and mesocoels and production of primary segments by subdivision of the metacoel, (c) formation of secondary segments by proliferation. (From Remane, 1950.)

subdivision of 4d, and that later segments are formed in a different manner when a pygidium has been formed, is derived from Iwanoff's (1928, 1933) theory of primary and secondary segments, which he considered to be of universal application (Iwanoff, 1944). This has now been shown conclusively to be incorrect in both arthropods and annelids (Manton, 1949; Anderson, 1959). The precocious formation of segments is a functional necessity in animals with small eggs and planktotrophic larvae, but primitively, all segments are formed from a subterminal proliferation zone or pygidium.

A third version of the cyclomerism theory has recently been proposed by Lemche (1960). It is close to Sedgwick's theory in that metamerically segmented animals are held to be derived from anthozoans, but differs from other variants of the cyclomerism theory in that a segmental sub-division of the coelom is not regarded as an essential feature of metamerism, although Lemche apparently accepts the enterocoel theory in principle. He also places

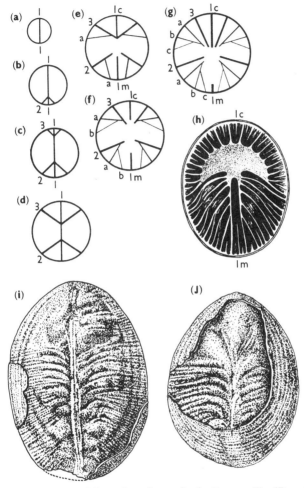

FIG. 13. (*a–g*) Sequence of formation of septa in the Tetracorallia, (*h*) septa of the Ordovician tetracorallian *Menophylla*, (*i*) muscle scars of the monoplacophoran *Cambridium nikiforavae*, and (*j*) of *Cambridium cernysevae*. (All from Lemche, 1960; *a–g* after Carruthers, 1906; *h* after Zittel, 1915; *i, j* after Horny, 1957.)

some reliance upon the existence of primary and secondary segments. The Stenothecoidea, a group of Cambrian monoplacophoran molluscs, have septa-like muscle scars which show a similar arrangement to the growth pattern of the septa of Ordovician corals of the order Tetracorallia (Fig. 13). Lemche suggests that both may have had a pre-Cambrian common ancestor, and he homologizes the counter region of the corals with the anterior part of the shell of stenothecoids. Thus far, segmentation is manifested chiefly in the arrangement of the dorso-ventral muscles, and this, together with the associated seriation of some other organs, such as gills and blood vessels, can be seen in the modern monoplacophoran, *Neopilina* (Lemche and Wingstrand, 1959). Molluscan segmentation is incipient; true metamerism is acquired in annelids and arthropods by the proliferation of secondary segments from a posterior budding zone and these contain segmented coelomic cavities. These 'secondary' segments correspond to the cardinal region of the Tetracorallia. The 'primary' segments correspond with the counter region and probably with the whole body of *Neopilina*.

Although, in Remane's (1950, 1952, 1958) hands, the cyclomerism theory has been persuasively elaborated in great detail, it has been vigorously attacked, particularly by Hyman (1951a, 1959). It is objected that the Anthozoa cannot be regarded as the most primitive coelenterates, and that anthozoan mesenteries grow inwards, instead of as outpocketings, as the enterocoel theory implies. Furthermore, the gastric pockets are arranged around a central cavity whereas the coelomic compartments are arranged along a longitudinal axis. Hyman suggests that there is no justification, apart from the needs of the theory, for this postulated change of axis of the animals. It may also be objected that the complete loss of the proto- and mesocoels in annelids militates against the theory (Sarvaas, 1933).

In addition to these objections to the cyclomerism theory and the enterocoel theory with which it is necessarily associated, the phylogeny of the Metazoa that is implicit in the theory has proved widely unacceptable. There is considerable reluctance to removing the Turbellaria, notwithstanding their obvious specializations, from the position of the most primitive triploblastic Metazoa and replacing them by the oligomerous phyla or the Mollusca (Steinböck, 1937, 1958; Hadži, 1953, 1958), just as there is a reluctance to accept the view that unsegmented coelomates are not more primitive than segmented ones.

The corm theory

Among alternatives to the pseudometamerism and cyclomerism theories, by far the oldest is the hypothesis that metameric segmentation resulted from incomplete asexual reproduction in which a chain of zooids was formed but the process was arrested before the zooids became fully individuated and separated from the parent body. Such a theory was supported by many early and mid-nineteenth century zoologists, including Quatrefages, Cuvier, Owen and Geoffroy St.-Hilaire, Haeckel and Hatschek, and was expounded in detail by Perrier (1882). An analogy was seen in the formation of chains of individuals in rhabdocoels and triclads, the proglottids of cestodes, and the strobilae of the scyphozoan scyphistoma.

The advantage of this over the two previous theories is that it derives segmented coelomates from unsegmented ones and so accounts for the existence of the latter (Naef, 1931). The objection most commonly raised to it is that the sequence in which the zooids are formed in platyhelminths or in which strobilae are formed in cnidarians, bears little relation to the sequence of segmentation in metameric animals. In turbellarian asexual reproduction the zooids are never serially arranged and the first plane of fission occurs in the middle of the worm, not sub-terminally (Fig. 14). The time of

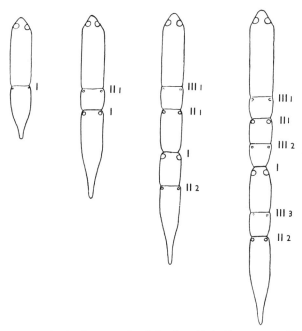

FIG. 14. Stages in the development of a chain of zooids in *Stenostomum*. Numbers indicate the sequence of appearance of fission zones. (From Child, 1941.)

appearance of subsequent fission planes is determined by the degree of development of the nearest fission plane anterior to them: the more differentiated the plane, the further away the new zone of fission occurs. Cestode proglottids are serially arranged, but in the reverse order to the metameres of chordates and annelids. The strobilation of the scyphistoma does present some analogy with metamerism in that the strobilae are produced in an oral-aboral sequence. However, any non-sessile animal that reproduced asexually in this way would be under the serious disadvantage of losing its head at each division and, as Hyman (1951a) points out, this method of reproduction appears to be particularly related to the sessile habit and is unlikely to have been rich in evolutionary potentialities.

Despite the weight of these objections, they are not conclusive. There is no reason to suppose that the fact that growth patterns in unsegmented animals differ from those in metameric animals, necessarily precludes the evolution of metamerism by incomplete fission. Many oligochaetes and some polychaetes reproduce asexually in a similar manner to turbellarians, yet also have a proliferation zone (Herlant-Meewis, 1958). It may be claimed that the theory is based on an analogy which cannot be sustained and in this sense fails for lack of evidence, but a more cogent argument against it is that segmentation appears to be primarily a result of the subdivision of mesodermal and associated tissues without interfering with the integrity of the individual, whereas strobilation is primarily the result of epidermal activity leading to the subdivision of the individual into a number of units each with the properties of an individual organism (Berrill, 1955, 1961).

Embryological theory

The theories of His (1874) and Caldwell (1885) are based on embryological considerations. According to these authors, metameric segmentation originated as an accident of embryology. Differential growth rates of ectoderm and mesoderm set up mechanical stresses in the latter during the elongation of the embryo or the larva, and this resulted in a fragmentation of the mesoderm. This fragmentation is manifested in the adult as a meristic repetition of all the mesodermal derivatives.

This theory has never been analysed in detail, nor was it supported by factual evidence when it was first proposed. One major obstacle to its acceptance is the lack of segmentation in nemerteans and some turbellarians, which are as long and slender as any segmented animal. Although some organs of these worms may be repeated at intervals along the body, the body-wall muscles are never divided in this way, although the muscle accounts for the greater part of the mesoderm.

This view of the origin of metamerism attracted scant attention when it was first proposed, possibly because it provided so little ground for phylogenetic speculation, but it has been revived in a modified form by Berrill (1955). He emphasizes that metamerism is acquired by the subdivision of the mesoderm in the development and growth of living animals, and that this rhythm or pulse in mesodermal growth along the longitudinal axis of the animal is but a reflection of a general property of growth processes, exaggerated in the mesoderm of segmented animals, but recognizable also in such phenomena as hydroid growth. A sudden change in the growth pattern of the mesoderm, or an exaggeration of an already latent pulsative quality, represents a slight initial modification, if one with profound structural consequences. Berrill therefore regards the evolution of segmentation as an event which could easily have been accomplished in a single step as a result of a relatively trifling mutation whenever the selective advantages of a metameric organization warranted such a change in morphology.

Berrill clearly recognizes three distinct 'problems which have generally been confused by other writers: the manner in which metamerism was acquired, the structure of animals in which it first appeared, and the selective advantage of metamerism that accounts for the appearance and survival of this type of structural organization. The embryological theory, as he has restated it, relates only to the first of these. Berrill has been concerned chiefly with chordate evolution, and suggested that metamerism appeared in the tail muscles of an ascidian tadpole larva as an adaptation to swimming. Analysed thus into its component parts, the problem of the evolution of metamerism begins to assume manageable proportions but we are still left with the unanswered question of the origin of annelid segmentation.

The locomotory theory

Several supporters of the pseudometamerism theory have suggested that swimming by undulatory movements resulted in metamerism of the musculature, and that this completed the process of segmentation already begun by the piecemeal repetition of other organ systems in the body (Snodgrass, 1938; Goodrich, 1946; Hyman, 1951a). This theory has little to commend it. Among invertebrates there is no correlation between the ability to perform undulatory swimming motions and the possession of a segmented musculature. And of all the animals that might be expected to offer the best evidence in support of this theory, the annelids fail miserably. It is clear that the theory demands in the first instance the segmentation of the longitudinal musculature, yet it is likely that in no existing annelid are the muscle fibres of the longitudinal body-wall muscles organized on a strictly segmental basis; instead, they usually extend through two or three segments (see below, p. 123). In fact, only the circular muscles are segmentally disposed, but they are irrelevant to sinusoidal swimming. Thus in most leeches, the circular muscles are relaxed and the body flattened when the animals swim (Gray, Lissmann and Pumphrey, 1938). The polychaete *Glycera* is normally able to swim only in a clumsy fashion by performing involved coiling movements, but spawning *G. dibranchiata* are able to perform typical undulatory movements despite the fact that the circular muscles are almost completely histolysed in this epitokal form (Simpson, 1962). In fact, there seems to be very little reason for supposing that undulatory swimming movements performed by pseudometameric worms could lead to metamerism of the musculature, and there is certainly no evidence to support such a view.

As an alternative, the segmentation of the musculature may be regarded as the primary event in the evolution of metamerism, and the associated or subsequent segmentation of other organ systems as only a subsidiary consequence of this. Such a view, as we have already seen, was proposed by

Berrill (1955) to account for the origin of chordate segmentation. He suggested that as the larval life of the ascidian tadpole was prolonged and it swam in stronger currents, it increased in size and strength. With the increase in the length of the tail, the developing muscle blocks revealed or acquired a tendency to discontinuous growth and this resulted in a segmentation of the caudal musculature, with the consequent benefit of greater swimming power. A similar view of the evolution of chordate metamerism was also suggested by Garstang (1928), except that he regarded the process as a more gradual one than Berrill has suggested.

Zenkevich (1945) attempted, with much less success, to extend a similar sort of argument to the evolution of metamerism in annelids. Like Berrill, he regarded the segmentation of the musculature as the most important feature of metamerism. He claimed that for the preservation of its shape, an animal required some sort of elastic recovery mechanism to oppose the effects of muscle contraction. Such 'shape stabilizers' took the form of the notochord in primitive vertebrates, the mesenchyme of turbellarians and nemerteans, the septa of annelids and the cuticle of nematodes. The existence of these stabilizers was essential for undulatory swimming, as for other muscular activity, and in an early stage in the evolution of this method of progression, there can have been no secondary body cavity because such a weakening of the parenchyma would have resulted, in the absence of other 'shape stabilizers', in the body breaking rather than flexing when the muscles contracted. Subsequently the worms acquired a stronger integument and this necessitated numerous flexible infoldings at which bending took place. The body-wall muscles became associated with these infoldings and so became segmented. The greater strength of the integument of annelids, or the existence of the 'shape stabilizer' in the form of the notochord in chordates, relieved the parenchyma of this function and so permitted the development of coelomic cavities. Zenkevich appears to have considered all unsegmented coelomates as being secondarily derived from segmented animals.

Whether or not annelid metamerism has been evolved as a similar adaptation to swimming as that of chordates, it is unlikely that such a view can be justified in Zenkevich's terms. While his 'shape stabilizers' may help in varying degrees to preserve the form of animals, they do not, except in the special case of the nematode cuticle, antagonize the body-wall muscles. The interruption of a strengthened integument by numerous flexible infoldings would not have resulted in a segmentation of the musculature as in annelids, with the muscles inserted into the infoldings, but in an arthropod type of musculature in which the muscles cross the flexible joints and are inserted on the sclerites. Numerous other objections can be raised to Zenkevich's theory, but the misconception about the role of supporting tissues and of antagonistic muscles, alone renders it untenable, at least so far as it relates to the origin of the coelom and segmentation.

The important function of the coelom as a hydrostatic organ, and of the segments in various types of locomotion has long been recognized (Sarvaas, 1933), but the similarity of swimming methods of unsegmented and segmented animals has always been taken to indicate that only swimming by undulatory flexions of the body can be considered as a possible factor in the evolution of segmentation. While the origin of chordate metamerism may be related to swimming, no convincing case has been made that this is true of annelid metamerism.

THE PRESENT POSITION

None of these theories to account for the evolution of the coelom and of metameric segmentation is entirely satisfactory, and three general criticisms can be made of the whole trend of discussions of the subject during the latter part of the last century when the current theories were first established.

First, most of the theories deal particularly with the structures from which the coelom might have been derived and the intermediate stages through which they passed in the course of their evolution, but hardly at all with the selective advantage gained from the possession of either a coelomate or a metameric type of organization. The advantages must have been considerable to judge from the success of the coelomates and, particularly, of the segmented coelomates. This failure to consider the adaptive significance of the coelom and of segments resulted in much of the speculation degenerating into a mere juggling with morphological types in order to arrange them in a tidy sequence. Furthermore, the preoccupation with the morphology of existing animals that has characterized much of the discussion, has tended to result in speculation about the relationships between modern phyla, and consequently the debate has become obscured with details of morphological similarity which might indicate an affinity between different groups of animals, but has little to do with either the secondary body cavity or metamerism. Some of the authors, too, have been unduly influenced by Haeckel's biogenetic law, and much of the discussion has turned on far too literal an interpretation of the embryological origin of coelomic cavities and of metameres.

A second confusing aspect of much of the discussion has been the failure to distinguish the evolution of the coelom from the evolution of metamerism, or indeed, to discuss whether or not any such distinction should be made. If the coelom originated from a series of metamerically disposed cavities, as many of the theories suggest, then as a logical consequence, either the unsegmented coelomates have had a previous history of metamerism, or, alternatively, a secondary body cavity has been independently evolved more than once. However, one of the few points of agreement between all the writers is that the coelom is homologous throughout the animal kingdom. Similarly, none of the earlier authors, with the notable exception of Bateson (1886) appears to have considered the possibility that metamerism may have been independently evolved in annulates and chordates. This is now generally accepted, but, even so, the implications of this possibility have received little consideration.

A third reason for a great deal of the confusion that has arisen is that during the period of the most intense speculation about the origin of the coelom, it was still not clear what in fact constituted a coelom. Initially, any cavity other than the gut, was regarded as coelomic (Haeckel, 1872). The Hertwigs (1882) excluded from the concept the blood spaces of molluscs, which they regarded as a pseudocoel, similar to the body cavity of rotifers and ectoprocts, and to the intercellular spaces in the parenchyma of platyhelminths. In 1884, Lankester drew attention to the fact that the mollusc pericardium is not a blood space, and it followed that the reno-pericardial canals and the genital ducts were coelomic in nature; thus the molluscs were reinstated to the rank of coelomates from which they had been excluded by the terms of the Hertwigs' definitions. During the next ten years the arthropods came to occupy a similar position. The blood spaces were thought not to be coelomic, but the coelom was represented by such cavities as the coxal gland of crustaceans (Gulland, 1885; Sedgwick, 1887; Weldon, 1890, 1891; Allen, 1893; Lankester, 1893). Subsequently it has been argued that these cavities represent nothing more than dilations in the genital or excretory ducts and can only be regarded as coelomic if it is assumed that they were once very much more extensive; if they are primitively small, then even in terms of the gonocoel theory, they are at best only incipient coelomic compartments (Sarvaas, 1933). This argument, of course, ignores the embryological evidence of the formation of incipient coelomic pouches in a number of arthropods, and the subsequent obliteration of the pouches with their incorporation into a myxocoel.

The present position is rather less richly confused and controversial, since there are now only two surviving major schools of thought.

One maintains that the coelom originated from enlarged and cavitated gonads, and that metamerism evolved as a swimming adaptation in pseudometameric animals. This is approximately the view of such writers as Goodrich (1946), Hyman (1951a) and Beauchamp (1959). This interpretation of the origin of the coelom and of metamerism is not without its difficulties and appears to have been reached by default rather than by analysis. Hyman has raised doubts about the validity of the gonocoel theory and considers metamerism to have been independently evolved in cestodes, annelids and vertebrates, but it is not clear from her writings how the three different types of metamerism could all have been evolved as an adaptation to swimming by serpentine movements. Goodrich evidently considered metameres to be homologous in annelids and chordates, and his interpretation is very close to Lankester's (1900) compromise between the gonocoel and enterocoel theories.

The chief alternative to this interpretation, also widely accepted, is the enterocoel-cyclomerism theory skilfully developed and argued by Remane (1950, 1952, 1954, 1958) and his supporters. This theory does not suffer from a lack of analysis, rather, it is the phyletic consequences of the theory that have proved an obstacle to its acceptance, although there is also substantial opposition to the enterocoel theory with which it is intimately associated.

Consideration of the evolution of the coelom and of segments has generally been conducted from the points of view of comparative anatomy and embryology. Both disciplines are notoriously liable to yield evidence that is susceptible of many different interpretations, and a weakness of most of the suggested theories has been a lack of serious discussion of the function of the coelom or metameres. It is also clear that the disinclination to consider the possibility of a polyphyletic origin of the coelomate and metameric conditions has introduced many difficulties. Before these evolutionary problems are set aside as insoluble, therefore, they should be considered from a functional viewpoint and also without preconceived notions that the coelom or metameres are homologous structures wherever they occur.

It is a fairly simple matter to give an acceptable morphological definition of the coelom. It is a cavity, originally bilateral and within the mesoderm, filled with liquid and bounded by an epithelium in which the gametes often originate. The cavity occupies the space between the ectoderm and endoderm layers and generally communicates with the exterior by coelomoducts and nephridia. Included within the definition is any cavity, as, for example, the coelomic canals of leeches, which may not occupy the whole space between the ecto- and endoderm, but which can be considered to be derived, phylogenetically or ontogenetically, from a coelom (Sarvaas, 1933). Metameric segmentation is not so easily defined, but it may be noted that one essential feature of it is that the body-wall musculature and the coelom are involved in the serial repetition of parts, although the coelom of many segmented animals may secondarily have lost its compartmented organization. The nervous system, excretory organs, gonads and other organs, may all show a comparable seriation, but unless the musculature also is segmentally arranged, or shows signs of having been so at a previous stage in phylogeny, the animal is regarded as pseudometameric.

The fact that it is the segmentation of the musculature that is so essential a feature of metamerism immediately suggests that the locomotory theories of the origin of metameric segmentation should be given special consideration, the more so because although they are beginning to win a vague acceptance, they have never been analysed in detail, and, as they stand at present, are obviously defective.

Segmented animals occur from the start of the fossil record, and their unsegmented ancestors must have been for the most part soft-bodied and unlikely to be fossilized, except under unusual circumstances. At all events, we cannot rely on direct evidence such as is provided by a palaeontological record, and it is unlikely that we shall ever be able to do so. We are therefore forced to use indirect and circumstantial evidence based on the principles of comparative morphology. Fortunately in examining the locomotory theory of metazoan evolution we are concerned chiefly with the dynamics of locomotion. From a knowledge of this we can predict the minimal structural requirements of an animal for it to move in a certain way. In this, we assume that the fundamental construction of organisms is based on sound mechanical principles and that physical and mechanical considerations bulk largely in determining the form of an organism. Such an assumption is implicit in most modern morphological studies and owes its inception above all to the influence of D'Arcy Thompson's great work, *Growth and Form*.

In considering locomotion and its influence on the form of animals, we shall be concerned, in the first instance, with the action of the physical and mechanical world upon the organism, rather than with purely biological influences. This makes our task much easier. The principles of dynamics do not change, so that an examination of the mechanics of locomotion in an existing animal provides information which is immediately applicable to any other organism which moves in the same way and has fundamentally the same body structure. So long as the argument rests on mechanical considerations, therefore, we shall be dealing with certainties which can be verified by experimentation, and not, as is the case with most of the classical theories, with probabilities or possibilities.

The analysis that I shall present is based primarily upon an investigation of the changes of shape of triploblastic animals, the factors controlling, limiting, and determining them, and the locomotory abilities conferred upon an animal by virtue of the possession of a particular body structure. This rather theoretical discussion can be checked at every stage by reference to the structure and locomotory abilities of existing animals. Following this, we shall be in a position to consider the sequence in which more and more efficient grades of body form have been evolved, and to describe the broad environmental situations in which these animals lived and moved.

2

THE ACOELOMATE CONDITION

THE HYDROSTATIC SKELETON

OF all the countless shapes and forms that triploblastic metazoan animals take, the least complicated and, in essence, the least specialized, is vermiform. We can define a worm in a set of postulates and justify them by reference to the structure and behaviour of living worm-like animals. Stripped of inessentials, a worm consists, from a mechanical point of view, of a flexible, muscular body wall enclosing an incompressible, but deformable medium which has a constant volume and in which fluid pressures can be transmitted. This is an economical definition and in living worms, of course, we often find considerably greater complexity of structure than we have postulated, but initially, we need only consider the potentialities of a hypothetical organism with this minimal structure.

With this apparatus, the hypothetical worm is able to change its shape by contraction of parts of the muscular body wall. But if, as we have postulated, the volume of the worm remains constant, contraction of muscles in one part of the body must be compensated by the dilation of other parts, and compensatory local volume changes of this sort are fundamental to reversible changes of shape in all soft-bodied animals.

This principle can be clearly illustrated by considering a long, cylindrical worm in which the body-wall musculature is organized in circular and longitudinal layers (Fig. 15). Contraction of the circular musculature reduces the diameter of the worm and causes a compensating increase in the length of the animal. Recovery from this extended position is possible only by contraction of the longitudinal muscles. Muscles, in this case those of the circular layer, are incapable of active elongation and, once contracted, external restoring forces must be applied to them to restore them to their original length. For this reason, negative fluid pressures can never be generated within the body of the worm and there is no way in which activity of the circular muscles can cause the body to shorten (Fig. 16). Thus extension of the circular muscles is produced by contraction of the longitudinal muscles and *vice versa*, and these muscles behave as antagonists in exactly the same way as flexor and extensor muscles of the limb of an animal with a rigid skeleton. The difference in this case is that the restoring forces which extend a contracted muscle are transmitted to it, not through the bone, but through the fluid contents of the body which, in a dynamic sense, can be regarded as a skeletal

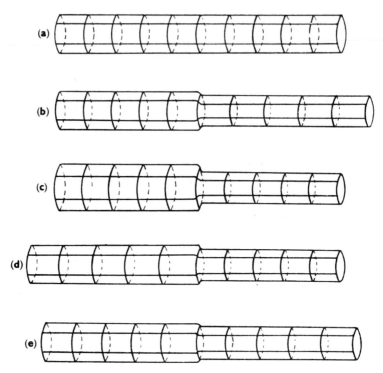

FIG. 15. Some consequences of the contraction of the circular muscles in the right-hand half of a cylindrical worm. (a) Initial state with all the muscles relaxed. (From Chapman, 1950.)

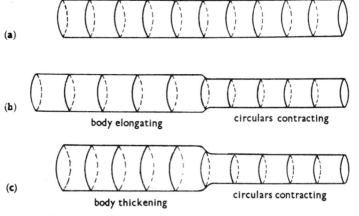

FIG. 16. Possible changes of shape of a cylindrical worm with only circular muscles in the body wall. (a) Initial state. Note that the contraction of the muscles of the left-hand half of the animal in (c) would restore the animal to its initial shape, but that this is impossible in (b). (From Chapman, 1950.)

system and are therefore called a fluid, or hydrostatic skeleton (Chapman, 1958).

More complicated changes of shape than mere extension or contraction of the whole body are possible even in the extremely simple, hypothetical animal that we have postulated. Local contractions of circular muscles may cause extension of other parts of the circular musculature without any change in length of the animal. One part of the circular musculature then antagonizes the rest. In the same way, contraction of the longitudinal muscles of one side of the body causes lateral flexure of the worm, and recovery from this position is achieved by contraction of the longitudinal muscles of the opposite side of the body. In all these changes of shape, the net effect of contraction of one part of the musculature depends entirely upon the behaviour of the remaining muscles. The pressure changes in the fluid contents of the body, produced by muscle contraction, are transmitted freely and without loss to all parts of the body wall. If the body-wall muscles remain in tonic contraction and do not extend when part of the musculature contracts, there will be no change of shape; instead, the hydrostatic pressure of the contents of the worm will increase. Extension or dilation of part of the body will follow upon contraction of some muscles only if the muscles in that region are relaxed and permit elongation. In some, possibly many, animals the whole body-wall musculature is nearly always partially contracted, maintaining a slight positive hydrostatic pressure in the body contents; under these circumstances relaxation of any part of the musculature results in its dilation (Wells, 1961). As a consequence of this dynamic interrelationship between all the muscles of the body wall, a considerable variety of localized lumps, constrictions and elongations of the body can be produced (Fig. 17).

Arrangement of the body-wall musculature other than in circular and longitudinal layers is mechanically possible; the fibres could be randomly orientated, for example. Such a system would be capable of the same changes of shape as one in which the muscles were disposed in circular and longitudinal layers, for the longitudinal fibres would antagonize all the other fibres to varying extents, depending upon their inclination to the longitudinal axis of the animal. However, such a system would be extremely uneconomic for anything but very localized adjustments of shape, and does not, in fact, occur in any worm. The chief advantage of a system in which the antagonistic muscles lie in circular and longitudinal directions, is that co-ordination of their activity makes relatively slight demands upon the nervous system. Co-ordination of contractions of a randomly orientated muscle system would call for an impossibly complicated nervous system.

The mechanical requirements of incompressibility and deformability which are demanded of the hydrostatic skeleton are best met by a liquid such as water which is virtually incompressible and, having a low viscosity, offers practically no resistance to indefinite deformation. It is not essential that the hydrostatic system should be a fluid, however—a loosely constructed parenchyma will serve for some purposes since it is practically incompressible and has a moderate, but for relatively small changes of shape, sufficient ability to undergo deformation.

Nearly all worms possess longitudinal and circular muscles and changes of shape are accomplished by their antagonistic contractions about a fluid

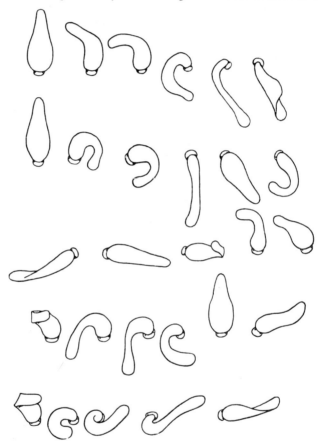

FIG. 17. Changes of shape of *Malacobdella*, a nemertean provided with a posterior sucker. (From Eggers, 1935.)

skeleton in the manner we have already considered in the model worm. In some respects, the system can be seen functioning in its least complicated form in the free-living acoelomate worms such as planarians and nemerteans. They have no cuticle which might be expected to restrict changes of shape and, although specialized in various ways, include many forms which are, from a mechanical point of view, among the simplest of all triploblastic metazoans. An examination of the functioning of the body wall in these worms therefore provides us with information about the mechanical capabilities of animals at this level of structural organization, and this information applies equally to any ancestral form that we may envisage, provided only that it has fundamentally the same type of organization as modern acoelomate worms. By this means we can gain some impression of the way in which a hydrostatic skeletal system may have been evolved in the triploblastic metazoans.

LIMITATIONS OF CHANGE OF SHAPE SET BY THE BASEMENT MEMBRANE

A worm with the idealized structure that we have postulated is theoretically capable of elongating and shortening indefinitely. In practice, of course, limits are set to the change of shape of any living animal, if only by the amount of passive stretching that the longitudinal muscles can withstand as the worm becomes longer and longer, or by the thickness of the body wall, which we have so far neglected, as the worm becomes thinner and thinner. Turbellarians, and more particularly the nemerteans, are noted for their remarkable extensibility, but even they cannot stretch or contract beyond certain limits. They do not possess a cuticle, but the inelastic fibres which make up the subepidermal basement membrane have a restraining influence upon changes in the length and circumference of the worms.

Indeed, at first sight it is surprising that a body bounded by a basement membrane composed of inextensible fibres should be capable of any change of shape at all, the more so since the basement membrane of many nemerteans and turbellarians is very thick. The fibres run in left- and right-handed spirals about the body of the worm in alternate layers (Fig. 18). They do not slip over each other where they cross, but, as the worm changes in length, the inclination of the fibres to each other changes. As the worm elongates, the angle between the fibres decreases and the fibres become more nearly longitudinal. Conversely, as the worm contracts, the fibres tend to become circumferential and the angle between them increases. This lattice of fibres, though composed of inelastic members, permits changes of length of the whole system (and of the worm), and in its behaviour resembles the extension and retraction of lazy tongs or a garden trellis.

We can best consider how the fibre system in the basement membrane influences and limits changes of shape in the worms by studying, in the first instance, the properties of an extremely simple and hypothetical system composed of a single inextensible fibre making a single turn of a geodesic spiral about a cylinder (Fig. 19a). We assume the cylinder to be circular and so simple relationships exist between the length of the cylinder, its volume, the length of the fibre, and the angle it makes with the longitudinal axis (Figs. 19b, c). Since the length of the fibre is constant, the cylinder has a characteristic length and volume for each particular inclination of the fibre. The same general condition obtains whatever the cross-sectional shape of the cylinder, though the volume corresponding with a particular orientation of the fibre will not be realized unless the cross-section of the cylinder is circular; if it is elliptical, the actual volume will be less than the theoretical capacity of the system.

The relationship between the inclination of the fibre and the capacity of the system is illustrated graphically in Fig. 19c. The calculation has been based on the assumption that the cross-section remains circular, and so the

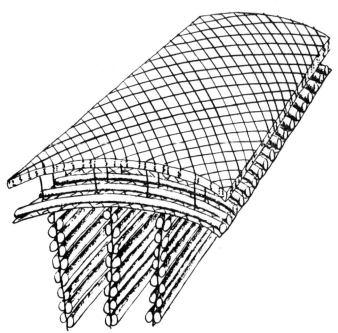

FIG. 18. Connective-tissue fibre system and musculature of the body wall of the nemertean *Amphiporus lactifloreus*. The thin layer of diagonal muscles between the circular and longitudinal muscle layers has been omitted. (From Cowey, 1952.)

volume plotted is the greatest volume the system can contain at any particular orientation of the fibre. The theoretical limit of extension occurs when the fibre is completely pulled out into a long thread and the volume it encloses is zero (D); the limit of contraction occurs when the fibre forms a complete circle, the 'worm' is in the form of a disk and the volume is again zero (E). Between these two extremes the volume reaches a maximum when the inclination of the fibre to the longitudinal axis is about 55° (R).

In one respect, the hypothetical worm which forms the model for these calculations has been simplified to the point of falsification. An essential feature of a worm in which the body-wall muscles act antagonistically about a fluid skeleton is that the volume remains constant. The horizontal line drawn in Fig. 19c represents the constant volume of the worm, or more precisely, of a slightly more realistic model than the hypothetical model we have considered so far. This line cuts the curve at two points, F and G, and these represent the theoretical limits of extension and contraction of our

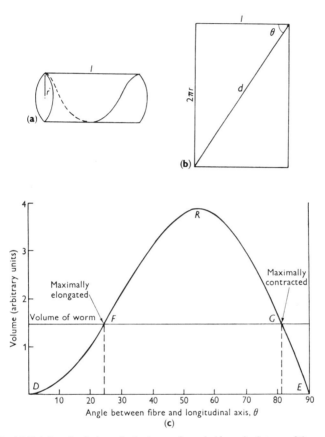

FIG. 19. (a) Unit length of a hypothetical worm bounded by a single turn of the geodesic fibre, (b) the same length of worm slit open and spread out, (c) relation between the volume enclosed by the fibre system and the inclination of the fibres to the longitudinal axis of the cylinder. The horizontal line represents the (constant) volume of a real worm. (From Clark and Cowey, 1958.)

model worm, as distinct from the limits of extension and contraction of the fibre system. Extension beyond F or contraction beyond G are impossible, for then the volume of the worm would exceed the capacity of the fibre system. Between these limits, the volume of the worm is less than the capacity of the fibre system and so the worm is elliptical in cross-section. The greatest degree of flattening occurs when the discrepancy between these two volumes is greatest, that is, where the inclination of the fibres to the longitudinal axis is 55°. Obviously, if the volume of the worm were equal to the greatest capacity of the system, i.e. the horizontal line in Fig. 19c cut the curve at R, no change of length would be possible because any change would demand a reduction in the volume of the worm. These conditions almost obtain in nemerteans that are turgid with eggs. Their cross-section is then circular and they are capable of almost no change in length.

The length at which the theoretical capacity of the fibre system is maximal is one of some significance. It represents an equilibrium position and one which the worms assume when they are anaesthetized and all the body-wall muscles are relaxed. This is because the semi-fluid contents of the worms, unrestrained by muscle tonus, tend to flatten out to the limit of the fibre system.

The extensibility of the worm, conveniently measured by the ratio of its greatest to its smallest length, depends upon the difference between its volume and the maximum capacity of the fibre system, which, in turn, is related to the degree of flattening of the cross-section of the worm when it takes up the relaxed, equilibrium position. It is therefore a simple matter to compare the actual performance of a worm with predictions based on a theoretical analysis of the mechanical properties of the fibre system. Cowey (1952) did this, and found a tolerable, though not exact agreement between the actual and theoretical ability of the nemertean *Amphiporus* to change its length. Subsequently, the theory was tested against a number of nemerteans and turbellarians, all of which are able to adopt a considerable variety of body forms (Clark and Cowey, 1958) (Figs. 20 and 21).

The turbellarian *Rhynchodemus* and the nemertean *Geonemertes* are both terrestrial and both have low theoretical extensibilities (Table 1) and in both, the actual extensibility agrees closely with the theoretical value. The fibre system therefore limits both extension and contraction of these worms in the way we have already explained. The difference between the volume of the worm and the greatest capacity of the fibre system is small (Fig. 20) so that the worms are almost circular in cross-section at all lengths. Terrestrial worms of this sort, living in forest litter, are liable to a considerable loss of water by evaporation; they secrete mucus on which they move and this also makes demands upon their water supply. Water loss can be reduced to a minimum if the surface area is small in proportion to the volume of the worm, and this condition is achieved if the cross-section is circular and not elliptical. Great extensibility, which necessarily involves a considerable flattening of the body, is therefore precluded in these terrestrial worms.

The nemerteans *Amphiporus*, *Lineus gesserensis* and *L. longissimus* are all marine, and water conservation cannot present a problem to them. The reasons advanced to account for the low extensibilities of *Rhynchodemus* and *Geonemertes* cannot apply and, in fact, these worms have large actual and theoretical extensibilities, but in no case does the actual extensibility quite

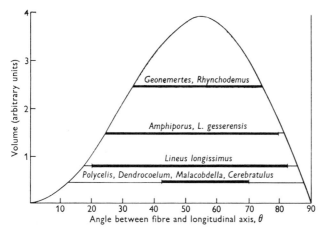

Fig. 20. The performance of a number of nemerteans and turbellarians, compared with the predictions of the theoretical analysis of change of shape. (From Clark and Cowey, 1958.)

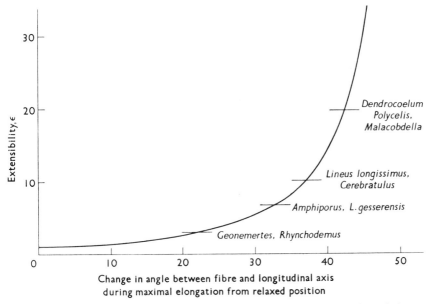

Fig. 21. Relation between extensibility and the changes in the orientation of the fibres in the basement membrane. (From Clark and Cowey, 1958.)

reach the theoretical value. *Amphiporus* and *L. gesserensis* do not reach the limit of contraction, while *L. longissimus* fails to reach the theoretical limits of either extension or contraction. This is shown by the worms not becoming quite circular in cross-section when fully extended or contracted. Now, a failure to become circular in cross-section at either limiting position indicates

TABLE 1
Extensibility of Nemerteans and Turbellarians
(from Clark and Cowey, 1958)

Species	Extensibility	
	Theoretical	Actual
Rhynchodemus bilineatus	3	3
Geonemertes dendyi	3–4	3–4
Amphiporus lactifloreus	6–7	5–6
Lineus gesserensis	6–7	5–6
Lineus longissimus	10	9
Cerebratulus lacteus	10+	2–3
Dendrocoelum lacteum	20+	2–3
Polycelis nigra	20+	2
Malacobdella grossa	20+	2–3

that the fibre system is not acting as a restraining influence and that some other agency is responsible. In these worms it is the compression of the epidermis. The surface area varies with the length, and so with the inclination of the fibres, in the manner shown in Fig. 22, and it can be seen that the surface

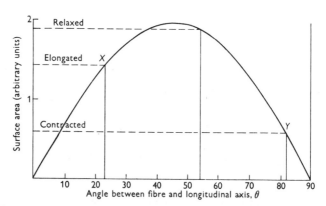

FIG. 22. Change in the surface area of a hypothetical worm as it changes length. (From Clark and Cowey, 1958.)

area is much smaller when the worm is fully contracted than when it is elongated. The available surface is apparently insufficient in this position to accommodate all the epidermal cells, and the limit of contraction is set not by the fibre system, but by the minimum surface area that is possible in these animals. To some extent the compression of the epidermal cells can be mitigated by folding of the basement membrane, and in all three species it is thrown into transverse folds when the worms are fully contracted, but evidently it is insufficiently folded to counteract this effect altogether. The same factor also operates in *L. longissimus* when the worm is fully extended, but in this case the folds run longitudinally because the circular muscles are contracted and pressures act transversely around the worm.

An extensibility of 9 or 10 appears to be near the limit for nemerteans, whatever the structure of the fibre system may theoretically permit, because the epidermis cannot be indefinitely compressed. This figure, based on theoretical considerations, is in good agreement with Coe's (1943) generalization, based on many years' practical experience of nemerteans, that they 'may often be contracted to one tenth [the length] of the fully extended worm'. Further effective contraction is possible only by throwing the body into coils as *Lineus socialis* frequently does (Fig. 23), but the mechanisms involved

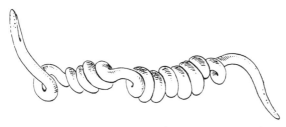

FIG. 23. The nemertean *Lineus socialis* showing extreme contraction produced by coiling. (From Coe, 1943.)

in such behaviour are, of course, entirely different from those in actual shortening of the body. The coiling is produced by the further contraction of the longitudinal muscles on one side of the worm.

The third group of worms, comprising the nemerteans *Cerebratulus* and *Malacobdella*, and turbellarians *Dendrocoelum* and *Polycelis*, all have very high theoretical extensibilities, but in fact show very limited abilities to change their length. They are all very much flattened and never even approximate to a circular cross-section. The agencies responsible for the limitation of their change of shape are:

(1) longitudinal, inextensible fibres in the longitudinal muscle layer, which limit the extension of the worm;

(2) similar fibres in the circular muscles which limit the circumference, and so the ability of the worm to shorten;

(3) dorso-ventral muscle fibres which preserve the flattened body-form and therefore resist changes of shape.

Clearly the structure of these species differs in several important respects from the simple model with which this discussion began. In fact, there is considerable variation in structure and ability to change shape in the Turbellaria and, to a lesser extent, in the Nemertea, and this is reflected in the variety of locomotory techniques employed by these worms, for in many of them changes of shape are, to a considerable degree, related primarily to movement.

LOCOMOTION IN TURBELLARIANS AND NEMERTEANS

The epidermis of turbellarians and nemerteans is generally heavily ciliated and the majority of species make some use of the cilia for locomotion. A ciliated epithelium is capable of performing relatively little work and, although it is not subject to fatigue as muscles are, provides too little power for locomotion except in small animals (Gray, 1928). Larger animals make use of the very much greater power of muscle contraction for locomotion. The size range of turbellarians is from less than 1 mm to several inches long. The smallest species are thus about the same size as the larger ciliates and they move in much the same way, while the larger turbellarians move almost exclusively by muscle contraction. Consequently, we can study within this group of worms the transition from ciliary to muscular locomotion and observe how changes in body shape which reflect the activity of the body-wall muscles, acquire a considerable, and often very specialized importance in locomotion.

The Acoela are small, inconspicuous animals and rarely exceed 1–2 mm in length. In many respects they are closer to the idealized worm than any we have considered hitherto, since they lack a sub-epidermal basement membrane and changes of shape are presumably limited only by the amount of passive stretching the muscle fibres can withstand. Most acoels move by swimming in the manner of ciliates and from the few observations on record of the behaviour of these very small worms, the impression is gained that the body-wall muscles are used only in very minor changes of shape, except in some species of *Convoluta*. Members of this genus are rather larger than most other acoels and they are somewhat sluggish swimmers. *Convoluta convoluta* rolls its body into a funnel with the lateral margins of the body almost touching ventrally, but diverging anteriorly (Gamble, 1893). Presumably this results in a more efficient displacement of water by the cilia inside the funnel and certainly, under these conditions, the velocity of water emerging from the narrow end of the funnel would be increased (Gray, 1928). The rolling up of the body therefore appears to be a locomotory adaptation which permits this relatively large animal to swim by cilia. The only acoelan in which the muscles play a dynamic role in locomotion is *Convoluta saliens*. The margins of the body are particularly extensive in this species and they are flapped violently up and down, so propelling the animal forwards (Gamble, 1893).

Rhabdocoels are generally somewhat larger than acoels and have a better developed musculature; nevertheless, changes of shape appear to be relatively trivial and locomotion is still by means of the cilia. The main difference which presumably results from the larger size but unchanged method of locomotion, is that rhabdocoels tend to swim far less often, and they creep over the substratum more than do the acoels. There are no specializations associated with swimming or creeping, though in *Ependytes* the body-wall muscles are modified to form a food collecting apparatus (Picken, 1937). Two flaps cover the mouth of this animal. When the worm feeds, these are suddenly opened, causing a swirl of water to rush into the space between them, carrying the food organism with it to the neighbourhood of the mouth. The flaps are then closed and the food swallowed.

The larger worms of the orders Tricladida and Polycladida show a greater

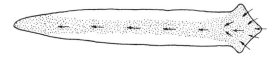

FIG. 24. Distribution of cilia on the ventral surface of *Planaria*. The arrows indicate the direction of ciliary beat. (After Pearl, 1903.)

variety of structure and locomotory habit than the acoels and rhabdocoels. All triclads and polyclads appear to be too large to swim by ciliary action, but aquatic triclads generally crawl on the substratum by cilia, and the ciliation of the epidermis is better developed on the ventral than the dorsal surface of the worm and may even be confined to the ventral surface (Fig. 24). A copious secretion of mucus attaches the animal to the substratum and is the medium in which the cilia beat to provide the propulsive force (Fig. 25).

FIG. 25. Ciliary locomotion in *Planaria*. Diagrammatic lateral view of the animal creeping on a secreted mucus trail. (After Pearl, 1903.)

A number of planarians are able to creep on a mucus trail on the underside of the surface film in still water.

Ciliary forces are too weak for the transportation of large animals even on the substratum and, as the flatworms increase in size, are at some stage replaced by the much more powerful forces of muscular contraction. A variety of factors permit turbellarians to postpone this change of locomotory method. The mass of most animals is proportional to the cube, the surface area to the square of the linear dimensions. Since the locomotory efficiency of animals that move by cilia is related to the surface area, there is a critical size above which the mass of the animal becomes too great to be transported by cilia. But if the thickness of the animal is the same whatever its size, both its mass and its surface area increase in proportion to the square of the linear dimensions and no limit is set by size to the effectiveness of the ciliated surface in locomotion. Triclads and polyclads are nearly all very much flattened, partly as a locomotory adaptation, and partly due to their lack of a blood vascular system and the slow rate of diffusion of oxygen through the tissues, making it necessary for all organs to be close to the surface of the body.

Most aquatic triclads are very much flattened and creep by ciliary action. Although most of the species for which data are available move slightly more slowly than the smaller rhabdocoels (Table 2), they are not much slower, and a large specimen of *Planaria maculata* 11 mm long is, in fact, able to crawl faster (1·49 mm/sec) than a specimen of the same species half the size (6 mm, 1·23 mm/sec). In other words, the increase of surface area more than com-

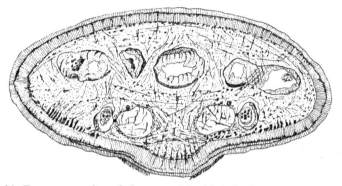

Fig. 26. Transverse section of the terrestrial triclad *Bipalium diana*, showing the restriction of cilia to part of the ventral surface. (After Moseley, 1874.)

pensates for the increase in mass. This is possible only if the animals present a relatively large surface area to the substratum and, as we have seen, such triclads as *Polycelis* and *Dendrocoelum* have very high theoretical extensibilities, which are indicative of great flattening when the body-wall muscles are relaxed. Furthermore, since they have low actual extensibilities, they remain flattened during all changes of length and breadth (Fig. 20).

A different situation obtains in terrestrial triclads and in polyclads. Since the former suffer a constant water loss, it is imperative that they should preserve a circular cross-section and so present a minimum surface area to the environment. This is illustrated by the fact that they have low extensibilities. In consequence of this, they present a small surface area to the substratum in proportion to their mass, and in some the ciliation of the ventral surface is correspondingly limited (Fig. 26). Furthermore, since they live in air instead of water and the effective weight of the animal is greater, the work that the cilia have to perform is also increased. It is not surprising therefore that the movements of terrestrial planarians should be relatively slow, nor that they move by muscular activity more often than by cilia.

TABLE 2
Rates of Locomotion of Turbellarians

Species	Rate of Locomotion mm/sec			Authority
	Min.	Max.	Ave.	
	CILIARY LOCOMOTION			
Rhabdocoela (aquatic)				
Mesostomum tetragonum			2·66	Lehnert, 1891
Tricladida (aquatic)				
Planaria polychroa	2·16	2·50		Lehnert, 1891
(exceptionally)		3·33		Lehnert, 1891
Polycelis tenuis	1·66	1·83		Lehnert, 1891
Planaria maculata			1·34	Pearl, 1903
Dendrocoelum lacteum	0·75	1·33		Lehnert, 1891
Planaria gonocephala	1·04	1·12		Parker & Burnett, 1900
Tricladida (terrestrial)				
Bipalium kewense	1·00	1·33		Lehnert, 1891
(exceptionally)		1·83		Lehnert, 1891
Geodesmus lineatus	0·50	0·66		Lehnert, 1891
	MUSCULAR LOCOMOTION			
Tricladida (aquatic)				
Planaria maculata			1·66	Pearl, 1903
(repeated stimulation)				
Polycladida (aquatic)				
Planocera californica		3·9	2·1	Olmsted, 1922
Phylloplana littoricola		4·3	4·0	Olmsted, 1922
Leptoplana lactoalba		5·0		Crozier, 1918

Polyclads are all aquatic, but the dimensions of many of them seem to exceed the size that can conveniently be transported by ciliary action, and in all of them muscular activity supplements or even replaces that of the cilia (Stringer, 1917; Crozier, 1918; Olmsted, 1922). According to Stringer, no planarian (her account obviously refers to polyclads) moves by cilia, for exposure of a number of species to magnesium chloride solution, which paralyses the muscles but leaves the cilia unaffected, stops movement, and exposure to lithium carbonate, which prevents the cilia beating but does not impair muscle contraction, has no effect on the locomotory abilities of the worms.

Three patterns of muscular activity may be employed by turbellarians and nemerteans for movement over the substratum. In no case do the worms enjoy exclusive rights to these locomotory techniques; indeed, all three are more effectively developed by other animals. But although imperfect and rudimentary, the muscular locomotory activities of turbellarians and nemerteans are of the utmost importance, because in them we see the origins of the dynamic functions of a hydrostatic skeleton. One locomotory method, employed by some turbellarians and a nemertean, involves looping in the manner of leeches. A second method, practised by many turbellarians, is by the use of pedal locomotory waves, a method that reaches its peak of complexity in gastropods. The third method, peristaltic locomotion, is used occasionally by nemerteans, though not by turbellarians, but is perfected only in coelomates with a true fluid skeleton.

Pearl (1903) observed that when the posterior end of *Planaria* is stimulated in some way, the tail is retracted and the head extended. The anterior end is then attached to the substratum and the longitudinal muscles contract, drawing the posterior end of the worm forwards. The posterior end may then be attached to the substratum and the head extended, again largely by contraction of the circular muscles, and the cycle repeated three or four times before the animal reverts to ciliary creeping, though by repeated stimulation the animal may be made to engage in this sort of progression for long periods. These movements are not invariably used only as an escape mechanism, and some species of *Planaria* and *Polycelis* often display them spontaneously. *Procerodes* (Ijima, 1887) and *Sabussowia* (Gamble, 1893) also crawl in this way and in the latter, this is the usual method of locomotion when the animal is out of water. This worm can move forwards or backwards, although, as the posterior tip never secures as good an attachment to the substratum as the anterior end, backward progression is relatively slow and inefficient. There is some doubt about the way in which planarians become attached to the substratum when engaging in this behaviour. It may be by adhesive mucus, or possibly part of the ventral body wall is raised to form a small, temporary sucker.

A nemertean, *Malacobdella*, also moves in a manner comparable to that of these planarians (Eggers, 1935). It lives as a commensal in the mantle cavity of lamellibranchs and has a permanent sucker at the posterior end of the body by which it remains attached to its host. When the nemertean crawls, the sucker is attached and the circular muscles contract, extending the worm forwards. The sucker is then detached and drawn forwards by the contraction of the longitudinal muscles (Fig. 27).

This type of movement is familiar from the behaviour of leeches (Herter, 1928, 1929; Gray, Lissmann and Pumphrey, 1938). Leeches have the great advantage of possessing permanent anterior and posterior suckers and so are able to move more efficiently than nemerteans or turbellarians. In almost all these animals, locomotory movements involve more complicated muscular activity than simple alternate and antagonistic contractions of the circular and longitudinal muscles. Very often, the body is thrown into a dorsal flexure during the phase of longitudinal muscle contraction (Fig. 28). This is produced by the ventral longitudinal muscles contracting more than the dorsals, but in addition, contraction of the dorso-ventral muscles in the middle part of the body causes flattening and so facilitates bending of the body in this

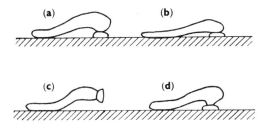

FIG. 27. Crawling with the aid of the posterior sucker by the nemertean *Malacobdella*. (From Eggers, 1935.)

region (Chapman, 1958). A consequence of this flexure is that the posterior sucker can be brought alongside the anterior one, and the length of a single 'stride' is increased and is almost equal to the length of the extended worm. Looping in this manner represents a specialized, if common, modification

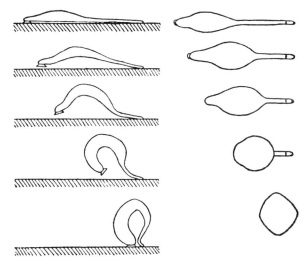

FIG. 28. Stages in the locomotory cycle of the leech *Helobdella stagnalis*. Lateral and dorsal views. (From Herter, 1929.)

of creeping by means of alternate extension and shortening of the body, and the extent of the dorsal flexure varies considerably in planarians and leeches (Mann, 1962). Dorso-ventral muscles are not essential to it and, indeed, the marine leech *Pontobdella* lacks dorso-ventral muscles and always has a circular cross-section (Chapman, 1958); it nevertheless engages in looping (Herter, 1929).

A number of triclads and the majority of polyclads do not employ the discontinuous movements of looping but instead, creep by means of waves of contraction that pass along the ventral longitudinal musculature. These are known as pedal locomotory waves by analogy with gastropods in which this manner of locomotion is most conspicuously developed, and make minimal demands upon the fluid skeleton. The result is a relatively smooth, continuous progression similar to the gliding observed in animals that move by cilia and, in fact, it is not always easy to distinguish ciliary creeping from movement by pedal muscular waves.

An approach to pedal locomotory waves can be observed in the terrestrial

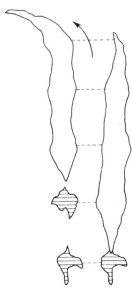

FIG. 29. Muscular creeping by the terrestrial triclad *Geoplana notocelis*. Dotted lines connect regions in which the longitudinal muscles are contracted. Points of contact between the body and the substratum are indicated by the patches of mucus left behind. (From Bresslau, 1928–1930.)

triclads *Geoplana* and *Bipalium*, in which this type of progression was first described by Moseley (1877b). Parts of the longitudinal musculature contract, causing slight local swellings of the body wall at these points and providing temporary points of attachment to the substratum by means of secreted mucus (Fig. 29). Anterior to these points of attachment, the longitudinal muscles are relaxed and are extended, presumably partly by contraction of the circular muscles and partly by the contraction of other longitudinal muscles anterior to relaxed ones. In this way the body is thrust and drawn forwards. Three or four such waves of contraction can be seen along the body of the worm at the same time and they pass backwards along the musculature at such a rate that the points of anchorage remain fixed relative to the ground and the worm leaves behind a dotted trail similar to that of a snail. In *Geoplana*, the longitudinal muscles of the ventral surface of the body, which are chiefly responsible for this type of locomotion, are much thicker

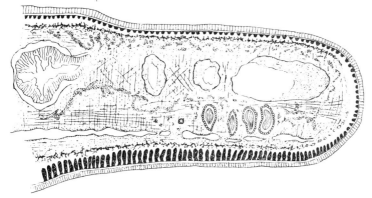

FIG. 30. Transverse section of the terrestrial triclad *Geoplana traversii*, showing the hypertrophy of the longitudinal muscles on the ventral side of the animal. (After Moseley, 1877b.)

than those of the dorsal surface (Fig. 30). A similar method of locomotion is sometimes employed by the shorter nemerteans (Gontcharoff, 1961).

The behaviour of the musculature of the polyclads is comparable to that of terrestrial triclads, except that the contractile waves tend to be more

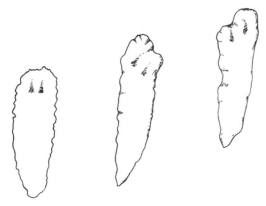

FIG. 31. Muscular creeping by the polyclad *Leptoplana*. (After Crozier, 1918.)

numerous and are generally confined to the edges of the worm instead of involving the whole of the ventral longitudinal musculature. Generally, the movement is slow and rather irregular. *Euryleptotes cavicola*, *Phylloplana littoricola* and *Leptoplana lactoalba* can achieve speeds of 4 or 5 mm/sec by this means (Table 2). A more regular and quite remarkable version of this behaviour was studied by Child (1904) and Crozier (1918) in *Leptoplana tremellaris*, in which regularly spaced contractions on opposite sides of the body alternate (Fig. 31), resulting in a kind of stepping, which resembles nothing so much as someone shuffling along with his feet in a sack, but recalls also the movements of such animals as polychaetes and arthropods which use appendages in fundamentally the same manner.

There is also a tendency in most turbellarians for the locomotory waves to be confined to the anterior part of the body and for the posterior part to be dragged along passively. This was particularly well shown by the

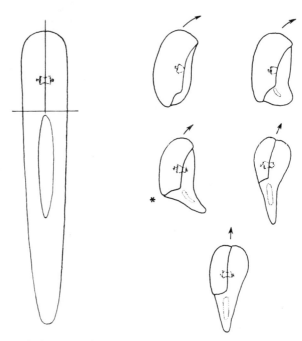

FIG. 32. Gradual change in the direction of locomotion of an anterior quarter of *Leptoplana* as it regenerates. Folds appear at the position marked. (From Child, 1904.)

experiments of Child (1904) on regenerating *Leptoplana*. He considered that the elongated form of this polyclad was a direct consequence of its method of locomotion. When an anterior quarter of the worm regenerates, the regenerated part initially grows out postero-laterally. Since, at this stage, traction is provided only by the original quarter, the animals tend to move in circles (Fig. 32). Later, when the posterior end has grown larger, folds appear on one side and the opposite side is stretched. This is held to result from the mechanical stresses set up in the regenerating, posterior portion of the worm as a consequence of the locomotory forces being confined to the anterior end. Subsequently the folds disappear and the body becomes linear.

Most nemerteans creep over the substratum, at least in part by ciliary action, but generally when the animals are not submerged, the propulsive force of the cilia is augmented by muscular contractions (Fig. 33). The characteristic peristaltic waves that pass along the body of nemerteans have been studied by Eggers (1924) and Friedrich (1933) and have something in common with the type of locomotion observed in terrestrial triclads. The peristaltic waves also foreshadow a type of locomotion more commonly found in coelomates and will be considered in more detail later. The peristaltic bulges

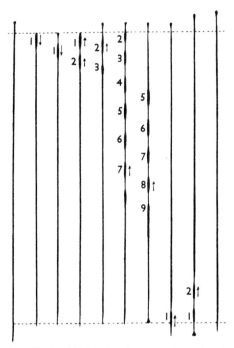

FIG. 33. Locomotion with the aid of peristaltic contractions of the body-wall musculature in the nemertean *Emplectonema*. (From Eggers, 1924.)

are produced by alternating contractions of circular and longitudinal muscles and, as in the terrestrial triclads, the regions where the longitudinal muscles are maximally contracted form temporary points of attachment to the substratum. Locomotory behaviour of this type is nearly always irregular and relatively uncoordinated in nemerteans, and frequently in very long worms the peristaltic waves of one part of the body bear no relation to those in other parts.

The fresh-water nemertean *Stichostemma* is unusual because peristaltic movements are observed only when the worm is moving backwards (Child, 1901). The cilia are then stationary and are directed forwards. Three or four peristaltic cycles may be observed along the body at the same time and, although the process appears to be surprisingly well co-ordinated, progress over the ground is slow. Reversed movements can generally be elicited by mild stimulation at the anterior end of the worm, but if stationary worms are violently stimulated at the anterior end, they do not crawl backwards but instead turn and creep off in another direction. Anterior locomotion, which is relatively fast and efficient, is by the combined action of cilia and muscles. Often the body is thrown into sinusoidal curves which pass backwards along the body. Since the worms live among algae, the crests of the locomotory waves, as they pass along the body, exert a backthrust against the vegetation and the worm is pushed forwards.

A feature of the anatomy of both nemerteans and turbellarians that we have so far not considered, is the layer of diagonal muscles in the body wall.

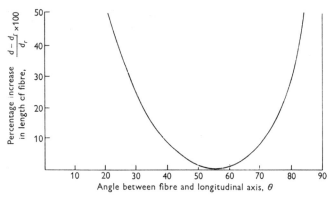

FIG. 34. Percentage increase in the length of the fibres composing the basement membrane of a hypothetical worm as the inclination of the fibres changes (cf. Fig. 19). The volume of the worm is assumed to be constant, and the fibres to be extensible. (From Clark and Cowey, 1958.)

This layer is conspicuous in most turbellarians (Bresslau, 1928–1930), is variously developed in nemerteans (Bürger, 1895, 1897–1907) and is also found in the body wall of leeches (Bhatia, 1941) and a few other worms. The diagonal fibres follow the same course as the spiral fibres in the basement membrane, and we are able to deduce something of their probable function by considering once more the behaviour of the fibre system in an idealized worm.

If we assume the fibres to be extensible, as the muscles are, and that the volume of the cylindrical model remains constant, the length of a single fibre must vary with its inclination to the longitudinal axis of the cylinder in the manner shown in Fig. 34. The contraction of a fibre restores it to its minimum length, which is at the equilibrium or resting position of the fibre system. The effect of contraction of the diagonal muscles in the real worm would therefore be to restore it to the length at which the flattening of the worm is greatest. The diagonal muscles are best developed in turbellarians and those nemerteans which preserve a flattened form, and the diagonal muscles in addition to supplementing the remaining body-wall muscles, as Mann

(1962) suggests, also assist the dorso-ventral muscles, the development of which is restricted by the presence of the gut and gonads between which they must penetrate.

Another function of the diagonal muscle may be to increase the turgor pressure of the animal. Clearly, if the diagonal muscles contract but the circular and longitudinal muscles resist elongation, the result will be not a change of shape, but an increase of turgor pressure making the worm rigid. Aquatic triclads, when they are creeping forwards, normally do so with the anterior end of the body raised from the substratum (Fig. 25). An exaggerated form of the same behaviour is seen when the animals rear up and execute searching movements while the posterior end of the body remains attached to the substratum (Lehnert, 1891). This searching movement is reminiscent of that of leeches which also possess diagonal muscles and in which the body remains turgid while this behaviour is in progress. Although there is not yet proof, there is every likelihood that the diagonal muscles provide the turgor necessary for these activities to take place.

GASTROPOD LOCOMOTION

The methods of creeping employed by turbellarians, in which pedal locomotory waves provide the propulsive force, foreshadow those of chitons and gastropods. But while in turbellarians, ciliary locomotion is sometimes supplemented, but rarely completely supplanted by muscular activity, this is not so in gastropods in which locomotion is generally through the agency of muscle contraction and cilia are employed in only a few species. Gastropods, being larger and more amenable to experimentation than turbellarians, have been studied very extensively, and their locomotory methods are accordingly understood in much more detail than those of turbellarians. Since the principles of pedal locomotion appear to be the same wherever it occurs we can amplify our knowledge of turbellarian movements by reference to the investigations of molluscan locomotion.

In spite of the fact that molluscs are usually held to be coelomate and turbellarians are not, the foot of polyplacophorans and gastropods may be regarded as analogous to the whole body of the flatworm. The turbellarian body is of course a closed system and its volume remains constant; the gastropod foot is usually an open system in that it is composed of erectile tissue and may be inflated by blood which is pumped from the rest of the body into the lacunar spaces of the foot. If the lacunae are then sealed off, as they are in some molluscs (Jordan, 1901), the foot, once inflated, may behave as a closed system while the animal is actually crawling.

The vascularity of the molluscan foot varies enormously. In the chitons and limpets, the foot is muscular and solid and can hardly be dilated at all; in others, such as pulmonates and some species that move over soft substrata like *Natica, Bullia, Polinices* and *Conus*, it can be inflated to an enormous extent (Schiemenz, 1884, 1887; Pelseneer, 1906; Morris, 1950; Brown, 1961; Brown and Turner, 1962). From a functional point of view, the gastropod foot bridges the gap between the acoelomate and coelomate conditions. The species with a relatively solid foot are comparable to acoelomates, those with an extensive pedal lacunar system may approach the coelomate condition, although in almost all of them the muscular contractions are very largely confined to the sole of the foot and the fluid skeleton plays, at most, a very minor role. However, there are exceptions, particularly among such forms as *Aplysia* in which the sole of the foot is relatively thin and flexible, and the foot itself is hollow, fluid-filled and not clearly separable from the visceral mass. The locomotion of these forms may be essentially that of coelomates with a fluid skeleton, and so depart radically from that of the majority of gastropods.

Muscular locomotion of gastropods, as in turbellarians, consists essentially of raising a part of the foot from the substratum, moving it forwards, and reattaching it. But here the similarity between gastropods and turbellarians almost ends, for gastropods have evolved a great variety of pedal locomotory techniques and although in principle they are all the same, in practice, the sequence of events varies considerably from species to species.

The various types of pedal locomotion have been described and categorized by Vlès (1907, 1913), Parker (1911) and Olmsted (1917a). In the great majority of gastropods, waves of muscle contraction can be observed to pass along the foot while the animal is in motion, and several such waves may be visible at once. These locomotory waves may pass along the sole of the foot in the same direction as the animal is moving, i.e. from posterior to anterior, or in the opposite direction to that of movement of the animal. These are known as direct and retrograde waves, respectively.

Parker (1911) and Lissmann (1945a) have analysed the sequence of muscular contractions during the passage of a direct locomotory wave along the molluscan foot (Fig. 35a). The point associated with the muscle fibre *2* is detached from the ground as the locomotory wave reaches it (B). The contraction of the longitudinal muscle fibre draws it forwards while it is still detached and then, as the locomotory wave overtakes the fibre, the point is reattached to the substratum (D) and remains in contact with it until the arrival of another locomotory wave. Each point on the sole of the foot is stationary except when a locomotory wave reaches it, and then it is moved forwards. Retrograde locomotory waves have not been analysed in detail, though it is supposed that essentially the same thing happens: a point on the sole of the foot is drawn forwards by contraction of the longitudinal pedal muscles at the time that it is detached from the substratum (Fig. 35b), but there are several important differences in the sequence of events. If the waves are direct, each muscle fibre is extended immediately after it has completed its contraction, and the stationary parts of the foot are those where the underlying longitudinal muscles are relaxed. If the waves are retrograde, the fibre is extended immediately before it starts to contract, and the stationary parts on the foot are those where the underlying muscles are contracted.

To some extent this analysis oversimplifies. Lissmann (1945b), using an apparatus in which successive regions of the foot of creeping snails (*Helix aspersa* and *H. pomatia*) passed over a narrow, swinging bridge, recorded

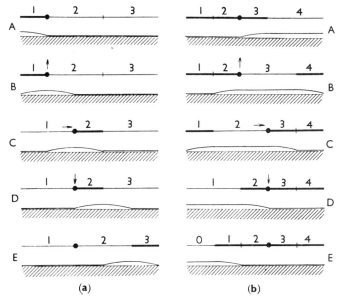

FIG. 35. Behaviour of the musculature during the passage of (*a*) a direct, and (*b*) a retrograde locomotory wave. (Partly after Parker, 1911.)

relative movements of different regions of the foot and their reactions with the ground. He found that static thrusts are developed between successive areas of fixation and the anterior end of the foot, and tensions between the fixation points and the posterior end of the foot. Thus the foot anterior to any fixation point tends to be pushed forwards, and that part of the foot posterior to a fixation point to be passively dragged forwards. The foot cannot be regarded as a series of mechanically balanced and independent, though co-ordinated, units; instead it must be considered as a mechanical whole. Nevertheless the simple analysis will serve as a first approximation to the dynamics of muscular creeping in snails, and it can be applied to any animal that moves by locomotory muscular waves in a manner similar to gastropods.

In his original classification of the various types of locomotory behaviour of the gastropod foot (Table 3), Vlès (1907) considered, in addition to the direction of movement of the locomotory waves, the simultaneous appearance of more than one independent series of waves. Thus both direct and retrograde

TABLE 3
Types of Muscular Pedal Locomotion in Gastropods and Chitons

I. Direct
    (a) Monotaxic   Most pulmonates (Vlès, 1907; Parker, 1911; Olmsted, 1917a; Crozier and Pilz, 1924; Lissmann, 1945a, b).
    (b) Ditaxic
        (i) Opposite:
        (ii) Alternate: *Haliotis* (Parker, 1911; Crofts, 1929), *Trochus* (Crozier, 1919), *Pomatias* (Lissman, 1945a, b).
    (c) Tetrataxic   Small species of *Littorina* (Vlès, 1907).

II. Retrograde
    (a) Monotaxic   Chitons (Vlès, 1907; Parker, 1911, 1914; Crozier, 1919), *Fissurella nodosa, Halcinia convexa* (Olmsted, 1917a), *Dolabrifera virens* (Parker, 1911), *Aplysia californica* (Parker, 1917b), *Tethys dactylomela* (Olmsted, 1917a).
    (b) Ditaxic
        (i) Opposite: *Nerita tessellata* (Parker, 1911).
        (ii) Alternate: *Nerita nodosa* (Parker, 1911), *Turbo* (Crozier, 1919), *Tectarius misricatus* (Olmsted, 1917a).
    (c) Tetrataxic   *Tritonidea tincta, Columbella mercatoria* (Olmsted, 1917a).

III. Diagonal
    (a) Monotaxic
    (b) Ditaxic

IV. Lateral   *Cypraea exanthina* (Olmsted, 1917a).
    (a) Monotaxic
    (b) Ditaxic

V. Composite

waves may extend across the whole width of the foot or they may form two or four parallel systems. These he termed monotaxic, ditaxic and tetrataxic waves, respectively. Later, Parker (1911) observed that ditaxic waves may be in phase on opposite sides of the foot (opposite) or out of phase (alternate). The latter type of locomotion can be particularly clearly observed in *Haliotis*

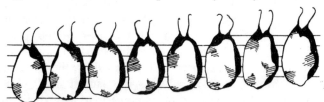

FIG. 36. Successive stages in the passage of muscular locomotory waves over the foot of *Haliotis*. Shaded areas represent regions of the foot in which the underlying longitudinal muscles are contracted. (From Lissmann, 1945a.)

(Fig. 36) since there are never more than one complete, and two partial waves on the sole of the foot at the same time (Lissmann, 1945a). The result is a type of stepping motion comparable to that which occurs also

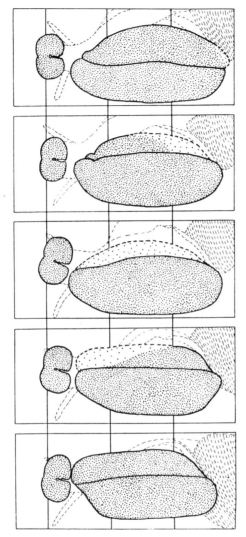

Fig. 37. Successive stages in the locomotory cycle of *Pomatias*. (After Lissmann, 1945a.)

in the polyclad *Leptoplana* (Child, 1904; Crozier, 1918), though in that animal, the waves of contraction of the longitudinal muscles appear to be much less regular and co-ordinated than those of *Haliotis*.

What is probably an exaggerated type of alternate ditaxic locomotion has been described by Lissmann (1945a) in *Pomatias elegans* (Fig. 37). One side of the foot becomes detached from the ground, peeling off from the posterior end. While this is in progress, the attached half of the foot dilates laterally and medially. The detached half is then swung forwards and

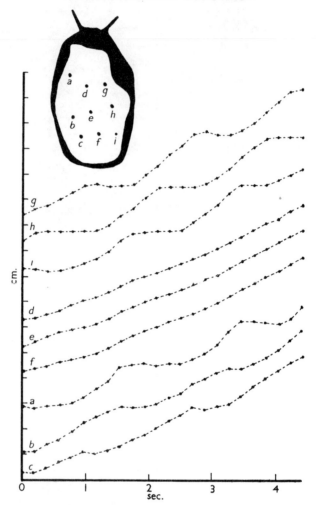

Fig. 38. Rates of progress of different points on the foot of *Pomatias*. (From Lissmann, 1945a.)

reaffixed to the ground, after which the cycle is repeated by the other half of the foot. This remarkable method of creeping, which is nothing less than bipedal locomotion, is aided to some extent by the proboscis which is stretched forwards and fixed to the ground as the locomotory wave of the foot begins and is then retracted, giving some forward traction while the locomotory cycle is in progress. It will be observed that in *Pomatias*, as in *Haliotis*, points on the sole of the foot are in motion only while the underlying longitudinal muscles are contracted; for the remainder of the cycle they are stationary. The mid-line of the foot, however, is in constant steady motion (Fig. 38).

GASTROPOD LOCOMOTION 59

Some extremely complicated types of locomotion have been discovered by Olmsted (1917a) in *Cypraea exanthina*. In this streptoneuran, locomotory waves may start at the posterior end of the foot as in normal retrograde

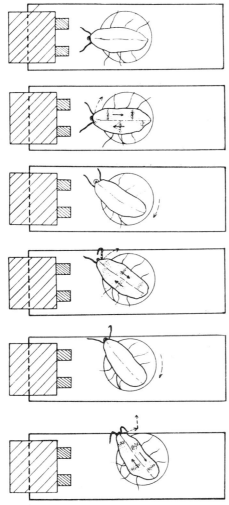

FIG. 39. *Gibbula* turning by rotating the visceral mass upon the foot and exhibiting muscular pedal waves which pass in opposite directions in the two halves of the foot. (From Gersch, 1934.)

locomotion, but then swing round and run laterally, generally from left to right, but sometimes in the opposite direction, as the animal is carried forwards. It is clear that when the pedal waves run from side to side of the foot, the longitudinal muscle fibres, instead of contracting in an anterior-posterior sequence, as in retrograde progression, or in the reverse sequence as in direct progression, must all contract simultaneously as the locomotory wave reaches them. The pedal waves may also cross the foot diagonally, from the postero-lateral corner to the antero-lateral. Further complication is introduced when several different sets of locomotory waves cross the foot at the same time. Composite wave patterns may be set up when, for instance, antero-posterior waves appear at the same time as the lateral waves and the two sets cross each other without interference. These waves may run the whole length and width of the foot or they may be confined to different regions of it, and often different wave patterns can be seen in the anterior and posterior regions of the foot. *Cypraea* rarely moves directly forwards for

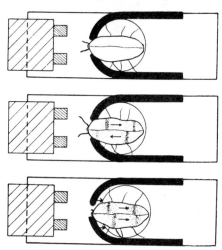

FIG. 40. Reversed pedal locomotory waves on the foot of *Gibbula* when the animal is forced to crawl backwards. (From Gersch, 1934.)

an appreciable distance, instead it tends to move in circles to left or right, changing direction frequently. When it is turning rapidly, the locomotory waves in the anterior third of the foot move laterally in the same direction as the direction of movement of the head, while the waves of the posterior third of the foot move in the opposite sense.

Different locomotory behaviour of opposite sides of the foot has also been observed in *Gibbula* (Gersch, 1934). Forward progression is by direct monotaxic waves, but when the animal is turning rapidly, locomotory waves on opposite sides of the foot move in opposite directions (Fig. 39). These movements, alternating with rotations of the visceral mass on the foot by contraction of the columella muscle, result in the animal being able to turn almost in its own length.

Gibbula is also one of the few gastropods that can crawl backwards. If it reaches an impasse where, even with its special powers of rotation, it is unable to turn round, the locomotory waves on both halves of the foot are reversed and the animal moves backwards until it has sufficient space in which to turn (Fig. 40). Similar behaviour can be observed in *Helcion pellucidum*. Both species have only very limited capabilities in this respect and distances of the order of 5 cm appear to be as far as they can crawl in reverse (Gersch, 1934).

Ciliary creeping, directly comparable to that of planarians, is found in only a few gastropods (Table 4). Surprisingly, the incidence of this habit

TABLE 4
Ciliary Locomotion in Gastropods

Species	Rate of Locomotion mm/sec	Authority
Aspidobranchiata		
Marginella arena	4·2	Olmsted, 1917a.
Taenioglossa		
Polinices draconis	1·9	Copeland, 1922.
Stenoglossa		
Nassarius obsoleta[1]	1·7–2·1	Copeland, 1919.
Nassarius trivittata[2]	2·8–3·8	Copeland, 1919.
Tectibranchiata		
Bulla occidentalis	0·4	Olmsted, 1917a.
Haminoea antillarum	0·4	Olmsted, 1917a.
Nudibranchiata		
Melibe leonina	—	Agersborg, 1919, 1921, 1923.
Pulmonata		
Lymnaea palustris	—	Walter, 1906.

[1] Described under the name *Ilyanassa*.
[2] Described under the name *Alectrion*.

does not bear any relation to the systematic position of the gastropods, nor to the size of the species, though within a species, large specimens may tend to rely more upon muscular than ciliary activity for locomotion.

This is so in *Polinices*, a gastropod with an enormously inflatable foot (Brown and Turner, 1962), which may reach a considerable size. Small specimens move chiefly by cilia, but large specimens rely almost entirely upon muscular activity for crawling. The cilia beat continuously even in those animals that are moving by muscular pedal waves, and so the transition from ciliary to muscular locomotion may be gradual. Muscular locomotion is complicated in *Polinices* by the use of the propodium, a tongue-like anterior organ, to supplement the forces generated by the pedal waves. In large specimens, a wave of contraction of the longitudinal muscles of the foot begins at the posterior end and moves forwards. As it reaches the anterior tip of the foot, the propodium is stretched forwards and attached to the ground ahead of the animal. The retractor muscles of the propodium then contract and draw the animal forwards as a new wave of contraction starts at the posterior end of the foot (Copeland, 1922).

Large specimens of *Polinices* can attain a speed of 5·9 mm/sec by this means, nearly four times the maximum rate when these gastropods move by ciliary action. All gastropods that rely solely upon cilia achieve modest speeds comparable to those of planarians, but even this is a surprising achievement when it is recalled that the locomotory surface of gastropods must transport a much greater load, including the visceral mass and shell, than that of a planarian. Undoubtedly, this is an important contributory factor to the almost universal use of muscular forces for locomotion in gastropods. Even so, ciliary locomotion may be commoner in gastropods than we suspect, for it is not always easily detected. Parker (1911) was unable to detect cilia beating on the foot of *Nassarius* (*Ilyanassa*) *obsoleta*, nor could he see signs of rhythmical contractions of the foot muscles. He concluded, *faut de mieux*, that locomotion must be by uncoordinated contraction of the longitudinal muscles and called it 'arhythmic' muscular locomotion. Crozier (1919) reported a similar type of locomotion in *Conus agassizi* from Bermuda. But on re-examining *Nassarius*, Copeland (1919) discovered that in fact it does move by ciliary action, though since the cilia do not beat regularly unless the foot is in contact with the substratum, it is easy to overlook this fact. This discovery raises some doubts about the existence of 'arhythmic' muscular locomotion. *Conus* has not been re-examined, so that whether it moves by ciliary or muscular forces is not certainly known.

The forward thrust available for transporting the animal, generated by the locomotory waves of the foot, cannot be greater than the thrust exerted against the substratum. If slipping occurs between the foot and the substratum, the forward thrust is obviously decreased. It follows that some parts of the foot must be stationary with respect to the ground while a thrust is exerted against it, and these parts of the foot, or, indeed, comparable parts of any organism, are known as fixed points or *points d'appui*. Lissmann (1945b) pointed out that in all animals that creep by muscular locomotory waves, each part of the body should undergo minimum deformation while it serves as a *point d'appui*, and if the changes of shape are smooth and sinusoidal, the best positions are at the turning points of the sine curve, when the longitudinal muscles are fully extended or fully contracted. As we have already seen, both alternatives may be found in gastropods.

Adhesion to the ground is obviously of paramount importance, for without it the locomotory waves do not result in forward progression of the animal, but merely slip over the substratum. The temporary fixation of the *points d'appui* may be by the secretion of adhesive mucus or by suction, though the relative importance of these two methods is not clear. Pulmonates and a number of other gastropods secrete copious quantities of mucus on which they move and which is left behind as a mucus trail. Those parts of the foot of *Helix* where the underlying longitudinal muscles are relaxed, the *points d'appui*, adhere by this mucus (Parker, 1911). This can be seen particularly clearly when the snails move in such a way that the foot is thrown into a series of standing waves and a dotted trail is left behind. Detaching a section of the foot that is fixed one moment, moving the next, may be aided by the extrusion of mucus from the scattered pedal glands; possibly the contraction of the longitudinal muscles immediately in front of the fixed point is responsible for squeezing the mucus from the glands (Lissmann, 1945b).

Not all gastropods move on a mucus trail however, and in those that do not, adhesion is produced by suction. A number of gastropods, including *Patella, Acmaea, Haliotis, Diodora, Crepidula*, nudibranchs and chitons, cling to the rocks limpet-fashion; the edges of the foot remain firmly in contact with the ground while the centre of the foot is drawn up to form a sucker. Of course it is impossible for the animal to move while the foot is behaving in this manner, but chitons and probably others are able to adhere with a third or less of the foot in contact with the ground and any portion of the foot can serve as an anchor (Crozier, 1919). When they are moving, however, chitons adhere by 80-90 per cent of the foot, that is the entire surface except in the region of the locomotory waves (Parker, 1914). In *Aplysia* particles of sand and gravel stick to the *points d'appui*, apparently by local suction caused by the contraction of the oblique muscles (Parker, 1917b). Since the smallest adherent particles are not more than 2 mm in diameter, the areas of the foot that can be independently raised to form a sucker must be quite small (Fig. 57, p. 96). It is also possible that some gastropods adhere by suction produced by the concavities of the locomotory waves themselves, that is, in the regions between successive *points d'appui*. Olmsted (1917a) fixed the open end of a manometer to a small hole pierced in a glass plate over which a variety of gastropods were made to crawl. He observed that the pressure under the foot was reduced as the locomotory waves passed over the mouth of the manometer, and concluded that this provided for the attachment of the foot to the substratum.

Adhesion by one method does not necessarily preclude the use of another. It is also possible for different parts of the animal to adhere by different methods. The anterior edge of the propodium of *Polinices* adheres principally by suction, but the entire surface of the foot is covered with mucus which also helps stick the animal to the substratum, and small specimens of *Polinices* are able to move upside down on the surface film of water when the propodium is useless, and the animal then creeps by ciliary action on a mucus track (Copeland, 1922).

Another matter which remains to be verified is the means by which the longitudinal muscles are extended. Early investigations of the mechanism of locomotion in *Limax* and *Helix* by Simroth (1879) were invalidated by his supposition that since contractions of oblique and transverse muscles cannot cause the elongation of the foot, the longitudinal muscles must be capable of active extension. This misconception was disposed of by Carlson (1905); and, at about the same time, Jordan (1901, 1905) and Biedemann (1905) provided the basis for a correct explanation in terms of the fluid skeleton. Jordan found that vesicles containing body-fluid under pressure protruded from the foot of *Aplysia* when isolated pieces of it were stimulated and the muscles contracted. The vesicles disappeared and the foot elongated when the stimulation ceased. However, if the vesicles were punctured, the foot did not extend when the stimulation was removed. He concluded that contraction of the transverse and oblique muscles at the posterior end of the foot increased the fluid pressure and so brought about the extension of the anterior end. While some gastropods, including *Aplysia*, certainly use a true fluid skeleton in this way during particular types of locomotion which will be discussed in the following chapter, this is not generally true of gastropods. As we have already seen of turbellarians, it is not necessary for a true fluid skeleton to exist for the locomotory movements exhibited by the majority of gastropods to occur, and, furthermore, no gross changes of shape are involved in these movements; the locomotory waves are usually confined to the sole of the foot and do not involve the rest of the body at all. Nevertheless, although longitudinal muscles in the middle of the foot may be extended by the longitudinal muscles immediately in front of them, the anterior tip of the foot must be extended by fluid pressure. The source of the increased fluid pressure which results in the extension of the anterior part of the foot has not been investigated. Undoubtedly it varies in different gastropods, but whatever specializations and refinements may have been evolved, the least complicated situation is one in which the longitudinal muscles alone undergo cycles of contraction and extension. The contraction of a part of the longitudinal musculature inevitably increases the hydrostatic pressure in the tissues of the foot, and automatically brings about the extensions of any muscles that are not in a state of tonic contraction. By such means the longitudinal muscles at the anterior tip of the foot can be extended. It is difficult to conceive of any alternative mechanism operating in turbellarians and there is every likelihood that it is employed by a number of gastropods.

PEDAL LOCOMOTION IN ANEMONES

Some actinians are able to creep over the substratum in essentially the same manner as polyclad turbellarians and gastropods, though the movements of anemones are extremely slow. Generally there are not more than two locomotory waves on the foot at the same time, and more often only one

(McLendon, 1906; Parker, 1917a) and these move slowly over the foot in the direction of movement of the animal (Table 5), that is, the waves are invariably direct. All the species that have been studied appear to be able to crawl in any direction with respect to the axis of the mouth, and in most, the foot becomes elongated in the direction of locomotion.

TABLE 5
Pedal Locomotion in Anemones
(from Parker, 1917a)

Species	Diameter of pedal disk	Frequency of movement	Rate of movement of waves across pedal disk (mm/sec)	Rate of locomotion (mm/min)
Actinia bermudensis	30 mm	Very frequently	0·15–0·17	1·1–1·5
Metridium marginatum	30 mm	Rarely	0·5	—
Sagartia luciae	4 mm	Frequently	0·027–0·067	0·4
Condylactis passiflora	130 mm	Frequently	0·54–0·72	1·25–5·6

Although the mechanics of locomotion in these anemones has not been studied experimentally, there is no suggestion in descriptions of it that the mouth remains closed throughout the period of activity. If this is so, the coelenteric cavity cannot function as a fluid skeleton, although it can do so under certain circumstances, and it follows that the restoring forces for the pedal musculature must be generated within the pedal disk itself. In this respect, therefore, actinian creeping is comparable to that of turbellarians and of some gastropods.

LOCOMOTION OF SOME BRYOZOANS

Most phylactolaematous Ectoprocta (Bryozoa) are sessile animals that remain attached to such substrates as the leaves of aquatic plants throughout their adult life. But a few, including *Lophopus*, *Lophopodella* and *Pectinatella*, are able to creep over the substratum as young colonies, becoming more sedentary as they grow older and larger. *Cristatella* has superior powers of locomotion and even old colonies are able to creep on the substratum. Unlike other phylactolaemes, colonies of *Cristatella* are provided with a muscular pedal disk reminiscent of the gastropod foot (Allman, 1856), which secretes copious amounts of mucus (Brien, 1960) and is more muscular than other parts of the body wall (Hyman, 1959).

The speeds that these animals are able to achieve are modest and are reported to range between a few millimeters and a few inches per day (Table 6). The method of locomotion is uncertain and has never been analysed in detail. Young colonies of *Pectinatella* have been observed to creep on a secreted mucus strand, and Allman's (1856) comparison of the base of a *Cristatella* colony with a gastropod foot, suggests immediately that creeping is accomplished by pedal locomotory waves (Hyman, 1959). Certainly, other suggestions that the colony is pulled along by contraction of the lophophore retractor muscles (Verworn, 1888; Wilcox, 1906), or by the combined effect of ciliary currents produced by the lophophores (Wesenberg-Lund, 1896; Brooks, 1929) are mechanically improbable, to say the least,

TABLE 6
Rates of Locomotion of Phylactolaematous Bryozoa

Species	Rate of Locomotion	Authority
Cristatella	1–15 mm per day	Marcus, 1926b.
	13 mm per 8¼ hr	Harmer, 1896.
	25 mm per day	Odell, 1899.
	Several inches per day	Allman, 1856.
Lophopodella carteri (young colony)	80–100 mm per day	Dahlgren, 1934.
Pectinatella (young colony)	8 mm per day	Brooks, 1929.
Lophopus (young colony)	6–8 mm per day	Harmer, 1896.

and have been disproved by the observations of Marcus (1926b) who showed that such colonies with only one polypide extended, and that laterally, creep as well in an anterior as a posterior direction.

Although coelomic cavities are contiguous with the ventral surface of these animals, it is unlikely that they are of significance in locomotion. Rather, the situation in phylactolaematous bryozoans appears to be remarkably similar to that in actinians and gastropods. Hyman's (1959) suggestion that the circular as well as the longitudinal muscles of the basal disk are employed in producing locomotory forces is, of course, an unnecessary assumption, although until the mechanics of creeping in these remarkable colonies has been investigated in detail, the precise origin of locomotory forces must remain problematical.

THE PARENCHYMA OF TURBELLARIANS AND NEMERTEANS

By far the commonest method of locomotion that involves the musculature in the animals we have considered is the use of pedal locomotory waves, and this, as we have seen, makes minimal demands upon the hydrostatic skeleton. For the most part, particularly in worms, the fluid skeleton is implicated only in slow changes of shape or in the maintenance of body turgor. It is not surprising, therefore, that the space immediately adjacent to the locomotory or body-wall muscles should be more or less occluded by connective tissue or parenchymatous tissue and not be filled with fluid. However, the looping movements practised by leeches and some triclads, and the peristaltic locomotion of nemerteans involve the entire body-wall musculature and consequently cause much greater and more rapid deformation of the hydrostatic skeleton than do pedal locomotory waves.

The parenchyma of turbellarians, which constitutes the hydrostatic skeleton in these animals, may vary in its constitution, though to what extent is uncertain because of our poor understanding of its nature. For many years it was thought to consist of a syncytial network with numerous fluid-filled interstices and free, wandering cells (Hallez, 1879; Prenant, 1922; Hyman, 1951a), but doubt has now been cast on this interpretation and in at least some turbellarians the parenchyma is not syncytial, but consists of several types of gland cell, muscle cells, and rounded, undifferentiated cells (Pedersen, 1959a, b; Skaer, 1961). The undifferentiated cells appear to correspond to the wandering cells of previous authors and to the neoblasts which play an important part in regeneration (Stéphan-Dubois, 1951; Brøndsted, 1955).

The space between the body wall and the gut of leeches is very largely occluded by botryoidal tissue, although this is a secondary condition and is not as complete as in turbellarians because coelomic sinuses ramify throughout the leech body (Gratiolet, 1862). In the looping movements which are occasionally practised by triclads and are regularly used by leeches, the ends of the body form suckers or adhesive organs and are alternately attached to the substratum while the circular and longitudinal muscles undergo cycles of contraction. The chief feature of this type of locomotion, and the one in which it differs from others involving the body-wall musculature, is that the whole body behaves as a single unit. When a wave of contraction passes along the circular muscles, the entire circular musculature becomes contracted before longitudinal muscles begin to contract. As a result, the stresses in the hydrostatic skeleton act in the same sense throughout the whole length of the animal at any stage in the cycle. The stresses must be much smaller than those in the hydrostatic skeleton of nemerteans during peristaltic locomotion. In this type of movement, the contracting longitudinal muscles exert thrusts in one direction and in one part of the body, while the contracting circular muscles exert thrusts in another direction in neighbouring regions of the worm. Even so, looping makes some dynamic use of the hydrostatic skeleton, and it is significant that in the leeches, in which this locomotory technique is regularly employed, fluid-filled coelomic sinuses should exist (Fig. 41). The movement of the fluid in them must permit the pressure changes caused by the contraction of some of the muscles to be more readily transmitted to all other body-wall muscles. The triclads, which employ this means of progression infrequently, lack such a communicating system of fluid-filled spaces.

Nemertean peristaltic movements entail considerable deformation of the hydrostatic skeleton and foreshadow the locomotory techniques of coelomate worms that possess a fluid-filled body cavity. While the nemertean parenchyma can never form as plastic a hydrostatic skeleton as the coelomic contents of higher worms, the extent of the parenchyma is often considerably reduced, and in many, is very largely gelatinous and contains few cells (Hubrecht,

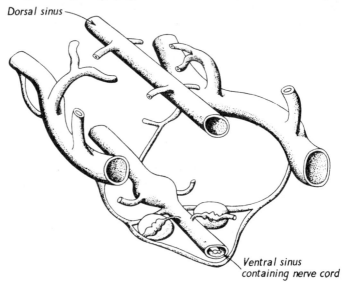

FIG. 41. Reconstruction of the coelomic canal system of a single segment of the leech *Hirudo*. (From Mann, 1962.)

1883, 1887; Montgomery 1897; Prenant, 1922). Although no studies have been made of its mechanical properties, it is likely that a virtually structureless material of this sort would offer much less resistance to deformation than a tissue layer such as occurs in turbellarians. Nemerteans may approach the coelomates in their possession of a plastic and deformable hydrostatic skeleton as well as in their use of peristaltic movements in locomotion, but even so, muscular locomotion of this sort is poorly developed and remains subsidiary to ciliary creeping in all but the largest species (Eggers, 1924; Friedrich, 1933; Coe, 1943).

CONCLUSIONS

The muscular body wall of acoelomate worms permits, and is responsible for reversible changes of shape during which the mesenchyme tissue occupying the space between the body wall and the gut serves as a hydrostatic skeleton,

CONCLUSIONS

and the longitudinal muscles antagonize the circular muscles. The changes of shape that these animals undergo are considerable and are limited only by the extent to which the body wall, or more particularly, the sub-epidermal basement membrane, can be deformed. Despite the versatility of these worms, their remarkable ability to change shape is not exploited to any appreciable extent. The changes of shape are rarely rapid enough or sufficiently powerful to perform external work.

Creeping by ciliary action is undoubtedly the fundamental and primitive method of locomotion in turbellarians and nemerteans. The smallest turbellarians are also able to swim by this means. However, the forces generated by beating cilia are small in absolute terms and are inadequate to provide a means of locomotion in large animals. Consequently, we find that a great many turbellarians and nemerteans employ muscular contractions in both swimming and creeping. The great variety of locomotory devices in these animals can be resolved into three main types. Crawling over the substratum is accomplished by pedal locomotory waves in triclad and polyclad turbellarians, involves peristaltic movements in nemerteans, and may be carried out by leech-like movements in a few members of both groups. Swimming by undulatory movements, and burrowing, further types of muscular locomotion, will be considered in later chapters.

Creeping by means of pedal locomotory waves foreshadows a locomotory technique common in gastropods and chitons and practised also by a few anemones and ectoproct bryozoans. A distinguishing feature of this type of movement is that waves of muscular contraction are confined to the longitudinal muscles of the surface of the body wall that is in contact with the ground. Particularly in some of the more specialized types of locomotion in gastropods, other muscles may be involved, but in most, the antagonist of the longitudinal muscles is the longitudinal musculature itself. The remaining body-wall musculature at most maintains a turgor pressure so that on relaxation the longitudinal muscles are extended. That the greater part of the body-wall musculature can play only a static role during locomotion is indicated by the fact that the locomotory waves are confined to the ventral or pedal surface of the animals. Although the ultimate restoring forces must be transmitted through the fluid skeleton, they have to be transmitted only over short distances and no large displacements of the body contents are involved. The demands made on the fluid skeleton are slight and it is clear that this type of locomotion is ideally suited to animals with a solid body in which the hydrostatic skeleton is the least plastic and offers the greatest resistance to deformation, and is therefore the least efficient, of any we shall consider.

Other types of muscular locomotion make greater demands upon the fluid skeleton than do pedal locomotory waves, and they are less often employed by solid-bodied animals. Looping movements, characteristic of leeches and practised occasionally by some triclads, are less exacting than the peristaltic locomotion employed by some nemerteans. The turbellarian parenchyma is evidently sufficiently plastic for the transmission of fluid pressures from one end of the body to another, but it is significant that leeches, which creep by looping much more frequently than triclads, possess a series of coelomic sinuses that must facilitate the transmission of fluid pressures around the body. The peristalsis of nemerteans is an ill-coordinated and relatively inefficient method of locomotion which is perfected only in worms that possess a true fluid skeleton, and nemerteans rely chiefly upon ciliary activity for crawling. Nevertheless, peristalsis must be regarded as a very advanced type of movement in animals of the nemertean grade of construction, and is possible only because the parenchyma tends to contain few cells and to be largely gelatinous, and hence to offer less resistance to deformation than a tissue layer would.

We may therefore conclude that the parenchyma of turbellarians and nemerteans serves as a fluid skeleton only in a very limited sense, and that the muscular activities used in locomotion either make virtually no demands upon the hydrostatic skeleton, as in pedal locomotion, or are rarely and relatively ineffectively performed. This conclusion can be confirmed by considering other animals that employ similar locomotory techniques. It seems unlikely that the chitons, gastropods, anemones and bryozoans that move by pedal locomotory waves, make use of a hydrostatic skeleton in these activities. Leech-like looping makes more, but still relatively slight demands upon the hydrostatic skeleton, and leeches, which practise this method of creeping far more frequently than planarians, possess coelomic sinuses which must serve some of the functions of a fluid skeleton. Peristaltic creeping does demand the use of a hydrostatic skeleton and the nemerteans which employ this method of locomotion have a parenchyma that is largely gelatinous and therefore approximates to a true fluid skeleton. Peristalsis is perfected in coelomate animals, and these will be considered in the next chapter.

3
THE COELOMATE CONDITION

PRECISELY the same changes of shape, produced by contractions of the body-wall musculature, as those observed in turbellarians and nemerteans occur in worms with a secondary body cavity, though all these worms possess a cuticle which restricts changes of shape much more severely than the basement membrane does in the acoelomate worms. The most important difference between coelomate worms and those we have previously considered is that they possess a true fluid skeleton instead of a loose parenchyma; there is no barrier to the movement of fluid and changes of fluid pressure caused by contraction of part of the musculature are immediately transmitted without damping to all parts of the body wall. The coelomate condition represents a logical development from the acoelomate, in which a tissue has perforce to serve a function for which it is not ideally suited. The possession of a fluid-filled secondary body cavity, whatever its morphological nature, allows the full realization of the potentialities inherent in the muscular system of the body wall and, as we shall show, permits the worm to perform a substantial amount of external work by means of its changes of shape. The relative size of the body cavity is unimportant so long as it is continuous and is, at least mechanically, apposed to all parts of the body wall.

CONSTANCY OF VOLUME

One of the properties demanded of a fluid skeleton is that it should be of constant volume during the performance of changes of shape. This presents no difficulty in a solid-bodied animal, but to an animal with coelomic fluid or its equivalent, and open nephridia or coelomoducts communicating with the exterior, there is a danger that fluid will be lost when muscles contract and the internal pressure rises.

The danger of loss of fluid through the segmental organs is generally avoided by the provision of sphincter muscles at the nephridiopores or coelomoduct pores. In *Arenicola*, the nephridiopores are slit-shaped, and may be opened or closed by dilator and constrictor muscles (Chapman and Newell, 1947). The pore opens to release urine only for a second or two at a time and at intervals of a minute or more. When the internal pressure is artificially raised, the dilation of the pores is lessened and is briefer, apparently because the constrictor muscles are a specialized part of the circular body-wall musculature and exert greater tension when the worm is distended. However, the frequency of opening is increased if the internal pressure is raised and on balance the worm does lose coelomic fluid at a slightly higher rate than when the internal pressure is low (Strunk, 1930; Chapman and Newell, 1947). Chapman and Newell suggest that the worms have no difficulty in making good this slight loss by uptake of sea-water. The important thing is that there is no uncontrolled loss of coelomic fluid through the nephridiopores at working pressures, i.e. up to 50 cm sea-water, nor when the internal pressure is artificially raised to as much as 150 cm sea-water. The same is true of the earthworm *Lumbricus terrestris* which has both nephridia and dorsal pores

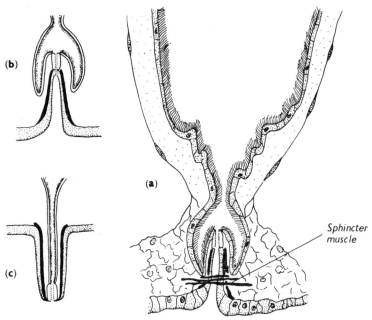

Fig. 42. Nephridiopore of *Sabella*. (a) The terminal part of the nephridial funnel and nephridiopore closed by a sphincter muscle, (b) and (c) the terminal portion inverted and everted.

in every segment leading from the coelom to the exterior (Newell, 1950). The dorsal pores are closed by a simple sphincter and after the muscles have been paralysed with a narcotic, coelomic fluid spurts from them at internal pressures of only 20 cm sea-water, although in an active worm, the working pressure may reach 35 cm sea-water. The unnarcotized worm can withstand pressures of up to 150 cm sea-water without uncontrolled leakage of coelomic fluid from the nephridiopores or the dorsal pores.

Experimental investigations have not been made of other worms, but it is likely that comparable mechanisms prevent uncontrolled leakage of coelomic fluid in most of them. Polychaetes with short, wide mixonephridia, such as *Sabella*, generally possess a sphincter muscle at the nephridiopore (Fig. 42), but species such as *Nereis* and *Lumbrinereis* appear to rely upon the great length of the very narrow nephridial duct to restrict the flow of coelomic fluid through them, since the nephridiopores of these worms are not provided with sphincter muscles. Worms such as *Nephtys*, with protonephridia and no coelomoducts, lack nephridiopore sphincters, since there is no direct communication between the coelom and the exterior.

Most coelomate worms are able to avoid the consequences of loss of coelomic fluid after amputation of part of the body by the contraction of the circular body-wall muscles tightly about the gut at the wound (Clark and Clark, 1962). Chapman and Newell (1947) found that after half the body of *Arenicola* was amputated, there was very little loss of coelomic fluid from the anterior part of the worm, and burrowing efficiency was not materially impaired. This is not always the case, however, particularly among short, rather globular coelomate worms. If part of the body of the polychaetes *Lipobranchius* and *Sternaspis* is amputated, for example, the worms cannot retain coelomic fluid in the anterior half and they die. *Priapulus* is also reported to die if it is transected anterior to the sphincter muscle that separates the caudal processes from the rest of the body (Hyman, 1951b).

The effect of loss of coelomic fluid can be easily understood in theory and can be seen in practice in the functioning of sea-anemones. In these animals, the coelenteric contents act as a fluid skeleton and, since the coelenteron communicates with the exterior by way of the mouth, the muscular-hydrostatic system can be an open or closed one. In *Calliactis* and *Metridium*, and probably other actinians, there is little, if any, uncontrolled loss of water through the mouth, and a constant volume is maintained during changes of shape (Chapman, 1949; Batham and Pantin, 1950), but over a long period the volume of the animal varies greatly (Fig. 43), so that the muscles must function at many different extensions. The muscle fibres of *Metridium* are, in fact, extremely extensible and may stretch from 200–300 μ to 1–2 mm in length. The limit of extension of the column is reached when the circular muscles form a single, unfolded layer of muscle fibres. As the column is contracted, the circular muscle layer buckles and at very great contraction second order buckling of the whole body wall occurs, and this ultimately sets a limit to the shortening of the column (Batham and Pantin, 1950). Buckling of the muscle layers is permitted by the peculiar viscous-elastic properties of the mesogloea, which may also set some limit to changes of shape, especially in anemones like *Calliactis* in which it is thick (Chapman, 1953a).

Within these limits, if the volume of the anemone is halved, it is still capable of the same changes of length as before, but only by a much greater change in diameter, and it can accomplish the same changes of diameter only by much greater changes of length. Inevitably, both a reduction in volume and excessive distension limit the total range of changes of shape available to the animal. Furthermore, each muscle has its optimal working length and a reduction or increase of volume entails some, or all of the muscles working outside this range. Batham and Pantin (1950) showed that if *Metridium* is inflated to two or three times its normal volume, the muscles tend to act isometrically against the internal pressure and produce very little change of shape, so that the anemone appears sluggish and immobile. The loss of

FIG. 43. Changes of shape of *Metridium senile*. All drawings are of the same individual on different occasions, and all are to the same scale. (After Batham and Pantin, 1950.)

mechanical efficiency of the muscles resulting from a decrease in the volume of the hydrostatic skeleton has been demonstrated by Chapman and Newell (1947) in *Arenicola*: the time taken for the worm to burrow into sand is increased two- or three-fold when a few tenths ml of coelomic fluid are withdrawn by hypodermic syringe.

Coelenterates are almost unique in their ability to vary their body volume over such a wide range. Actinians may lose water through the mouth and they can recover their original volume by pumping themselves up by means of the siphonoglyphal currents, and, accordingly, are capable of an enormous variety of shapes and sizes. Batham and Pantin (1950) have demonstrated that the body wall of *Metridium* is capable of extension by about 400 per cent, although during normal changes of shape, muscle contractions are usually of the order of 30 per cent of the body length. The muscles are therefore capable of working over a considerably greater range of lengths than those of most other animals. The only worm-like animals that may also be subject to substantial changes of body volume are holothurians. Pantin and Sawaya (1953) suggested that there may be some exchange of water between the respiratory trees and the perivisceral cavity of *Holothuria grisea* and that this accounts for prolonged changes of volume in this animal.

The functioning of anemones illustrates a further point of some importance. Theoretically, the gut can serve as a fluid skeleton in any animal, providing there is insufficient tissue between it and the body-wall musculature to act as a shock absorber and damp pressure changes generated by contractions of the muscles. However, this carries with it two disadvantages. First, unless the mouth and anus are tightly closed during changes of shape, there is a likelihood of loss of volume (and also of gut contents). Second, the use of the gastric cavity as a fluid skeleton prevents morphological specialization of the gut. The fluid forming the skeleton must move freely within the animal as parts of the body-wall musculature contract. If the gut contents serve this function, the orderly progression of food from mouth to anus is precluded and a sequential secretion of digestive enzymes is impossible. Coelenterates are the only animals in which the gut does serve this function and, in them, digestion is for the most part intracellular and there is no anus. We may therefore conclude that a body cavity entirely separate from the gut is essential if the body-wall muscles are to act antagonistically about a fluid skeleton and the gut is to be freed from the influence of changes of shape of the body wall.

THE CUTICLE

All coelomate worms possess a cuticle, and the fact that planarians and nemerteans have no such protective covering of the epidermis is instrumental in permitting them to undergo very large changes of shape. But in the higher worms, the cuticle imposes an important restraint on change of shape. How significant a factor the cuticle is in this respect, it is difficult to estimate, but certainly, none of the higher worms has anything like the extensibility of sea-anemones or nemerteans. An interesting and extremely specialized extension of the function of a cuticle is found in nematodes. In these worms, the restraint imposed by the cuticle is actually instrumental in producing the changes of shape that these worms show.

So far as chemical analyses of cuticles have gone, it appears that most are composed largely of collagens or chitins. Two classes of chitin have been reported. That of arthropod cuticle is α-chitin and a similar substance has been found in the coenosarc of hydrozoans, the pneumatophore of siphonophores, and the coenosteum of millepores. The secretion of α-chitin appears to be incompatible with the synthesis of collagen, and animals with a cuticle of α-chitin have little collagen in the layer immediately beneath the epidermis (Rudall, 1955). Onychophorans have a very thick collagen 'dermis' (basement membrane) but the cuticle, composed of α-chitin, is extremely thin (Lotmar and Picken, 1950; Rudall, 1955). The cuticle of the annelids *Aphrodite* and *Lumbricus* is predominantly of collagen (Rudall, 1955; Watson, 1958), though that of earthworms differs in important respects from vertebrate collagen (Gustavson, 1957; Watson, 1958). The chaetae, which are of course secreted by cells derived from the epidermis, are of β-chitin (Lotmar and

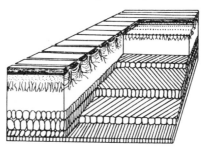

Fig. 44. Stereogram showing the structure of the cuticle of *Ascaris*. (From Bird and Deutsch, 1957.)

Picken, 1950), the secretion of which is not incompatible with the production of collagen, and indeed, the brachiopod, *Lingula*, has a thick cuticle of β-chitin and a thick basement membrane of collagen, both secreted by the same cells (Rudall, 1955).

Except in the nematodes, little is known about the mechanical properties of the cuticle, nor how it permits or restricts changes of shape of the animal. The nematode cuticle is both structurally and chemically complicated. Chitwood and Chitwood (1950) and Bird and Deutsch (1957) distinguish nine layers in it (Fig. 44), conspicuous among which are the three layers of crossed fibres forming a spiral basket-work about the body of the animal (Picken, Pryor and Swann, 1947; Harris and Crofton, 1957), recalling the structure of the basement membrane of nemerteans and turbellarians. The fibres are collagenous (Fauré-Frémiet and Garrault, 1944; Picken, Pryor and Swann, 1947) and are regarded as inextensible; and in the same way as the fibrils of the nemertean basement membrane, they permit anisometric extensions of the cuticle. Collagen, possibly of a different type, also occurs in a fibrillar layer and in the inner layer of the cortex of the cuticle

(Chitwood, 1936). The external cortical layer contains keratin and a fibrous protein, 'matricin'. Glycoproteins and albumins have also been identified in the cuticle, but have not been referred to any of the structural layers (Chitwood, 1936; Bird, 1956). The annelid cuticle is structurally far less complicated than that of nematodes, but it resembles the latter at least insofar as spirally arranged collagen fibres occur in it (Picken, Pryor and Swann, 1947; Reed and Rudall, 1948). Although a change in the inclination of these inextensible fibres permits a change of shape, they obviously cannot do so unless the matrix in which they lie is plastic and deformable. Most chemical and physical analyses have been made of whole cuticle and the properties of the matrix have not been characterized, but Gansen-Semal (1960) has identified a non-fibrillar elastin in the cuticle of the gizzard of *Lumbricus*; curiously, it does not occur in the epidermal cuticle.

The nematode cuticle, although it has striking parallels with the nemertean basement membrane, functions in a highly specialized way (Harris and Crofton, 1957). The nematodes and the nematomorphs are set aside from all other worms by the possession of only longitudinal muscles in the body wall. Obviously reversible changes of shape must be achieved by some method different from that prevailing in other worms. The hydrostatic pressure in the pseudocoel of *Ascaris lumbricoides* is considerably higher than that recorded in the body cavity of any other animal (Table 7, p. 109). As a consequence of this high internal pressure, the worm is distended and the cross-section is always circular. In *Ascaris*, the spiral fibres in the cuticle are inclined at an angle of 75° to the longitudinal axis of the worm and at that inclination, any contraction of the longitudinal muscles tends to increase the angle between the fibres and so to reduce the volume enclosed by the fibre system (Fig. 45). However, the circular cross-section indicates that the volume of the worm and that enclosed by the fibre system are the same, so that contraction of the longitudinal muscles involves an increase in internal hydrostatic pressure rather than a change of volume, and it is this which antagonizes the longitudinal muscles and provides the restoring force. Relaxation of the muscles results in a decrease in the internal pressure and an elongation of the worm to the elastic limit of the cuticle, or towards the maximum capacity of the fibre system where the inclination of the fibres to the longitudinal axis of the worm is 55°. In practice, *Ascaris* never reaches the latter limit, which would demand an increase in length of more than 50 per cent; such extensions appear to be well beyond the capabilities of the animal. The greatest changes of length reported by Harris and Crofton (1957) are of the order of 10–15 per cent. The fibres of *Oxyuris equi* are inclined at an angle of 60° (Bird, 1958), so that a much smaller increase in length would bring it to the limiting position of maximum volume, though whether it ever does so is unknown.

Clearly an animal with a high internal hydrostatic pressure and with a cuticle of the nematode type containing fibres orientated at any angle greater than 55° to the longitudinal axis requires only longitudinal muscles to undergo reversible changes in length. It may also be noted that were the fibres orientated at an angle less than 55°, only circular muscles would be required to produce reversible changes of length. No such animal exists, no doubt because it would be under the serious disadvantage of being unable to flex its body.

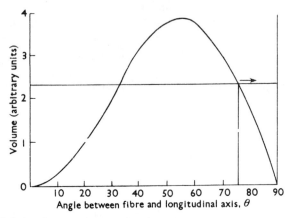

Fig. 45. Relation between the capacity of the fibre system of the nematode cuticle and the inclination of the fibres to the longitudinal axis of the animal. The horizontal line represents the volume of the worm (*Ascaris*). Contraction of the longitudinal muscles increases the inclination of the fibres and causes reduction in volume or increase in turgor pressure.

NEMATODE LOCOMOTION

Despite the apparent simplicity of their musculature, nematodes are capable of a variety of changes of shape. The common view that they are rather stiff, immobile animals (Hyman, 1951b) is incorrect and they undergo a considerable range of body movements which may be employed as means of locomotion.

The most familiar method of locomotion in nematodes is by a flexion of the body in the dorso-ventral plane, produced by waves of contraction passing along the dorsal and ventral parts of the longitudinal muscle coat (Fig. 46). No change of length is involved in this movement and the dorsal and ventral muscle bundles antagonize each other. Precisely the same movements are used by nearly all long, narrow animals when they swim, and the mechanism of this type of locomotion will be discussed later in more detail. Unlike leeches and nemerteans, nematodes are unable to present a flattened surface to the water when swimming, and so are at some disadvantage. However, nematodes are ideally suited to swimming in viscous media (Gray, 1951) and this type of locomotion is also very successful when they are moving between solid objects against which the body can exert a thrust, as, for example, when moving among water-weeds or through the gut contents of a host animal.

Essentially the same movements can be observed in snakes (Fig. 99, p. 181). A serpentine motion is displayed whenever the ground on which the snake crawls is sufficiently rough to provide a number of solid objects, vegetation, pebbles, etc., against which the body may be engaged (Mosauer, 1932a, b; Gray, 1946). Propulsion is effected by forces acting normally to the body at the points of contact with obstacles fixed to the ground. Other types of movement, including crotaline side-winding, the rectilinear crawling exhibited by boas, and concertina movements, are used only when these fixed points

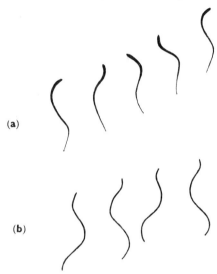

FIG. 46. Swimming movements of nematodes, (a) *Rhabditis marina*, a spindle-shaped worm, (b) *Plectrus palustris*, a filiform species. (From Stauffer, 1924.)

with which the body may make contact are lacking (Gray and Lissmann, 1950; Lissmann, 1950).

As Harris and Crofton (1957) emphasize, the great majority of nematodes are structurally almost identical, they function on the same principles as *Ascaris*, and they move in the same way. Only a small number of aberrant free-living nematodes have adopted alternative methods of locomotion (Stauffer, 1924). *Hoplolaimus*, for example, burrows in a manner comparable to that of earthworms (Fig. 47). The longitudinal muscles show regional contractions that pass backwards along the body, and the body wall undergoes compensatory swelling in the regions where the longitudinal muscles are contracted. Providing the spaces through which the nematode is burrowing are small, the swollen part of the body can be wedged while the more anterior parts are extended; the anterior half of the worm then becomes dilated in turn and serves as an anchor while the posterior region of the body is drawn forwards. *Hoplolaimus* is also capable of crawling through spaces larger than the greatest diameter of the body. Those regions of the body that are undergoing contraction are then thrown into waves, so making contact with the sides of the cavity, against which a back-thrust may be exerted. This is a

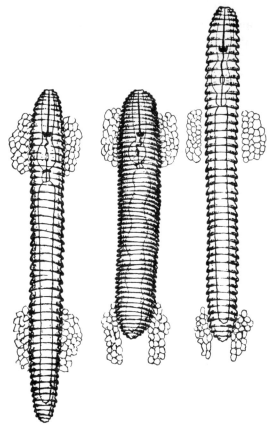

FIG. 47. Locomotion of *Hoplolaimus* through densely packed material. (From Stauffer, 1924.)

compromise between the peristaltic type of locomotion and the sinusoidal movements shown by swimming nematodes. The success of both, and particularly of peristaltic burrowing, depends upon the frictional force between the cuticle and the substratum remaining sufficiently high to prevent slipping. Species that move by peristaltic waves usually have strong annulations, spines, or scales on the cuticle, apparently as an adaptation to this end.

Stauffer (1924) suggested that members of the free-living, marine families Epsilonematidae and Draconematidae, by virtue of the possession of ambulatory or stilt bristles, are capable of moving over solid surfaces, caterpillar-

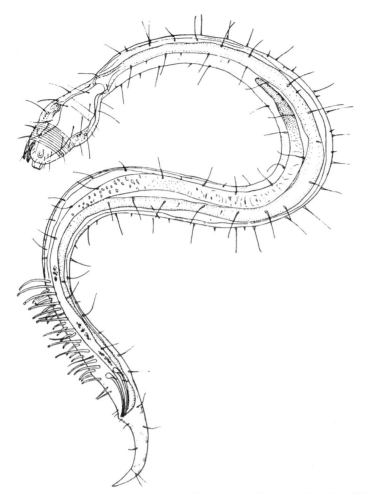

FIG. 48. *Chaetosoma haswelli*, a draconematid nematode. (From Irwin-Smith, 1918.)

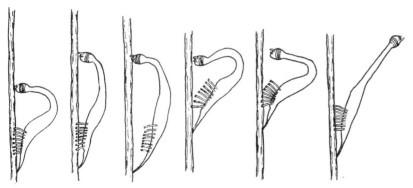

FIG. 49. Stages in the locomotory cycle of a draconematid nematode. (From Stauffer, 1924.)

fashion. Draconematids (Figs. 48 and 49) attach by the stilt bristles, extend the body to its fullest extent and then adhere to the substratum by the adhesive bristles in the cephalic region; the posterior bristles are reattached close to the head. The cycle is then repeated. The Epsilonematidae (Fig. 50) creep in essentially the same manner, though, lacking cephalic adhesive bristles, they move by alternate attachment of caudal adhesive glands and the stilt bristles, which in these worms are in the middle region of the body.

The movements of draconematids and epsilonematids do not involve

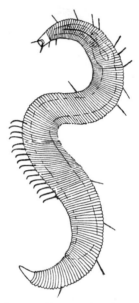

FIG. 50. An epsilonematid nematode. (From Hyman, 1951b.)

very complicated reactions of the musculature. No more is demanded than that the body should be flexed by contraction of the ventral longitudinal muscles and extended when they relax, and that these movements should be co-ordinated with appropriate reactions of the stilt and adhesive bristles, though how these function and how the co-ordination is achieved is unknown. Hoplolaimid locomotion, on the other hand, does involve somewhat different principles from those that operate in other nematodes. Although *Ascaris* is capable of some dilation, Harris and Crofton (1957) found that contraction of part of the longitudinal musculature does not result in an increase in the diameter of the worm at that place, but causes a local decrease in volume which is compensated by elongation of the body elsewhere. In other words, while longitudinal muscle contraction in general is antagonized by the cuticle in conjunction with the high internal hydrostatic pressure of the worm, localized contractions of the longitudinal muscles are antagonized by the rest of the longitudinal musculature. This clearly cannot be so in *Hoplolaimus* if Stauffer's account of its movements is correct, for in this nematode local contraction of the longitudinal muscles does cause a local dilation of the body. This suggests that the properties of either the cuticle or the muscles of *Hoplolaimus* differ from those of the cuticle or muscles of *Ascaris*.

The changes of length of *Ascaris* are trifling compared with the extensibilities of other worms, even including those that possess a cuticle. This is partly because the cuticle of other worms is thinner and more flexible than that of nematodes, but such cuticles are incompatible with high internal hydrostatic pressures. Under these conditions, the cuticle cannot be instrumental in antagonizing the longitudinal muscles, so that an additional layer of body-wall muscles, the circular muscles, is essential for the functioning of the system.

THE FUNCTIONING OF THE BODY-WALL MUSCULATURE IN COELOMATE WORMS

The muscular-hydrostatic system of nematodes and nematomorphs is highly specialized and unlike that of any other animals. When we consider the movements of coelomates, and the remaining pseudocoelomate worms, therefore, we return once more to the mechanical principles that we have already found to apply in acoelomates. Pseudocoelomates and coelomates are functionally equivalent, providing they possess both circular and longitudinal muscles in the body wall, and no distinction need be made for the present between animals with fluid-filled, secondary body cavities of different morphological natures.

If, in an unsegmented coelomate worm, the whole of the longitudinal musculature and most of the circular muscles are contracted, a bulge appears in that part of the body wall where the circular muscles are relaxed. If, now, a wave of relaxation passes along the circular muscles, the bulge travels along the body wall. This peristaltic behaviour of the musculature can be used for a variety of purposes, but it cannot produce forward movement of the animal so long as the longitudinal muscles play a purely static role, for the contraction of one part of the circular musculature, instead of extending the worm and providing a longitudinal thrust that can move the centre of gravity of the animal forwards, merely extends the relaxed parts of the circular musculature. Locomotion is possible only if the longitudinal muscles are employed in a dynamic sense. Indeed, locomotion is possible by the use of longitudinal muscles alone, with the circular muscles playing a static role, as we have already observed in pedal locomotion of gastropods and turbellarians, but the opposite is impossible. Locomotion in coelomate worms thus invariably involves the dynamic participation of both circular and longitudinal muscles to produce peristaltic locomotory movements.

The essential difference between peristaltic and pedal locomotory waves is that extension of part of the longitudinal musculature, with the consequent forward movement of that part of the body relative to the ground, is caused in the former by the combined contraction of antagonistic circular muscles and the longitudinal muscles immediately anterior to it, instead of, as in pedal waves, merely by the contraction of other longitudinal muscles. Thus, both circular and longitudinal muscles exert a forward thrust on each part of the body and, since the hydrostatic skeleton is implicated in peristaltic locomotion, the entire body-wall musculature contributes to the locomotory force, instead of only that on the ventral surface of the animal. The result is that far greater forces are available for locomotion.

The ability to develop peristaltic locomotory waves appears to be a consequence of the existence of a fluid medium as hydrostatic skeleton. We have already observed that in the apparently exceptional nemerteans, which do employ peristaltic waves to supplement the ciliary locomotory forces, the mesenchyme is largely gelatinous and not, as in planarians, a tissue layer.

THE BURROWING HABIT

Another consequence of the possession of a highly deformable fluid skeleton is that it permits the development and use of a proboscis. The basic mechanism of proboscis eversion is simple. Any weak, i.e. thin, portion of the body wall that is in direct communication with the general body cavity, is preferentially distended when the muscles of the body wall contract. This region may be prevented from distending and everting, and is withdrawn after it has been everted, by the contraction of proboscis retractor muscles. Alternatively, as in a number of worms, the proboscis may be retained within the body simply by keeping the mouth closed. Relaxation of the lips, which act as a sphincter, allows the proboscis to be everted by the internal hydrostatic pressure. Whatever the mechanism, the movement and distension of what is usually a relatively large structure necessarily involves a considerable displacement of coelomic fluid and demands far greater mobility of the hydraulic medium than can generally be provided by a tissue. Even so, most turbellarians possess an eversible proboscis despite their lack of a true fluid skeleton. While their proboscis is undoubtedly everted by hydrostatic pressure generated by contraction of the body-wall muscles, the exact mechanism of proboscis eversion in planarians is unfortunately unknown. It seems clear, though, that neither the speed nor the force of proboscis eversion in these worms can be compared with those of coelomate worms.

It is most likely that the eversible proboscis evolved and, as in planarians, still serves primarily as a feeding structure, but the organ has an important secondary function. Most coelomate worms live in burrows and many excavate the burrow with the aid of the proboscis apparatus. When the proboscis is used in this way to perform substantial amounts of external work, its eversion invariably entails the use of a true fluid skeleton.

Even nemerteans use a true hydrostatic skeleton for the eversion of the proboscis (Böhmig, 1929). The rhynchocoel is fluid-filled, and contractions of the circular and longitudinal muscles of its walls cause a sudden rise in the fluid pressure and the eversion of the organ. So effective is the apparatus that a proboscis two or three times the length of the worm can be shot out with great rapidity. The nemerteans use the proboscis for trapping prey and for burrowing, and, indeed, the only activity of these worms involving rapid and powerful muscular action is through the agency of the proboscis apparatus. Comparatively few nemerteans burrow, and, while very small animals like the acoelan *Convoluta* are able to creep through the interstices between the sand grains, cilia are obviously inappropriate for burrowing by large animals the size of most nemerteans. *Cerebratulus* is a rapid and efficient burrower and is one of the largest nemerteans. Its method of burrowing, which is probably similar to that of other burrowing nemerteans, has been studied by Wilson (1900). The proboscis is everted into the sand, making a cavity in it. The end of the proboscis is then dilated and crooked, and serves as an anchor while the body is drawn into the sand by the contraction of the proboscis retractor muscles. The circular body-wall muscles at the anterior end of the worm, are contracted so that the head is narrow and pointed as it is drawn into the burrow. Once inside, the head is dilated and the proboscis withdrawn, so that the head now forms an anchor while the wave of contraction of the longitudinal muscles spreads backwards along the body, drawing the rest of it into the hole. At this point the proboscis is everted again and the cycle is repeated.

A manner of burrowing broadly comparable to that used by *Cerebratulus*, but with an entirely different structural basis, is shown by the holothurian *Thione briareus*. This animal is able to burrow by peristaltic contractions of the body-wall musculature, but usually it attaches its oral tentacles to a stone buried in the sand and then, by contracting the tentacular muscles, draws its body down. If there are no stones in the sand, burrowing, though still possible, appears to present some difficulty to the animal (Pearse, 1908).

Sipunclids also employ similar methods of burrowing to nemerteans, but in these worms, as in most other unsegmented coelomates, the proboscis cavity forms a part of the general body cavity, so that the entire body-wall musculature participates in proboscis eversion and in burrowing movements.

The activities of the sipunculid proboscis and its use in burrowing have been studied in *Sipunculus nudus* by von Uexküll (1903) and Zuckerkandl (1950a) and, in much less detail, in *Dendrostoma zostericola* by Peebles and Fox (1933). The proboscis of these animals terminates in a small cluster of tentacles and a slightly swollen collar, but the former are not necessarily exposed during proboscis eversion, are generally protracted only when the animal is feeding, and appear to play no part in burrowing. When *Dendrostoma* burrows, the proboscis is everted by contraction of the circular body-wall muscles, particularly those of the posterior half of the body, and is thrust into the substratum. The posterior end of the worm is constricted, driving coelomic fluid forwards into the everting proboscis, and at the same time, the anterior part of the body remains dilated, so wedging the animal in the sand. When the proboscis is fully everted, the region immediately behind the collar is markedly dilated and forms a new anchor while the strong proboscis retractor contracts and draws the body of the worm forwards.

Burrowing *Sipunculus* execute approximately the same cycle of activities, though when they are burrowing vigorously, the worms may show some modifications of the cycle. Anchorage is not invariably provided by dilation of the body; in rapidly burrowing worms the middle region of the body is flexed to the limit permitted by the walls of the burrow and this flexure provides an anchor while the proboscis is everted. Besides providing an extremely solid anchor, the flexure of the body has an additional advantage since the concavity between the wall of the burrow and the body of the animal provides a space into which sand particles, displaced by the eversion of the proboscis, flow. The hydrostatic pressure of the coelomic fluid generally rises rapidly during the initial phases of the proboscis eversion and falls again as the proboscis becomes fully protracted and also during retraction. A rise in hydrostatic pressure is not essential to proboscis eversion, however, and during slow eversion in water, the coelomic fluid pressure may never rise above 0·6–0·7 mm Hg. Commonly, however, pressures of the order 15–30 mm Hg are recorded. During rapid burrowing in sand, the cycles of pressure change associated with successive proboscis eversions may overlap, so that the pressure rise in anticipation of the next eversion coincides with the proboscis retraction of the previous cycle. When this happens, the proboscis is shot out with explosive force as soon as the retractor muscles relax. Under these circumstances, and when the worm is buried in sand, maximum internal pressures of the order of 70 or 80 mm Hg have been recorded by Zuckerkandl (1950a). Since the greatest hydrostatic pressures achieved by both *Sipunculus* and *Phascolosoma gouldi* are very little greater than this and occur when the animals go into extreme contraction as a defence posture, it appears that for penetration into extremely dense substrates the worms can generate a very large thrust by bringing the forces of contraction of the entire body-wall musculature to bear on the proboscis. Such considerable thrusts must be matched by a suitably strong anchor if the animal is not to slip backwards as the proboscis is everted. Since the defence posture involves an extreme flexure of the body, which also appears during rapid burrowing, the worm remains firmly wedged in the sand. Furthermore, since no part of the body-wall musculature need remain relaxed in order to provide a symmetrical dilation of the body, such as forms an anchor during slow movements, all the muscles are free to contribute to the generation of a high internal hydrostatic pressure.

THE BURROWING HABIT

An extremely specialized method of burrowing in which the extrovert retractor muscles once more provide the main tractive force, may be seen in *Priapulus* (Fig. 51). According to Friedrich and Langeloh (1936), the sequence

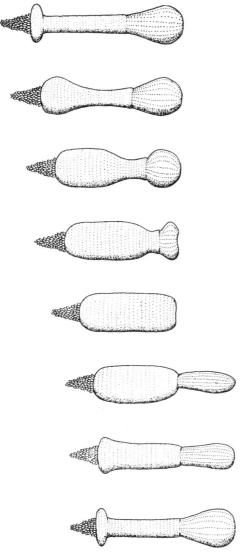

FIG. 51. Cycle of activity of a burrowing *Priapulus*. (After Friedrich and Langeloh, 1936.)

of events during burrowing is as follows. The relaxed worm lies with its proboscis fully everted, the anterior tip of it dilated into a balloon shape. The circular muscles of most of the body wall are slightly contracted except for those at the posterior end which are relaxed and the body wall forms an annular swelling in this region. As burrowing movements begin, the circular muscles of the body wall relax and the longitudinal muscles contract. The body of the worm is gradually drawn forwards over the proboscis which remains anchored in the mud by its distended end up to the time it is finally withdrawn into the body. Since the proboscis of *Priapulus* is equal to about one-third the total body-length, the net forward progression during this stage of the cycle is equal to that amount. When the longitudinal muscles are maximally contracted, and the proboscis fully inverted, the body is uniformly distended along its whole length and wedges the animal in the burrow during the next stage of the cycle. The circular and longitudinal muscles of the body wall

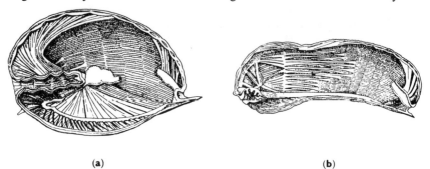

(a) (b)

FIG. 52. The aberrant polychaete *Sternaspis* with anterior end (*a*) inverted and (*b*) everted. (After Goodrich, 1897b.)

suddenly contract almost simultaneously, causing the proboscis to be shot out with great speed and force. As coelomic fluid flows into the proboscis, the body of the worm gets smaller in diameter and so ceases to anchor the animal, but by this time, the proboscis has dilated and this prevents slipping. The cycle is then repeated. A synergic contraction of both circular and longitudinal muscles is unusual in worms and, as in sipunculids, relatively high internal pressures must be generated, with the result that proboscis eversion is a very powerful action. The animal does of course progress in a series of jerks, but this appears to be a most successful method of burrowing through a soft substratum.

The aberrant polychaete *Sternaspis* (Fig. 52) lives in a similar habitat to *Priapulus* and its peculiar anatomy suggests that it may burrow in much the same way (Rietsch, 1882; Vejdovsky, 1882; Goodrich, 1897b), although this has not been confirmed by direct observation. The extrovert consists of the first seven segments which can be retracted within the rest of the body by the action of a complicated series of retractor muscles. Eversion is accomplished by contraction of the circular muscles of the posterior part of the worm. Segments 2, 3 and 4 are provided with a crescentic, lateral row of strong, spine-like chaetae. These are protracted as the anterior end of the worm is everted and, judging by their shape and disposition, must not only help provide a firm anchorage of the anterior segments in the substratum while the posterior segments are drawn forwards by contraction of the retractor muscles, but also assist in excavating the substrate in front of the worm as the extrovert is everted.

Although the proboscis may be used in a variety of ways in the formation of a burrow, the proboscis retractor muscles do not generally provide traction for the rest of the body as they do in the worms we have considered so far. The reason for this is that the proboscis is generally relatively short and if contraction of the proboscis retractor muscles alone were responsible for dragging the body forwards, progress into the burrow at each cycle of the burrowing activity would be small indeed. The use of the proboscis to excavate a hole into which the animal then crawls is the commonest way in which worms with a short proboscis burrow. This manner of burrowing is exemplified by the polychaete *Nephtys* (Clark and Clark, 1960b) and is probably characteristic of the burrowing activities of a number of other worms, unsegmented as well as segmented.

The proboscis of this worm is violently everted by the contraction of the longitudinal muscles of the first 30–35 segments aided by the dorso-ventral muscles of the anterior part of the animal. This punches a hole in the sand. The body is unable to slip backwards at the moment of impact of the proboscis with the sand, because the widest segments of the body, from segment 15 to segments 45–50, are wedged against the walls of the burrow and the parapodia and chaetae are extended from the sides of the body to oppose slipping by increasing the frictional forces between the worm and the sand. Segments 20–35 are the widest of the body since they house the very large, muscular proboscis when it is inverted. With the eversion of the proboscis, most of the coelomic fluid contained in them is driven forwards to inflate the proboscis and they are consequently reduced in volume. However, it is the longitudinal and dorso-ventral dimensions that are reduced and these segments do not become appreciably narrower, so that they are still able to perform the role of anchors. The proboscis is inverted by the contraction of the proboscis retractor muscles and the relaxation of the longitudinal muscles. Segments 15–35 are now dilated by the presence of the pharynx within them and segments 35–50 become less dilated as the body-wall muscles relax. The worm then crawls forwards into the hole that has been made by the proboscis. When the prostomium reaches the end of the hole previously made by the proboscis, the entire burrowing cycle is repeated, starting again with the eversion of the proboscis.

The proboscis of the echiuroid *Urechis* is employed in a different manner (Fisher and MacGinitie, 1928b). It is not very extensible but is extremely mobile. It excavates a cavity in the soft mud in which the animal lives not by violent eversion, but apparently by scraping movements. The anterior part of the worm is then elongated by the contraction of the circular muscles, thrust into the cavity made by the proboscis, and expanded by the relaxation of the circular and contraction of the longitudinal muscles. In this way it is wedged against the walls of the burrow while the remainder of the body is drawn forwards. The burrow is enlarged by the use of the chaetae. Material is scraped from the sides of the burrow by the two oral chaetae which are protracted for this purpose, it is worked backwards by the ring of anal

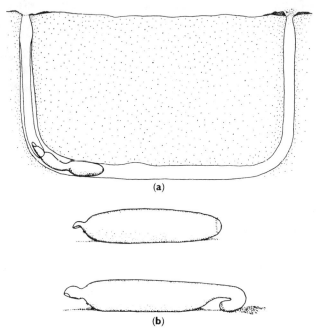

FIG. 53. The echiuroid *Urechis caupo* (a) in its burrow, and (b) the position adopted by the animal when using its posterior ring of chaetae for enlarging its burrow. (After Fisher and MacGinitie, 1928b.)

chaetae and finally ejected from the tube by a jet of respiratory water from the anus. The animal is able to rotate in its burrow and the walls are widened on all sides by the action of the ventrally placed oral chaetae. Removal of the debris is also achieved by the posterior end of the worm being doubled under and then vigorously straightened out (Fig. 53). The anal chaetae, which are curved forwards and scrape away the mud, are strongly everted during this process, and the mid-ventral chaeta of the series, which would otherwise be pressed against the ventral surface of the worm when its posterior end is flexed, is lacking and the two chaetae flanking this position are reduced in size (Fisher and MacGinitie, 1928a). Other echiuroids which possess a complete ring of posterior chaetae presumably do not engage in this behaviour.

The burrowing holothurian *Caudina* uses its oral tentacles to manipulate a cavity in the sand in front of the advancing body in much the same way as the echiuroid proboscis is employed (Gerould, 1896; Yamanouchi, 1929). Movement of the body in the sand is by means of peristaltic locomotory waves, but if the tentacles are amputated, the animal is unable to burrow. It is also reported that this and a number of other burrowing holothurians ingest a considerable quantity of sand while burrowing, but it has so often been claimed that burrowing worms essentially eat their way through the soil and later shown unequivocally that they burrow by muscular activity, that this observation must be accepted guardedly.

The burrowing of *Arenicola* is comparable to the initial phases of burrowing by *Urechis*, though a different function has been ascribed to the proboscis than merely digging a hole into which the animal then crawls. *Arenicola* is of course segmented, but the main part of the body is aseptate, so that functionally the coelom is equivalent to the body cavity of unsegmented coelomates. According to Chapman and Newell (1947), the repeated eversion of the proboscis of *Arenicola marina* agitates the sand, exploiting its thixotropic properties and rendering it more easily penetrable. In a series of experiments these authors found that the resistance to penetration of sand in which *Arenicola* lives was reduced by as much as 90 per cent by repeated application of a pressure of 600 g/cm^2. Within 30 sec of agitation, the sand returned to its normal consistency. They therefore proposed that the mechanism of burrowing is as follows: repeated eversion of the proboscis agitates the sand until it becomes sufficiently fluid for the anterior end of the worm to be thrust into it. The sand quickly sets around the buried part of the worm, so that when this is then dilated it forms a substantial anchor and a much greater thrust against the substratum can be generated by the proboscis when it is again everted. Burrowing, when part or all of the worm is beneath the surface, is by a combination of softening the substratum by agitation due to repeated proboscis eversion, and by strong forward thrusts of the anterior part of the worm. In support of this interpretation, they showed that a worm lying on the substratum cannot generate a forward thrust greater than its own body weight, say 2·5 g, representing a pressure of 16 g/cm^2, and probably much less than this, since the greater part of the weight is supported by the ground and cannot be applied to the substratum at the point of entry of the worm. This figure is much less than the 400 g/cm^2 necessary to penetrate unagitated sand. Sands exhibit thixotropic properties only if the particles composing them fall within a certain size range, and only if the water content is above a certain level. Chapman and Newell (1947) claim that *Arenicola* is unable to burrow into dry sand. Wells (1944) had previously suggested that the lateral movement of the diverging, chitinized, buccal papillae of *Arenicola marina* excavated a hole in the sand as the proboscis was everted (Fig. 54), rather in the manner a rabbit digs a hole, and has remained unconvinced by the arguments of Chapman and Newell; he has also shown that *Arenicola* can burrow into materials that are certainly not thixotropic (Wells, 1948, 1961).

The anemone *Peachia* burrows in what appears to be a very similar manner to *Arenicola* (Faurot, 1895). If placed on the surface of the sand, *Peachia* eventually bends its column into an arc so that the basal part is held vertically against the ground. The mouth and oral disk are tightly contracted and the body is turgid. Circular constrictions then appear on the column and pass towards the base, so that the enclosed fluid is carried towards the basal disk which is in the form of an inverted hemispherical bowl. As the shock wave reaches it, the disk is suddenly everted, becoming first cylindrical and then pointed in the process, and displacing sand grains centrifugally from beneath

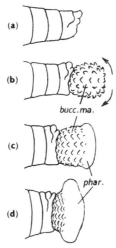

Fig. 54. Scraping action of the chitinized papillae on the buccal mass of *Arenicola marina* as the proboscis is everted. (*b*) Incomplete extrusion of the buccal mass, (*c*) and (*d*), eversion of the pharynx. (From Wells, 1954.)

the disk. The basal part is then introverted until the arrival of another wave some five or six seconds later. The effect of this repeated eversion of the disk against the sand is to make a depression with the sides of which the column remains in contact so that a tube of uniform diameter is constructed. Once part of the column is buried the movements of the anemone become more rapid and burrowing is accomplished by essentially the same means as in coelomate worms, by peristaltic locomotory waves. The oral disk remains closed throughout the burrowing process, which may take an hour, so that the coelenteron behaves as a true fluid skeleton.

Whatever the precise function of the proboscis in burrowing, it is evident that its chief use is in opening a way through the substratum into which the animal can move. The great advantage of using the proboscis for this purpose is also clear: a great part of the body-wall or equivalent musculature, and sometimes the whole of it, can be employed in the single function of everting the proboscis, so that a considerably greater pressure can be brought to bear against the substratum than if a localized region of the body-wall musculature is employed for the purpose. An eversible proboscis is by no means essential for burrowing. It is used by *Priapulus*, but not by *Halicryptus* (Friedrich and Langeloh, 1936), and the use of the proboscis of *Urechis* does not involve impacting it against the substratum (Fisher and MacGinitie, 1928b) and it is not eversible. Nor must it be supposed that the exploitation of the thixotropic properties of sand, which may sometimes contribute to burrowing by *Arenicola*, is necessarily a general phenomenon. Many substrata do not exhibit thixotropic characteristics, some in fact show the opposite property of dilatancy, becoming less easily penetrated when pressure is applied to them, and must be quite difficult to burrow into. Nevertheless, sands which do not possess a fauna of worms are a rarity, and are generally those which are subject to considerable disturbance by wave action (Clark and Haderlie, 1960, 1962; Clark, Alder and McIntyre, 1962). Whatever the nature of the substratum, the fundamental method of burrowing into it by soft-bodied animals is the same. Part of the body wall is dilated and forms an anchor while the anterior circular muscles contract and force the head of the worm forwards into the substratum. The anterior end of the worm then dilates, forming a new anchor while the rest of the body is drawn forwards by the contraction of the longitudinal muscles. The whole process is repeated and so the worm forces its way through the substratum. This method of burrowing of course makes it likely that if the worm does possess an eversible proboscis, it will be brought into play during the burrowing cycle. The proboscis is not necessarily everted when the body-wall muscles contract; it may be restrained by the proboscis retractor muscles or by the closed mouth, but it is easy to see how such refinements as its function in the burrowing of *Arenicola* may have been evolved.

LOCOMOTION IN THE BURROW AND OVER THE SUBSTRATUM

With the exception of nematodes, nearly all of which move by undulatory swimming motions, most unsegmented animals with a secondary body cavity are capable of forming a burrow and most are sedentary. The burrows are of a semi-permanent nature and their walls are lined and consolidated with mucus. The animals crawl up and down the burrows by peristaltic locomotory movements (Fig. 55) similar to those that are employed during the construction of the burrow. One part of the body is dilated by contraction of the longitudinal muscles, so wedging the animal against the walls of the tube, while the circular muscles anterior to it contract, elongate the worm and push it forwards, and several zones of muscle contraction may appear along the body simultaneously. Since the relaxation of a short length of the circular muscles and contraction of the associated longitudinal muscles cause bulging around the entire circumference of the body wall, contact is made with the walls of the burrow on all sides to form *points d'appui*. It is for this reason that locomotory waves in these worms are invariably retrograde. Direct locomotory waves, in which the *points d'appui* occur where the underlying longitudinal muscles are relaxed, are obviously inappropriate for animals

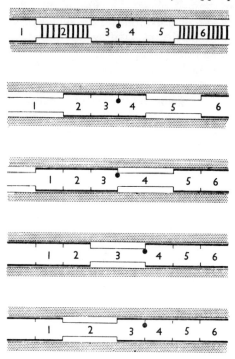

Fig. 55. Behaviour of the musculature of a cylindrical worm during the passage of a peristaltic locomotory wave along the body (cf. Fig. 35*b*, p. 55).

living in burrows since the diameter of the body would be narrowest at points where the thrust is exerted against the substratum.

Although these worms with a roughly circular cross-section are admirably adapted for living in, and moving up and down burrows, they are extremely ill-adapted for moving over the substratum for then only a small part of the body wall is in contact with the ground to provide a *point d'appui*, and the forward thrust, which cannot exceed the frictional forces opposing slipping, is correspondingly lessened. Nevertheless, a number of animals of this grade of construction are able to crawl over the ground. *Arenicola* is able to do so (Guberlet, 1933; Werner, 1956), notwithstanding claims to the contrary (Newell, 1948). Even the anemones *Halcampa chrysanthemum* and *Peachia* are both able to crawl if laid horizontally on the sand (Faurot, 1895) and the contractile waves pass along the column from mouth to basal disk, as the anemone moves oral end foremost (i.e. the locomotory waves are retrograde), as one would expect in burrowing anemones that are able to move up and down the burrow with some facility. *Aiptasia* is also able to move in a similar manner, but in this non-burrower the locomotory waves are direct (Fig. 56).

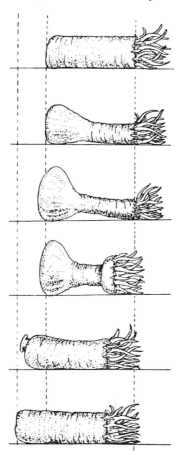

FIG. 56. Locomotion of the anemone *Aiptasia carnea*. (From Portmann, 1926.)

The movement of the anemones over the substratum may be facilitated if they have some means of adhesion, so increasing the frictional forces at their points of contact with the ground. An example of how this may be accomplished is furnished by *Aiptasia couchii* which, according to Stephenson (1935), is even able to climb by means of the adherent rugae of the column and can move at an astonishing pace.

The use of special means of adhesion to the substratum is, of course, a characteristic of animals that habitually move over it rather than through it.

Both gastropods and echinoderms do so, but most of the former make little, if any use of a fluid skeleton, and most of the latter do not move by means of the body-wall musculature as worms do. There are a few exceptions in both groups however.

Some holothurians move over the substratum by means of peristaltic locomotory waves of the body-wall musculature in a similar manner to worms. Their progress, though slow, is rendered more efficient by the use of the podia to provide temporary anchors. *Stichopus* creeps by means of a muscular wave of contraction that starts at the posterior end and sweeps towards the head (Parker, 1921). Initially all the podia of the trivium are attached, but as the wave of contraction begins, the posterior tube feet are detached and the hind end of the body is raised and lifted from the surface. The posterior

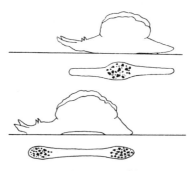

FIG. 57. Locomotory movements of *Aplysia californica*, lateral and ventral views. (From G. H. Parker, 1917b, *J. exp. Zool.* **24**, 141, fig. 1.)

end of the animal is contracted and set down again near the attached anterior end, then, as the contractile wave reaches this part of the body, it too is lifted from the ground and projected forwards. Not more than one contractile wave is visible at the same time and there is usually a pause before a second wave commences.

This type of locomotion is reminiscent of that in *Aplysia* (Fig. 57) and, on occasion, *Helix* (Fig. 58), save that in these gastropods, the locomotory waves are not discontinuous. The anterior end of *Aplysia* is raised from the ground and is extended forwards. As much as half the animal may become detached from the substratum in this way. The anterior end is then reattached while the rest of the body forms an arch and a wave of contraction of the longitudinal pedal muscles passes backwards along the foot. The posterior end is then detached and crowds forwards towards the head, when a second wave of contraction begins. A comparable method of locomotion has been reported in *Helix* by Carlson (1905) (Fig. 58). Adhesion to the substratum is by suction in *Aplysia* and by mucus in *Helix*. Obviously the locomotion of both animals is a development from the normal pedal locomotion of gastropods; the chief difference is that the locomotory waves in *Aplysia* and *Helix* are not confined to the longitudinal muscles of the foot, but involve the whole of the pedal musculature and there is no doubt that in these cases, the lacunar tissue of the foot serves as a fluid skeleton.

Adhesion to the substratum is not essential for locomotion. The echiuroid *Urechis* appears to be able to move over the substratum better than most acoelomate worms. The anterior part is elongated and the viscera and coelomic fluid from the posterior part of the animal are then squeezed forwards

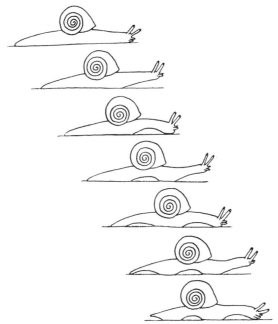

FIG. 58. Exaggerated locomotory waves sometimes employed by *Helix*. (From Carlson, 1905.)

into it by the contraction of the posterior circular muscles, moving the centre of gravity forwards. The posterior part is then drawn up and the cycle repeated (Fisher and MacGinitie, 1928b). It can crawl backwards or forwards with equal facility and, according to Fisher and MacGinitie, can do so as fast as the earthworm. While this type of movement bears some relation to peristaltic locomotion, in that the sequence of contraction of circular and longitudinal muscles in different parts of the body is the same, it differs in an essential feature. A swelling is not produced in the body wall which anchors the worm by increasing at that point frictional resistance to backward movement, the centre of gravity is heaved forwards within the sac formed by the body wall. It is a method of progression that is available only to worms that are capable of becoming extremely flaccid and which have mobile viscera

Another way in which the difficulty of providing adequate *points d'appui* can be solved appears in nemertean locomotion. The proboscis is everted and anchored in the substratum. It is then retracted, dragging the body passively forwards. The success of this method depends on the presence of a long proboscis which can be attached to the substratum in some way. This practice has not been reported in other worms but finds a parallel in the holothurian *Synaptula* (Olmsted, 1917b). It lives among algae and moves chiefly by means of the oral tentacles, although it is capable also to peristaltic locomotory movements of the body wall. A tentacle is extended and adheres to the substratum by mucus, it is then contracted, drawing the animal forwards, and another tentacle is extended and repeats the cycle. So efficient is this method that the animal is able to crawl up vertical surfaces with the body hanging free. Similar behaviour has been reported of another holothurian, *Rhabdomolgus ruber* (Becher, 1907).

PERISTALSIS OF THE BODY WALL OF WORMS

Locomotory peristaltic waves involve contractions of both the circular and longitudinal muscles, but simple peristalsis of the body wall, produced by waves of contraction passing along the circular muscles only, also occurs in tubicolous worms and is used to provide respiratory and feeding water-currents. All these worms are aquatic and, since they live in semi-permanent burrows, it is important that they should be able to prevent water stagnating in the burrow. A peristaltic bulge which occludes the burrow and travels along the body wall, carries with it a quantity of water. Continued peristalsis therefore results in a water current being pumped through the tube. In *Arenicola*, these waves of peristalsis produced by contraction of the circular muscles generally pass from tail to head, although reversed peristalsis is possible. In *Urechis* the water current passes from head to tail.

The production of a respiratory current in *Arenicola* is a periodic activity under the influence of a pacemaker in the central nervous system. The control of the activity cycles of *Arenicola* and the influence of environmental conditions upon them have been the subject of detailed investigations by Wells (1937, 1949a, b, 1950; Wells and Albrecht, 1951; Wells and Dales, 1951). Similar peristaltic waves producing respiratory water currents have also been observed in terebellids (Dales, 1955), and sabellids (Nicol, 1931; Fox, 1938; Wells, 1951, 1952a).

The respiratory current produced by the peristalsis of the body wall may be exploited in another way. *Urechis* secretes a sheet of mucus from the proboscis across the mouth of the tube, and then retreats down the burrow, leaving a mucus cone extending from the proboscis to the rim of the burrow entrance through which all the water drawn in is filtered (Fisher and Mac-Ginitie, 1928b). Periodically this filter becomes clogged by particles in the water. The worm then crawls forward and eats the mucus sheet and the trapped food material, and then secretes a new mucus net (Fig. 53a).

THE LOCOMOTION OF SOME ECHINODERMS

A major group of unsegmented coelomates to which we have so far directed little attention is the Echinodermata. In all but the Holothuroidea, the coelom, though often spacious, is contained within a rigid, or nearly rigid body wall and therefore cannot behave as a hydraulic system of the type with which we are familiar in worms. The locomotion of echinoderms is either by movements of the arms (in ophiuroids and crinoids), or by the use of podia (in most asteroids and echinoids, and many holothurians).

Our knowledge of the structure and functioning of the tube-foot/ampulla system is due largely to the investigations of Smith (1946, 1947, 1950) and Nichols (1959a, b, 1960, 1961). In asteroids, echinoids and holothuroids, the circumferential muscle fibres of the podial ampulla antagonize the longitudinal muscles of the tube-foot. Crinoids, the podia of which are used in food collection and respiration, have no ampullae, but muscle fibres crossing the radial water vascular canal are able to constrict portions of it and so cause distension of the podia (Nichols, 1960). Differential contraction of the podial longitudinal muscles inclines the distended podium in any appropriate direction and plays an obvious role in producing the stepping movements of the tube-feet by which many echinoderms are transported over the substratum (Fig. 59). At least in asteroids, the podia may move the animal by leverage or traction, depending upon the circumstances (Kerkut, 1953) and depending upon whether the podia are provided with basal suckers or not.

Chapman (1958) has pointed out that the increase in volume of the ampullae when the podia are retracted, must be accommodated by dilation of the body wall. This is possible in asteroids and holothurians which have a flexible body wall, but in echinoids the body wall is rigid and it is not clear how the volume is adjusted. The peristome region is flexible and may accommodate some of the volume changes of the interior, but whether or not this is adequate to account for the volume change resulting from the retraction of all the podia is uncertain.

As we have already seen, the holothurians approach the worms in both their fundamental structure and in the behaviour of the body-wall musculature which is organized in circular and longitudinal layers. There is a tendency within the class to a loss of radial symmetry and the evolution of a ventral locomotory surface. For example, *Thione briareus* has podia in all five radii and moves equally well with any side uppermost (Pearse, 1908; Mast, 1911), *Holothuria surinamensis* and *H. rathbuni* are intermediate (Crozier, 1914, 1915b), and in *H. captiva*, tube feet occur in only three radii (the **trivium**) and the animal always moves on this surface, mouth forwards (Crozier, 1915b). In most holothurians, the body-wall musculature plays very little part in locomotion which is similar to that of echinoids and asteroids, but

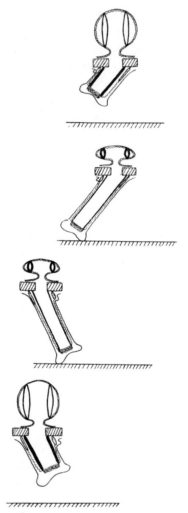

FIG. 59. Successive stages in the locomotory movements of a podium of the starfish *Asterias*. (After Smith, 1947.)

Stichopus and *Synaptula*, and presumably all apodous synaptids, move by means of locomotory waves of contraction of the body-wall muscles which approximate in varying degrees to peristaltic locomotory waves. In those species that crawl over the surface of the substratum, the locomotory waves are closer to pedal waves of gastropods than to peristaltic locomotory waves, and the podia provide adhesion between the holothurian and the substratum at the *points d'appui*, but in burrowing species, the whole body-wall musculature is usually employed in generating peristaltic locomotory waves, and the podia decline in usefulness. The most successful burrowing holothurians are, in fact, apodous.

The coelom of holothurians has hydrostatic functions other than those

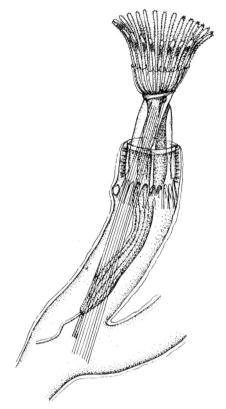

FIG. 60. The phylactolaematous ectoproct *Fredericella sultana*, with polypide everted. (After Allman, 1856.)

associated with locomotory movements of the body wall. Expulsion of water from the respiratory trees is, at least in part, brought about by contractions of the body-wall muscles when the oral tentacles are retracted and the buccal sphincter closed (Crozier, 1916; Budington, 1937). The internal pressure set up by the contraction of the body-wall muscles is transmitted to the respiratory trees through the coelomic fluid. The body-wall muscles are of course not responsible for filling the respiratory trees with water; they are inflated by pulsations of the cloaca.

OTHER MECHANICAL FUNCTIONS OF THE COELOM

Although there is a correlation between the burrowing habit and the presence of a large, patent coelom, the existence of a secondary body cavity is by no means invariably associated with locomotion. A number of animals that possess such a cavity, such as bryozoans, are sessile and permanently attached to the substratum, but even in these, the coelom or an equivalent body cavity subserves a mechanical hydrostatic function.

It is only proper that a consideration of non-locomotory, but mechanical functions of the coelom should begin with the ectoproct Bryozoa, since it

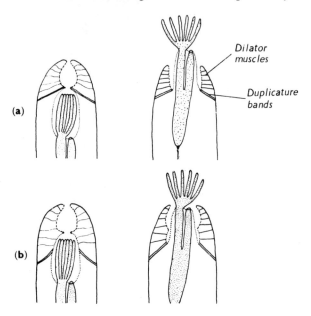

FIG. 61. Mechanisms of polypide eversion in phylactolaematous ectoprocts. (*a*) *Lophopodella*, in which the duplicature bands are inserted at the level of the diaphragm and permit the complete eversion of the polypide, (*b*) *Stoltella*, in which the duplicature bands are inserted on the tentacular sheath so that only partial eversion of the polypide is possible. (Modified from Rogick, 1937, and Marcus, 1941.)

was in these animals that the use of a coelom as a hydrostatic organ was first clearly appreciated (Farre, 1837; Allman, 1843). The mechanism of eversion and inversion of the polypides of bryozoans is exactly comparable to the mechanism of proboscis eversion in coelomate worms. Admittedly, the mechanisms become rather complicated, although in principle the same, in those bryozoans—the majority—which have a rigid, calcareous zooecium, but in the manner of eversion and retraction of the polypide, the soft bodied Phylactolaemata approach very closely the hydrostatic system found, for example, in the proboscis apparatus of a sipunculid. The various devices evolved in the calcareous ectoprocts to permit polypide eversion, have been elucidated by many students of this phylum over the last century. The outcome of all these investigations of bryozoan morphology was admirably

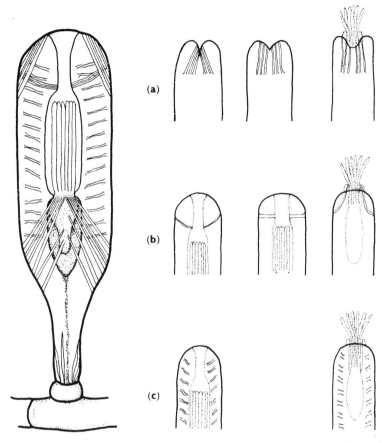

FIG. 62. A ctenostomatous gymnolaeme, *Farrella repens*, with the polypide inverted. The smaller figures show the behaviour of (*a*) the parieto-vestibular muscles (in lateral view), and (*b*) the parieto-diaphragmatic and (*c*) parietal muscles (in frontal view) during polypide eversion. (Modified from Marcus, 1926a.)

summarized by Harmer in 1930, and since that time, little has been discovered to warrant any re-evalutation of Harmer's views.

The Phylactolaemata (Fig. 60) have a flexible, sometimes gelatinous, muscular body wall which includes both circular and longitudinal layers of muscles. Contraction of these increases the fluid pressure within the coelom and is the chief agency in causing eversion of the polypide, as Allman (1856) clearly understood. As the fluid pressure in the coelom is increased, the dilator muscles of the vestibule and the duplicature bands are contracted, and the lophophore retractor muscles are relaxed. The polypide then emerges to the full extent permitted by the duplicature bands (Fig. 61). This varies somewhat: in *Stoltella*, for example, the bands are inserted on the tentacle sheath (Fig. 61*b*) instead of, as more commonly, at the distal end of the cystid wall (Fig. 61*a*), and the polypide is incompletely everted (Marcus, 1941). Retraction of the polypide is effected chiefly by contraction of the main lophophore retractor muscles aided, in some species in which they are appropriately disposed, by the muscles of the duplicature band (Brien, 1960). The existence of restraints to polypide eversion is important, since they permit the independent eversion and retraction of polypides which share a common hydrostatic skeleton, in that the coelom of each individual is in communication with that of its neighbours (Harmer, 1896). Relaxation of the vestibule dilator muscles and closure of the vestibule by contraction of its sphincter are additional factors which help prevent the eversion of the polypide during changes in coelomic fluid pressure caused by activity elsewhere in the colony.

Although the majority of the Gymnolaemata have a more or less calcified zooecium, in ctenostomes (Fig. 62) it is flexible and composed of chitin (Hyman, 1958). In these animals, eversion of the polypide is accomplished in much the same way as it is in the phylactolaemes, by contraction of the muscles of the cystid wall increasing the internal hydrostatic pressure and by the removal of restraints to eversion. The parietal muscles of ctenostomes do not form complete circular and longitudinal coats as they do in phylactolaemes, but are reduced to widely separated and incomplete circular elements (Farre, 1837; Allman, 1843; Marcus, 1926a, etc.). Opening and closing of the vestibule is effected by similar means in these ctenostomes as in the phylactolaemes.

In the calcareous cheilostomes, and also on some encrusting ctenostomes, the walls of the zooecium are attached on four sides to those of neighbouring individuals, and by the base to the substratum. Only one surface of the boxlike zooecium remains free to move appreciably, and the parietal muscles are correspondingly modified, running obliquely across the body cavity from the base to the flexible frontal membrane (Figs. 63 and 64). Contraction of these muscles depresses the membrane and causes eversion of the polypide (Nitsche, 1871). Additional complication is introduced in some anascan cheilostomes by the development of a calcareous plate, the cryptocyst, beneath the flexible frontal wall of the zooecium. The oblique parietal muscles are then reduced to a single pair which is inserted into the frontal membrane by way of ligaments running through pores in the cryptocyst (Harmer, 1930). In the ascophoran cheilostomes, the frontal membrane becomes completely calcified and rigid, and eversion of the polypide is accomplished by the contraction of parietal muscles inserted into the ventral wall of a compensation sac which communicates with the exterior. Dilation of the

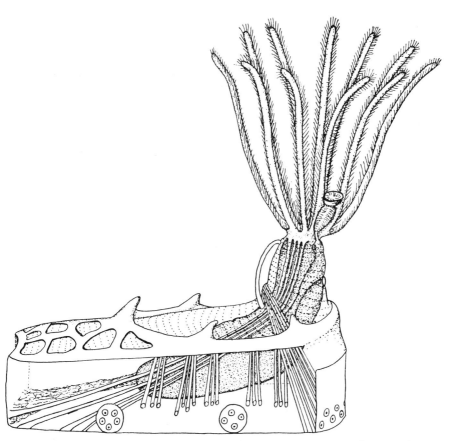

Fig. 63. The cheilostomatous gymnolaeme *Electra pilosa*, with polypide everted. (After Marcus, 1926a.)

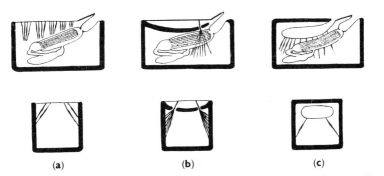

Fig. 64. Mechanisms of polypide eversion in cheilostomatous gymnolaemes; diagrammatic side and end views. (a) *Membranipora*, (b) *Micropora*, (c) the Ascophora. (After Harmer, 1930.)

sac, by contraction of the parietal muscles, causes an influx of water from outside, compresses the coelomic contents, and everts the polypide (Busk, 1884; Jullien, 1888a, b; Harmer, 1902).

The various mechanisms of polypide eversion in ctenostomes and cheilostomes are clearly related to one another, and Harmer (1930) discusses the possible course of evolution of the ascophoran condition from the clearly more primitive anascan condition, but the hydraulic system of cyclostomatous gymnolaemes is rather different. These have tall, narrow, cylindrical zooecia, which, save for the terminal membrane, are completely calcified (Fig. 65).

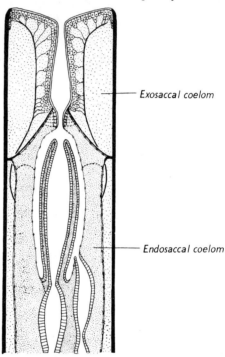

FIG. 65. The coelomic compartments of a cyclostomatous gymnolaeme. (After Borg, 1926.)

The terminal membrane is probably homologous with the frontal membrane of cheilostomes, but is of such a small area as to be insufficient to cause eversion of the polypide. The cyclostomes lack a compensation sac and therefore cannot function in the same way as other heavily calcified bryozoans. The method of polypide eversion has been inferred by Borg (1923, 1926) from a consideration of the morphology of these animals. The terminal membrane is continued inwards to line the long vestibule. The coelom is subdivided into exo- and endosaccal portions by a membranous sac which extends almost to the proximal end of the zooecium. At its distal end, the membranous sac has an annular insertion into the bottom of the vestibulum, and is also attached to the body wall by eight ligaments. Muscles, which Borg claims are modified parietal muscles, run from the outer surface of the membranous sac just below the ligaments, to the wall of the vestibule and to the terminal membrane. Contraction of these muscles dilates the vestibulum and coelomic fluid is displaced from the distal part of the exosaccal coelom into the more proximal parts. This, in turn, compresses the membranous sac and the fluid in the endosaccal coelom, and causes eversion of the polypide. Harmer (1930) complained that this inference of the mechanism of polypide eversion from the structure of the cyclostomes failed to account for the presence of the membranous sac, but, as Chapman (1958) has observed, this structure is necessary if the coelomic fluid displaced by the dilation of the vestibule is to be driven into the basal part of the zooecium where it can best effect protrusion of the polypide.

It is unnecessary to multiply examples of the use of hydrostatic systems to produce dilations or eversions of parts of animals. Mechanisms of this sort are extremely common and of widespread occurrence whenever large or powerful changes of shape are necessary (Chapman, 1958). The fluid skeleton need not necessarily be coelomic; a pseudocoel can serve precisely the same function. In the pedunculate barnacle *Lithotrya*, to cite but one example of the use of these mechanisms by arthropods, distension of the body and extension of the cirri is effected by pressure of the haemolymph, generated by contraction of the muscular walls of the peduncle (Cannon, 1947). The lamellibranch *Mya* presents a rather unexpected situation in which the hydrostatic system by which the siphons are extended is not contained within the body at all (Chapman and Newell, 1956). Sea water contained within the mantle cavity is forced into the siphons by contraction of the adductor muscles of the shell. The tips of the siphons remain tightly closed by a sphincter muscle during this operation, and the mantle edges are fused except at the anterior end where there is a small opening for the protrusion of the foot. The mantle edges here are closely applied to one another and are sealed by valvular flaps. So long as the mantle opening and the distal siphonal openings remain tightly closed, the mantle cavity and siphons behave as a single hydraulic system.

HYDROSTATIC PRESSURE OF THE BODY FLUIDS

In those animals in which it has been measured, there is great variation in the pressure of the fluid in the body cavity (Table 7). This variation is not only, as might have been expected, between members of different phyla engaged in different activities, but also among 'resting' individuals of the same species.

Sea-anemones, and possibly also holothurians, function at widely differing internal pressures and over a considerable range of body volumes (Batham and Pantin, 1950; Pantin and Sawaya, 1953) and, as Wells (1961) points out, all soft-bodied animals are probably subject to some variation in both. If the gut is distended with food, the fluid pressure in the coelom must rise, or the body wall be distended, or both may occur. In all animals except coelenterates, the gut is morphologically distinct from the hydrostatic skeleton (there are, as we have seen, great advantages to separating the digestive tract from the influence of movements of the body wall), but at the same time, the gut is still effectively part of the fluid skeleton, even if only as dead space.

A few animals like sea-anemones become enormously distended after a meal, but for the majority of worms, it is unlikely that variations in gut volume are large enough to account for the variations in the internal hydrostatic pressure that have been recorded. But even though variation in gut volume is unlikely to afford a complete explanation, it suggests a means by which internal pressure may be, in part, regulated (Wells, 1961). Water may be taken into the gut through the mouth or anus and in this way adjust the total volume and internal pressure of the animal. Pantin and Sawaya (1953), for example, noted a drop by about 30 per cent of the internal pressure of resting *Holothuria grisea* as water was expelled from the respiratory trees, and Ramsay (1953) has suggested that mosquito larvae regulate their body volume by adjusting the quantity of water voided through the anus as a result of peristaltic waves passing down the intestine. However, while this may be so in mosquito larvae, the mere fact that water passes into and out of the gut by way of the mouth or anus, need not necessarily imply that this is related to adjustment of internal hydrostatic pressure and body volume. Monro Fox (1952) concluded that anal respiration was not related to the maintenance of body turgor in crustaceans such as *Daphnia* which have a thin and fairly flexible exoskeleton, rather turgor is likely to be maintained in these animals by osmotic pressure, the body fluid being hypertonic to the external medium. It is clear, too, that all soft-bodied animals that experience osmotic stress are liable to variation of body volume and presumably also of internal hydrostatic pressure.

To a very great extent, however, variations in the resting pressure of the body fluids of soft-bodied animals are caused by changes in the state of tonic contraction of the body-wall muscles. Depending upon the mechanical strength of the tissues, some slight excess of internal over external hydrostatic pressure is necessary to prevent collapse of the body wall, but beyond this, excess internal pressure is gratuitous and reflects the general level of activity of the animal and the tonus of its muscles. Batham and Pantin (1950) record that the resting pressure of one *Metridium senile* averaged only 0·7 mm sea-water (0·05 mm Hg) on one occasion when it was quiescent, but on another day, when the animal engaged in frequent muscular activity, the resting pressure was 6·1 mm sea-water (0·45 mm Hg).

Excepting the acutely specialized nematodes, in which very high internal

TABLE 7

Hydrostatic Pressure of Body Fluids

Species	Pressure (mm Hg)			Activity	Authority
	Maximum	Resting	Working		
COELENTERATA					
Calliactis parasitica	10·0 ?	0·1–2·4	2·2–10·0	Peak pressure on contracting.	Chapman, 1949.
				Circular and parietal muscles of column.	Chapman, 1949.
Metridium senile	c. 1·0	0·05–0·45 (av. 0·19)	0·18–0·48	Change of shape by column muscles.	Batham and Pantin, 1950.
			2·94–7·34	Forcible contraction by mesenteric retractor muscles.	Batham and Pantin, 1950.
			5·0	Peak pressure on contracting.	Batham and Pantin, 1950. Parker, 1917a.
NEMATODA					
Ascaris lumbricoides	400–450	16–127 (av. c. 70)	to 225	Fluctuating basal pressure produced by local contraction and extension. Coiling.	Harris, 1962 (estimated from Bradley, 1959). Harris and Crofton, 1957. Harris and Crofton, 1957.

TABLE 7—Continued

Species	Pressure (mm Hg) Maximum	Resting	Working	Activity	Authority
SIPUNCULIDA					
Sipunculus nudus	60				Uexküll, 1903.
	73·5	1·47–2·2			Zuckerkandl, 1950a.
			2·4–29·1	Free proboscis eversion.	Zuckerkandl, 1950a.
			66·2–70·6	Burrowing.	Zuckerkandl, 1950a.
Phascolosoma gouldi (Golfingia)	79·4	2·2–2·9	8·4–18·4	Free proboscis eversion.	Zuckerkandl, 1950a, b. Zuckerkandl, 1950a.
ANNELIDA					
Nereis sp.		0·735	11·0 (max.)		Zuckerkandl, 1950b.
Glycera sp.		0·4–1·5	5·9 (max.)		Zuckerkandl, 1950b.
Arenicola marina	58·8–66·7	5·1–17·6 (av. 9·6)	12·5–30·1 (av. 19·9)	Burrowing.	Chapman and Newell, 1947.
			17·6–41·9 (av. 26·5)	Contraction in response to touch.	Chapman and Newell, 1947.
			1·5	Free proboscis eversion.	Wells, 1954.
			11·0	Calculated max. required for tube irrigation.	Wells, 1945, 1961.
			22·0	Calculated max. required for constructing head shaft.	Wells, 1945, 1961.
Lumbricus terrestris	25·7		2·9–14·0 (av. 9·9)	Active wriggling.	Newell, 1950. Newell, 1950.
Earthworm		4·4–5·5	> 9·3		Inada, 1950.

TABLE 7—Continued

Species	Pressure (mm Hg)			Activity	Authority
	Maximum	Resting	Working		
CRUSTACEA					
Carcinus maenas			3·7–14·1 (av. 9·6)		Picken, 1936.
Pomatobius fluviatilis			5·5–18·8 (av. 14·7)		Picken, 1936.
ONYCHOPHORA					
Peripatopsis spp.			2·2–14·7		Picken, 1936.
ECHINODERMATA					
Holothuria grisea	14·0	0·5–1·5 (av. 1·2)		Injection of acetylcholine chloride.	Pantin and Sawaya, 1953. Pantin and Sawaya, 1953.
			(0–4)	Drop in coelomic pressure on expulsion of water from cloaca.	
			2·2 (max.)	Locomotory movements.	Pantin and Sawaya, 1953.
			3·3–11·8	Contraction in response to mechanical stimulation.	Pantin and Sawaya, 1953.
Caudina chilensis	33·1–36·8	7·4–14·7	22·1–29·4	Burrowing.	Yazaki, 1930. Yazaki, 1930.
Thyone sp.	79·4	0–1·8	14·7		Zuckerkandl, 1950b. Zuckerkandl, 1950b.

fluid pressures are obligatory for the successful functioning of the body-wall musculature (Harris and Crofton, 1957), no animal requires to generate higher pressures than are necessary to overcome external forces. If these are small, as when the animal is in air or lying in shallow water, a very slight increase in internal hydrostatic pressure is adequate to produce a major change of shape. Indeed, if the animals execute movements in shallow water, the external forces may be negligible and only the resistance of the body wall to deformation has to be overcome. In *Metridium*, contraction of the height of the column by about 19 per cent causes only the very slight increase of internal pressure from 1·4 to 3·2 mm sea-water (0·1 to 0·24 mm Hg) because of the diaphanous nature of the body wall and the slight resistance it offers to deformation (Batham and Pantin, 1950). Similarly, an excess internal pressure of only 20 mm sea-water (approximately 1·5 mm Hg) is sufficient to permit proboscis eversion in *Arenicola marina* (Wells, 1954); in *Sipunculus nudus* the comparable figure is 4·5 mm Hg (Zuckerkandl, 1950a). Much greater forces may be generated, and sometimes are, but clearly they can be of any size to the limit of the power and speed of contraction of the musculature.

Wells (1961) has calculated that for most of its activities, *Arenicola* in fact requires only moderate internal pressures. Irrigation of the burrow, for example, demands a pressure equivalent to about half the head of water to be moved, that is, 15 cm sea-water (about 11 mm Hg) if, as in large worms, the burrow is about 30 cm deep (Wells, 1945). When it sets up its burrow initially, the worm digs down and then, by vigorous pumping, forms a head shaft to its burrow by blowing out the overlying sand. The pressure necessary to achieve this is equivalent to a head of water somewhat greater than the depth of sand, i.e. of the order of 20–25 mm Hg. Burrowing is another activity which demands rather high internal pressures, and Chapman and Newell (1947) have recorded values between 8 and 30 mm Hg in the coelom of *Arenicola* during this activity. However, such high internal fluid pressures are encountered only when the animal is setting up its burrow, and this is a relatively infrequent occurrence.

Batham and Pantin (1950), in an extremely useful discussion of hydrostatic pressure in relation to body movement, have stressed the need for detailed observation of the behaviour of the musculature and of the whole animal that is associated with fluctuations in the internal pressure. Without this information, pressure changes may be almost impossible to interpret. Nevertheless, some of the gross differences between the hydrostatic pressures generally recorded in different animals and shown in Table 7, can be partially explained, and some generalizations can be drawn.

The order of magnitude of pressures recorded—resting pressures, normal working range and maximum pressures—bears some relation to the habits and mode of life of the animal. Thus, the burrowing holothurian *Caudina chilensis* (Yazaki, 1930) has consistently higher internal pressures when 'resting' (7·4–14·7 mm Hg), burrowing (21·1–29·4 mm Hg) or on extreme contraction in response to electrical stimulation (33·1–36·8 mm Hg), than the non-burrowing *Holothuria grisea* (Pantin and Sawaya, 1953) in which the following pressures were recorded: average 'resting' pressure, 1·2 mm Hg; 'normal body movements', up to 2·2 mm Hg; extreme contraction caused by the injection of acetylcholine chloride, 14·0 mm Hg. Other burrowing animals like sipunculids and annelids also work at relatively high pressures of the same order of magnitude as those in *Caudina*. The 'resting pressure' which features in many of these measurements is probably something of a misnomer, since most animals are rarely completely inactive, as the name implies. Resting pressure may perhaps best be regarded as a tonic pressure that is maintained when the level of activity of the animal is low.

Fluid pressures in *Calliactis parasitica* (Chapman, 1949) appear, under comparable conditions, to be appreciably higher than those recorded in *Metridium senile* (Batham and Pantin, 1950), sometimes by as much as an order of magnitude. Resting pressures vary between 2 and 33 mm sea-water in *Calliactis*, but average around 2 or 3 mm in *Metridium* if the animal is quiet, or about 6 mm sea-water if it is active and muscle tonus is high. On sudden contraction, the internal pressure of *Metridium* may rise to 20 mm sea-water, although it is usually less, and on further stimulation may exceptionally rise to 50 or 100 mm sea-water. In *Calliactis* contraction commonly produces pressures of the order of 100–140 mm sea-water. As we have already observed, the chief resistance to change of shape in sea-anemones is offered by the body wall, and principally by the mesogloea. *Metridium* is capable of very much larger changes of shape than most other actinians, including *Calliactis*, and has a correspondingly thin mesogloea. That of *Calliactis* is very thick (Chapman, 1953a, b), and is likely to be chiefly responsible for the much higher pressures recorded in that animal.

Batham and Pantin (1950) showed that if *Metridium* is distended to near its maximum volume, either by feeding or by artificial inflation with sea-water, the parietal and circular body-wall muscles nearly reach their isometric limit. The internal pressure is then about 12 or 13 mm sea-water (approaching 1·0 mm Hg). The normal working range of these muscles corresponds to pressures ranging from about 2 mm, to an average maximum of about 6 or 7 mm during changes of shape, that is, about 20 to 50 per cent of their isometric pressure. In *Arenicola*, Chapman and Newell (1947) showed that the resting pressures are of the order of 10 mm Hg, burrowing entails pressures up to about 30 mm, while the maximum pressure that can be generated (quoted by Batham and Pantin) is about 59–66 mm Hg (800–900 mm sea-water). Again, the working range is below half the isometric pressure. In the nematode *Ascaris* also, resting pressures and those generated in normal, slight activity average 70 mm Hg, and extreme pressures generated when the animal coils may reach 225 mm Hg (Harris and Crofton, 1957). From calculations based on the maximum tension that can be developed in muscle preparations when they are electrically stimulated (Bradley, 1959), it appears that the maximum pressure that can be generated in the pseudocoel is of the order of 400–450 mm Hg (Harris, 1962). Batham and Pantin remind us that the range of optimum mechanical efficiency of frog skeletal muscle is just below 50 per cent of the isometric tension (Fenn, 1923; Hill, 1939) and suggest that it would not be altogether surprising if this range of working tensions proved the most efficient in a wide variety of animals. Certainly the working range quite often appears to be about half the maximum tension that can be generated (Table 7).

Sipunculus and the burrowing holothurian *Caudina* both appear to employ pressures much closer to the isometric limit of the muscles than annelids or coelenterates do. These high pressures are recorded when the animals are burrowing and imply either that the muscles of sipunculids and holothurians have very different properties from those of other animals, or, more likely, that burrowing is an extremely infrequent activity for which the animals can afford to employ pressures well outside the range for optimal mechanical efficiency, for the sake of penetrating the substratum. Burrowing for them would then be of a similar character to defence reactions in which animals sacrifice efficiency to the needs of survival in an emergency situation.

Metridium, though turgid to the point at which the body-wall muscles can no longer produce changes of shape, is still capable of sudden and forcible retraction by contraction of the powerful mesenteric retractor muscles, with considerable further increases in internal pressure. As Batham and Pantin (1950) point out, animals with a hydrostatic skeleton differ from those with a rigid skeleton in that changes of shape involve considerable changes in the length of the muscles and also cause considerable deformation to the whole of the musculature. It follows that rapid changes of shape, such as we find in defence reactions, are extremely wasteful of energy and animals can continue to behave in this spendthrift fashion only if periods of uneconomic use of the musculature are kept to a minimum.

This is certainly true of almost all defensive and escape mechanisms and is apparently true of burrowing in some of the unsegmented coelomates. Even *Arenicola*, in which burrowing is carried out at internal pressures probably well within the range of optimal mechanical efficiency of the muscles, usually inhabits the same burrow for weeks on end (Schwarz, 1932; Thamdrup, 1935), though some of Wells' (1945) observations of worms in the aquarium, suggest that the worm may often modify the shape of its head shaft without actually forming a new burrow. Possibly the explanation of the fact that burrowing demands the use of forces well within the capabilities of the muscular system in *Arenicola*, is that although it is a sedentary animal, it still engages in rather more frequent minor burrowing activity than other coelomate worms.

CONCLUSIONS

Worms which possess a true fluid skeleton in the form of a secondary body cavity are able to make use of changes of body shape produced by contractions of the body-wall musculature. The same changes of shape can be produced in nemerteans, but they are at best of subsidiary importance in locomotion and are generally too feeble and slow to perform mechanical work. It is particularly significant that the only muscular activity of nemerteans which involves overcoming substantial external resistance, burrowing into the substratum, is performed by the proboscis, the functioning of which is engineered by contraction of muscles about a true fluid skeleton in the form of the rhynchocoel. Many coelomate worms also make use of the proboscis when burrowing, particularly in order to make a cavity in the substratum into which the rest of the body may move. The eversion of the proboscis can be achieved with greater speed and force than the protraction of any other part of the body because the entire body musculature can, if necessary, contract at once and contribute to the sudden rise in fluid pressure causing the eversion of the proboscis.

The bodily changes of shape of coelomate worms that are of the greatest significance are peristaltic waves that pass along the body wall. These are of two types. The first, produced by waves of activity passing along the circular muscles alone, do not result in movement of the worm, but can be employed to produce water currents through the burrow, and these water currents generally are used as a source of food or oxygen. Peristaltic contractions involving the antagonistic contractions of both the circular and longitudinal muscles are employed in locomotion. This represents a considerable advance over pedal locomotory waves that occur in acoelomates and which are produced by contractions of the longitudinal musculature alone. Extension of the longitudinal muscles, in the former type of locomotory wave, is caused not merely by the contraction of other longitudinal muscles, but chiefly by the contraction of the antagonistic circular muscles. As a result, during the phase of the locomotory cycle when a portion of the body wall moves forwards, the forces causing that forward movement are produced by the combined contraction of both sets of muscles and greater thrusts can be achieved. The fact that the circular muscles are involved in this locomotion has important consequences. As the circular muscles contract, the cross-section of the worm must become circular. Hence the surface applied to the ground is minimal, unlike the situation in acoelomates, where a large surface may be applied to the substratum. Peristaltic locomotory waves are inappropriate in animals that move over the substratum, but are ideally suited to animals that move in a burrow, since the entire circumference of the body where the longitudinal muscles are contracted and the diameter is greatest, can be applied to the substratum, so providing a *point d'appui* against which considerable thrusts may be exerted before slipping occurs. Another consequence of the circular cross-section is that the entire body-wall musculature may be employed in locomotion instead of only that on the pedal surface.

Peristaltic locomotory waves involve a considerable deformation of the fluid skeleton and demand the rapid transmission of pressure changes, without damping, from one part of the body to another. A parenchymatous tissue is insufficiently plastic to be suitable and, with the exception of the nemerteans which are in any case quasi-coelomate, only worms with a true fluid skeleton, whether it be a coelom or a pseudocoel, show these types of movements. They are nearly all capable of burrowing and the majority live in permanent or semi-permanent burrows or tubes. None is capable of rapid locomotion over the surface of the ground, an activity for which they are ill-adapted. However, it will be observed that none of these worms, with the exception of *Priapulus* which burrows in a highly specialized manner, is capable of sustained locomotory activity, and all of them are sedentary.

Nematodes and holothurians both present interesting parallels with the situation in the coelomate worms, though both are highly specialized. The nematodes possess only longitudinal muscles in the body wall, which are antagonized by the high internal hydrostatic pressure in conjunction with a very thick cuticle. They are capable of a considerable range of body movements, but the majority swim by undulatory movements or progress by movements which are closely related to swimming by flexions caused by alternate contractions of longitudinal muscles on opposite sides of the body. Because the cuticle is necessarily extremely thick and tough, they are not capable of great dilation of the body wall, and in consequence almost without exception do not perform peristaltic movements. The most usual method of locomotion, in holothuroids, as in other echinoderms, is by the use of podia. Most are adapted for progression over the surface of the ground and only in very few cases does the body-wall musculature assist these movements. The apodous holothurians, most of which burrow, have a musculature organized on the same lines as that of the body wall of worms and are reported to move by peristaltic locomotory waves in much the same manner.

Although the use of a coelomic fluid skeleton is a very common locomotory adaptation in soft-bodied animals, this is not the only way in which fluid skeletons are employed. Almost any activity that involves considerable distortion of the body wall, as, for example, the eversion of the nemertean proboscis or of a bryozoan polypide, can best be effected by the use of fluid pressures generated by contraction of the muscles bounding the hydrostatic skeleton. If the distortion involved is slight, a tissue-filled cavity can serve, but when the distortion is considerable or must be accomplished with speed or power, the resistance to deformation afforded by a parenchymatous tissue

CONCLUSIONS

renders it inappropriate as a hydrostatic skeleton and it is replaced by a fluid than can be more readily deformed.

Changes of shape in animals with a hydrostatic skeleton have two important features. Since soft-bodied animals are unable to make use of levers as animals with a rigid skeleton can, changes in the lengths of muscles are necessarily great. Also, it is impossible for the effect of muscle contractions to be localized; the entire musculature is influenced by muscular activity anywhere in the body. Because of the magnitude of muscle contraction and the widespread deformation it entails, changes of shape are slow. Rapid movements that occur in defensive reactions are wasteful of energy and can be used only occasionally and in emergency.

Despite these fundamental differences between animals with hydrostatic and rigid skeletons, the muscles of soft-bodied animals, like those of vertebrates, appear likely to show their greatest mechanical efficiency when operating at tensions between about 20 and 50 per cent of their isometric limit. Most of the changes of shape in coelenterates take place within these limits. The same is true of *Arenicola* when it is burrowing and carrying on a variety of maintenance activities. However, in some unsegmented coelomates, the internal hydrostatic pressure generated when they are burrowing is proportionately very much greater and suggests that their muscles are working outside the limits of optimum mechanical efficiency. If this is true, it can only be explained as an adaptation to a truly sedentary existence in which burrowing is a very infrequent activity. Possibly this is a physiological concomitant of the fact we have already observed, that unsegmented coelomates tend to live sedentary lives, buried in the substratum.

4

THE SEPTATE CONDITION

THE significance of segmentation can best be understood by considering the differences in structure, and in locomotory and other abilities, between segmented and non-segmented worms. Peristaltic locomotory waves that are used rather ineffectively by some unsegmented coelomate worms when they are crawling, become perfected as a means of locomotion in annelids that have a segmented musculature and in which the coelom is subdivided by intersegmental septa. The earthworm, with its segmented circular body-wall muscles and an almost complete series of septa, represents the nearest approach to the structure of an idealized model annelid that we are likely to meet. The mechanisms of earthworm locomotion have been analysed by Gray and Lissmann (1938), and from their analysis we are able to gain some impression of the essential characteristics of locomotion in these segmented worms.

Not all annelids move like earthworms, however, and in the locomotion of some polychaetes, we find a radically different type of movement and a different morphology from anything we have considered hitherto. By considering the function and behaviour of the segmental musculature and the septa in different annelids with their varied structure and locomotory techniques, we may begin to understand the fundamental nature of segmentation, its functions, and the advantages it confers on animals that are constructed in this way. But we must not confine our attention solely to animals that show complete segmentation. Much can be learned from animals that are derived from segmented worms but have suffered a reduction or loss of their segmental organization. The body of oligomerous animals, at least in its primitive condition, can be considered to be divided into three segments, each separated from its neighbour by a septum. These animals are constructed on an entirely different plan from annelids and their segments function in a different way. They, too, must obviously be taken into account in any analysis of segmentation.

LOCOMOTION AND MUSCULATURE OF EARTHWORMS

According to Gray and Lissmann (1938), when an earthworm such as *Lumbricus* is at rest, both circular and longitudinal body-wall muscles are partly relaxed and the worm has approximately the same diameter from head to tail. The start of locomotion is marked by a contraction of the circular muscles in the anterior part of the body, so extending a number of segments. A wave of circular muscle contraction then passes backwards along the body, and when it has travelled some distance it is followed by a wave of longitudinal muscle contraction, also starting from the anterior end. A second wave of circular muscle contraction follows, and in a crawling worm there may be several regions of alternating circular and longitudinal muscle contraction along the body (Fig. 66). The effect of contraction of the circular muscles is, of course, to make the segments longer and thinner, while contraction of the longitudinal muscles shortens the segments in preparation for the next phase of circular muscle contraction.

The elongation of the body caused by contraction of the circular muscles

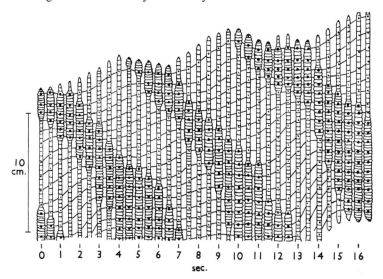

FIG. 66. Stages in the locomotory cycle of an earthworm. (After Gray and Lissmann, 1938.)

can be used for locomotion only if a thrust is exerted against the substratum. Those parts of the body in which the longitudinal muscles are maximally contracted are the fattest, and, when the earthworm is burrowing, they are wedged against the wall of the burrow. Resistance to slipping is increased by the protraction of the chaetae as the longitudinal muscles contract. *Lumbricus*, with the chaetae confined to the ventral half of the body, is adapted to crawling over the ground as well as through it, but many earthworms, including the very successful and numerous Megascolecidae, are adapted to a purely subterranean existence and in many of them, the chaetae are symmetrically disposed around the segments.

Segments in which the longitudinal muscles are maximally contracted thus form the anchor or *point d'appui*, and are instantaneously stationary with respect to the ground. In segments anterior to this anchor, the circular muscles contract, the segments elongate and, exerting a backthrust against the stationary segment, force the anterior end of the worm forwards. In segments posterior to the anchor, longitudinal muscles contract and draw the posterior part of the worm forwards, exerting a pull on the anchoring segment as they do so. A short time after the instantaneous situation represented in Fig. 67, the muscles of the segment immediately behind the fixed one become maximally contracted and this segment in its turn forms the new anchor while the circular muscles of the more anterior segment begin their phase of contraction. Each segment in succession forms a *point d'appui* and there may be several such stationary points along the length of the worm at the same time.

It is clear that in this type of locomotion there must be very precise co-

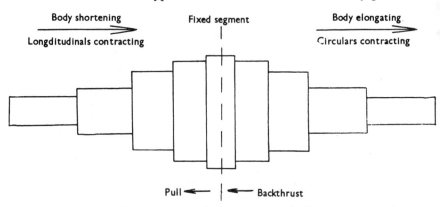

Fig. 67. Forces acting upon the stationary segment and adjacent segments of an earthworm performing peristaltic locomotory movements.

ordination between the activity of neighbouring parts of the body. Coordination is maintained in part by the contraction of the longitudinal muscles in one segment placing the longitudinal muscles of adjacent segments under tension and, presumably by a proprioceptive reflex arc, causing their contraction also. However, section of the entire body-wall musculature does not disrupt the passage of the locomotory wave, providing the ventral nerve cord remains intact, so that there is probably also an overall nervous coordination of locomotory activity in adjacent segments.

So far, we have treated each segment as though it were a hydrostatically independent unit, its activity unaffected by that of neighbouring segments. Mechanical independence of this sort would result in the contraction of the circular muscles in one segment antagonizing the longitudinal muscles of the same segment but being without effect elsewhere in the body, and the changes of shape of a single segment would be functionally equivalent to the changes of shape of a whole acoelomate worm (see Chapter 2).

For adjoining segments to be autonomous, the septa should be both water-tight and rigid, otherwise changes in the fluid pressure generated by the contractions of the body-wall muscles of one segment would be transmitted to the next either by leakage of coelomic fluid across the septum or by deformation of the septum. But in fact, annelid septa are neither complete diaphragms nor are they perfectly rigid and we must now consider to what extent they can function as hydrostatic isolating agencies.

Both the structure of the septa of *Lumbricus* and the manner in which they function have been studied in some detail by Newell (1950). All the septa are perforated by a ventral foramen through which the nerve cord and the ventral blood vessel pass (Fig. 68). However, this foramen is bordered by a

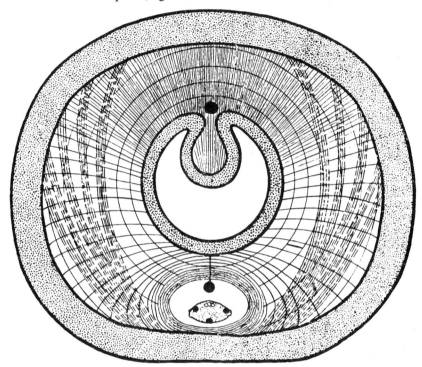

FIG. 68. Disposition of the muscle fibres in the septum of *Lumbricus*. (From Newell, 1950.)

sphincter muscle which effectively closes it when the worm is crawling or burrowing. Fluid injected into one segment does not pass into the next unless it is injected at such a pressure as to rupture the septum. But if the worm is narcotized, the sphincter muscles of the foramina do not contract and the injected fluid passes freely from segment to segment. The effectiveness of the septa as barriers to the passage of coelomic fluid across them has also been confirmed by X-radiography in *Lumbricus* by Newell (1950) and in *Tubifex* by Scully (1962).

As to the rigidity of the septa, it is very difficult to render a flexible diaphragm rigid by applying tensions at its edge when the distorting pressures are applied perpendicularly to its surface. The septa of *Lumbricus* are provided with circular, radial and oblique muscles in the plane of the septum, all of which help to keep it taut, but even so they are only partially successful in this. There are, in addition to the muscles in the plane of the septum, extrinsic muscles which run posteriorly from the body wall to a few anterior septa, but these are unimportant and cannot affect the argument as it applies to the rest of the body. The behaviour of the septa in a crawling worm cannot be observed directly in an animal such as *Lumbricus*, but they can in small oligochaetes with a transparent body wall. In *Tubifex*, there is a considerable distortion of the septa as pressure changes occur in the coelomic fluid of a segment, but the tension in the septal muscles offers some resistance to deformation and pressure changes are damped down so that they are not transmitted for more than a few segments in either direction. In fact, if saline is forced into the clitellar region of *Lumbricus*, the hydrostatic pressure in the anterior third of the worm does not rise appreciably (by less than 10 per cent of the normal pressure in these segments in active worms) unless the applied pressure is sufficient to rupture the septa (Newell, 1950).

Muscular septa, perforated by a ventral foramen and comparable to those of *Lumbricus* probably exist in all earthworms (Ribaucourt, 1901; Bahl, 1919; Stephenson, 1930) and presumably function in a similar manner. Apart from the modification of a few anterior septa in association with the genital system (a common feature in oligochaetes), the only substantial departure from this situation is found in the megascolecid *Pheretima*. In this Indian earthworm, all but the first few specialized septa have numerous perforations (Fig. 69); Bahl (1919) counted no less than 68 in half a septum from the middle region of the body of *Pheretima posthuma*, though there are somewhat fewer perforations in the more posterior septa. All the foramina are bordered by a sphincter muscle and can probably be sealed off in the same manner as those of *Lumbricus*. The situation in *P. posthuma* is extreme, and other species of *Pheretima* have only one or two additional perforations in each septum.

The arrangement of muscle fibres in the septa appears to vary considerably in different oligochaetes. Newell (1950) described circular, radial and oblique fibres in the septa of *Lumbricus*, but in the tubificid *Limnodrilus* the fibres run dorso-ventrally and laterally (Nomura, 1913), and in another tubificid, *Monopylephorus*, there are semi-circular fibres inserted on the ventral body wall and variously orientated fibres running from the dorsal and lateral parts of the body wall to the gut and to the semi-circular fibres (Nomura, 1915). Whatever the disposition of fibres in the septum, they serve a similar function of keeping the septum taut.

Since oligochaete septa are imperfectly rigid, no segment is completely independent of events taking place in adjacent segments. However, the damping of pressure changes by a succession of septa does permit pressure gradients to develop over short lengths of the body. The contraction of one set of muscles in a particular segment affects chiefly the remaining musculature of that segment; it has a diminishing effect in neighbouring segments, and none at all in segments a short distance away. This has two advantages: pressure changes produced by localized muscle contraction are not dissipated throughout the body of the animal, and the tonus of body-wall muscles in one segment does not have to be constantly adjusted in response to contractions

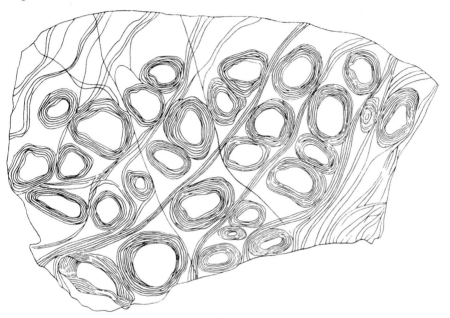

FIG. 69. Disposition of muscle fibres in a portion of the septum of *Pheretima posthuma*, showing also the numerous sphinctered foramina. (From Bahl, 1919.)

of muscles elsewhere in the worm, which are without relevance in that segment.

The imperfection of the septa as bulkheads is also reflected in the structure of the longitudinal musculature. It is often supposed that the longitudinal muscles of earthworms, if not all annelids, are organized on a strictly segmental basis. In fact, the longitudinal muscle fibres of *Lumbricus* are several millimeters long, that is two or three times the length of the segments (Schneider, 1908; Stephenson, 1930), and the same appears to be true of polychaetes. This is not surprising in view of the failure of the septa to provide hydrostatic isolation of the segments. From a mechanical point of view, the division of the body into segments is partial and approximate—an advantage, possibly, in that the interdependence of adjacent segments is likely to result in the smoother transmission of locomotory waves along the body. Although it is incorrect to regard individual segments of earthworms as totally isolated from adjacent segments, they have some degree of autonomy and, since in locomotion no segment behaves in a very different manner from its immediate neighbours, it is sufficient that a segment should be isolated from other segments some distance away where totally different events are taking place.

This is not to say that localized changes of shape are impossible in worms without a segmental musculature and septa—they clearly are not—but the subdivision of the body into mechanically nearly independent units makes for a far more efficient locomotory system than that of unsegmented worms. Manton (1961b) demonstrated that the lateral force that can be generated by contraction of the longitudinal muscles of the non-septate part of *Arenicola* is 130–200 g/cm^2. A specimen of the earthworm *Allolobophora* of about the same size and weight can generate forces of about 80–100 g/cm^2, less than *Arenicola* since only a small part of the body-wall musculature contributes to the thrust. However, whereas *Arenicola* is capable of making a maximum heave only once or twice, whereupon leakage of coelomic fluid and bursting of the capillaries in the gills impair its pushing ability, the earthworm is capable of repeating its maximum thrust many times over. This capability of sustained activity appears to be a major difference between segmented, septate worms which crawl rapidly and burrow frequently, and unsegmented, non-septate worms which, with the single exception of *Priapulus*, are relatively sedentary, crawl slowly, except possibly when moving up and down their tubes, and burrow infrequently.

LOCOMOTION OF ERRANT POLYCHAETES

Polychaetes form a diverse group of animals and have evolved a variety of methods of locomotion with corresponding changes in the musculature. The type of movement peculiar to them, is crawling by means of parapodia which are used as locomotory appendages or levers in a manner comparable to the legs of land animals. Not all polychaetes move in this way, but it is characteristic of errant worms such as *Nereis*.

During slow crawling movements by *Nereis*, the body-wall muscles play very little part and locomotory forces are provided by the parapodial musculature. The cycle of activity has been described by Foxon (1936) and Gray (1939) (Fig. 70). Initially, a parapodium is directed forward and its tip is applied to the substratum. Contraction of the parapodial flexor muscles exerts a forward pull on the body of the animal, which advances, and at the end of this power-stroke the parapodium is inclined obliquely backwards. It remains inactive until the rapid preparatory movement anticipating the next power-stroke, when it is lifted clear of the substratum and swung forwards again. During the back stroke, the aciculum and chaetae are protracted, helping to anchor the tip of the parapodium. During the preparatory stroke they are retracted. As one parapodium begins its effective stroke, the parapodium next anterior to it follows suit after a brief interval, and so on forwards along the body. Only a small number of segments separates adjacent active parapodia and since the parapodia of opposite sides of the body alternate with one another in performing their power-strokes, the worm is provided with a great many *points d'appui* along its body and must therefore experience a fairly constant and uniform forward thrust from the simultaneous contraction of numerous parapodial flexor muscles.

When it is crawling rapidly over the substratum, *Nereis* exhibits a more complicated type of locomotory activity. This involves the longitudinal body-wall muscles as well as the extrinsic muscles of the parapodium, and in consequence, much more powerful forces are available than when the

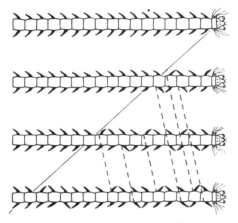

FIG. 70. Cycle of activity of the parapodia of *Nereis* during slow crawling. Broken lines connect successive parapodia which are about to execute a power stroke; the line marks the spread of excitation backwards along the body. (After Gray, 1939.)

parapodial muscles alone are employed. Zones of contraction alternating on opposite sides of the body pass along the longitudinal muscles from posterior to anterior, so throwing the body into lateral sinusoidal waves (Fig. 71). Fig. 72 shows the successive positions of two points on opposite sides of a segment during the passage of one complete locomotory wave. Each point, in turn, remains stationary with respect to the ground while its fellow rotates about it and is moved forwards. While a point on a segment is temporarily fixed, the underlying longitudinal muscles are relaxed and maximally extended. This is the converse of the situation in earthworms, in which the *points d'appui* are formed in segments where the longitudinal muscles are contracted. As a consequence of this, the locomotory waves of earthworms pass along the body in the opposite direction to the direction of motion of the worm, whereas in *Nereis*, they pass in the same direction. In fact, to apply to annelids the terminology developed to describe the pedal waves of gastropods, the locomotory cycles of earthworms are retrograde, those of

Nereis are direct; in *Lumbricus*, the locomotory waves are monotaxic, in *Nereis*, they are ditaxic and opposite.

Rapid ambulation in *Nereis* is complicated by the activity of the parapodia, which is superimposed upon that of the longitudinal muscles (Fig. 73). The parapodial power-stroke is performed as the parapodium passes over the crest of the locomotory wave, that is, when the underlying longitudinal muscles are extended. The backthrust exerted by the contraction of the longitudinal muscles of the opposite side of the body is thus transmitted to

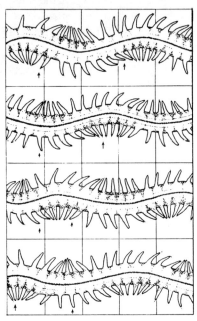

Fig. 71. Rapid crawling by *Nereis*. Arrows indicate successive positions of two parapodia. (Partly after Gray, 1939.)

the ground through the parapodia while they are in the act of performing their power-stroke.

During rapid ambulation, therefore, there are two sets of forces contributing to forward progression; one due to the contraction of the longitudinal body-wall muscles, the other to the contraction of the parapodial flexor muscles. The forces generated by the latter are evidently weak and by themselves are capable of transporting the worm only very slowly. For rapid locomotion the far more powerful forces available from contraction of the longitudinal muscles are necessary.

Although the parapodia may be relatively ineffective as appendages, their use in locomotion is of the greatest theoretical importance, for it represents a major departure from the type of locomotion observed in other worms

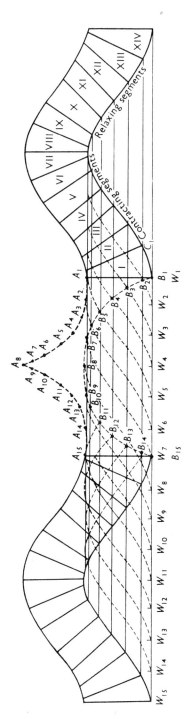

FIG. 72. Movement relative to the ground of two points on opposite sides of a segment of *Nereis* during the passage of one complete locomotory wave. (From Gray, 1939.)

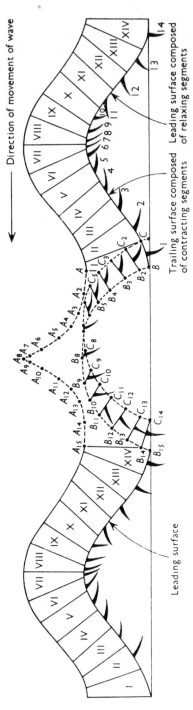

Fig. 73. Movement relative to the ground of one side of a segment of *Nereis*, with the corresponding activity of the parapodium superimposed. (After Gray, 1939.)

and foreshadows the use of legs rather than the body-wall muscles for locomotion, which is a feature of the arthropods.

The mechanism of crawling in *Nereis* is probably typical of other errant polychaetes, though undoubtedly on further investigation we shall find a number of variations in this pattern of activity, some of them important.

Nephtys, an errant polychaete superficially similar to *Nereis*, gives us an illustration of the type of variation upon the fundamental locomotory patterns in these worms that we may expect to find as investigations are extended to a greater number of animals (Clark and Clark, 1960b). *Nephtys* is rather unusual among errant polychaetes in that it is an active, burrowing worm. It probably never displays slow crawling for which parapodial muscles alone are responsible. When buried in the sand it displays burrowing movements and rapid ambulation. Movement over the substratum is by swimming rather than crawling.

As we have already observed (p. 89), its method of burrowing is essentially to make a cavity in the sand by suddenly and violently everting its proboscis. During this phase, the body of the worm is anchored by the widest segments of the body, between the 15th and 45th, which are the only ones wide enough to reach the sides of the burrow. The parapodia of these segments are inclined obliquely backwards and increase the frictional forces between the worm and the sand. The proboscis is then inverted and the worm crawls forward into the cavity it has just made. The proboscis of *Nephtys* is very large and muscular, and it has a considerably greater diameter than the first 5–10 segments of the body. Special anatomical modifications of these segments are necessary to permit the passage of this large organ through them (Clark, 1956, 1958), but the point of immediate concern is that these segments are too narrow to reach both sides of the cavity that the proboscis has made. An undulatory locomotory wave begins, as in *Nereis*, at the anterior end of the body and extends backwards along it. As the crest of the locomotory wave reaches the parapodia of the anterior segments, they are carried laterally and make contact with the walls of the burrow as they perform their power-stroke. The wider, succeeding segments also become involved in this activity, but since they are almost as wide as the burrow, the locomotory wave is very much reduced in amplitude by the time it reaches them and has very little effect. The tapering segments posterior to segment 50 remain passive throughout the cycle and are dragged forwards without contributing to the total locomotory forces.

The differences between the locomotion of *Nephtys* and *Nereis* are, thus, that *Nephtys* does not exhibit slow ambulation and the undulatory waves in rapid ambulation decrease in amplitude very rapidly as they pass from the anterior to the middle segments of the body. These changes are of course related to the different morphology and to the burrowing habit of *Nephtys*. An additional difference that we shall examine later, is that only the distal half of the parapodium, instead of the whole member, is moved when it performs its power-stroke in *Nephtys*.

A quite different modification of parapodial locomotory activity can be observed in *Sabella* when the worm crawls along its tube. In some ways, *Sabella* presents the exact antithesis of the situation found in *Nephtys*, for instead of the segments most actively engaged in locomotion being too narrow to reach the sides of the burrow so that the longitudinal muscles are necessarily involved with crawling, in *Sabella* the tube fits the worm so tightly that locomotory waves produced by contraction of the longitudinal muscles are virtually precluded. When the worm crawls along its tube, a parapodium is extended until its tip meets the wall of the tube, it is then swung into the appropriate direction, e.g. inclined backwards if the worm is moving forwards,

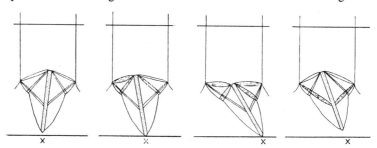

FIG. 74. Cycle of activity of a notopodium of *Sabella*, when the animal is 'poling' itself along its tube. (Original.)

and then protracted still further, so pushing the animal along (Fig. 74). The effect is an activity very like 'poling' a barge. At the end of the movement, when the parapodium has been protracted to its greatest extent, it is retracted into the body again. All the parapodia contribute to this locomotion, but are generally uncoordinated in their activity, except that parapodia adjacent to and opposite an active one are themselves inactive. *Sabella* is also capable of peristaltic movements in its tube, both for irrigation (Nicol, 1931; Wells, 1951, 1952a) and for locomotion, the latter generally when the worm is backing slowly down its tube.

THE MUSCULATURE OF ERRANT POLYCHAETES

The very different methods of locomotion employed by errant polychaetes from those of unsegmented worms, oligochaetes and even some sedentary polychaetes, demand a radically different organization of the segmental musculature. The most significant development is the appearance of an extrinsic parapodial musculature. This has profound consequences on the whole of the segmental architecture. The parapodia are moved by a series of muscles which run diagonally from the top and sides of the ventral nerve cord to insertions in the parapodial walls. These muscles cross the coelom and are incompatible with the existence of complete circular and longitudinal muscle coats. In most errant polychaetes, the circular muscle layer is very much reduced and is confined to the neighbourhood of the intersegmental boundary; in some, it is lacking altogether. The longitudinal muscles are confined to the dorsal and ventral surfaces of the segment. The lateral body wall, now very much less muscular than it is in worms with a complete muscle coat, is further weakened by the insertion of the thin-walled parapodia into it. It is supported by oblique muscles which run from the ventral nerve cord to the intersegmental body wall. It may also be supported by other structures.

This pattern of muscular organization is very much modified in many worms, particularly in the more sedentary species. A few have acquired a condition very similar to that of earthworms and many, that of unsegmented coelomates. The longitudinal and circular muscles then form almost complete muscle coats and there is generally an associated reduction of the oblique muscles which now become redundant. But these changes are possible only at the expense of the extrinsic parapodial muscles and ultimately of the parapodia themselves.

Thus we find that peristaltic locomotion is fundamentally incompatible with movement by means of parapodia. *Sabella* is one of the few polychaetes to have reached a compromise between these two locomotory techniques and is able to move by either means, and parapodial locomotion in this worm is through the agency of the intrinsic parapodial muscles. The extrinsic muscles have nearly disappeared and as a result it is possible for almost complete circular and longitudinal muscle coats to develop. But this represents a very unusual condition and the musculature of errant polychaetes must obviously be considered in a totally different light from that of oligochaetes or, indeed, of any other worms.

Although such movements play very little part in locomotion, errant polychaetes are capable of reversible changes of length. Indeed, a number of 'errant' worms, despite their name, live in burrows or tubes and display, among other features typical of tubicolous worms, a well-marked contraction reflex when they are stimulated. In some worms, recovery from the shortened state is undoubtedly achieved by contraction of the circular muscles, so that circular and longitudinal muscles antagonize one another just as they do in earthworms. But this is not always so. In *Nephtys* the circular muscles are lacking and dorso-ventral muscles in the proximal part of the parapodia antagonize the longitudinal set (Clark and Clark, 1960b). The animal is approximately rectangular in cross-section and changes in length are compensated by changes in height. The width of the segment remains constant. This is a far cry from the situation in worms in which the circular muscles antagonize the longitudinals and the cross-sectional shape is perforce approximately circular. Whether a similar system obtains in other errant polychaetes is unknown, but in view of the great reduction of the circular muscles in most of them, comparable modifications are quite likely.

The extrinsic parapodial musculature is generally complicated and various modifications of it exist in different polychaete families. Basically there appear to be four sets of muscles inserted into each parapodium, two dorsal and two ventral. The dorsal muscles arise at the outer edges of the dorsal longitudinal muscles, cross one another diagonally in the root of the noto-

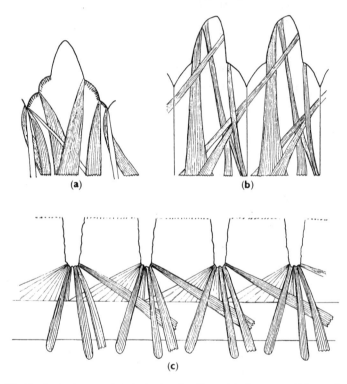

FIG. 75. Tendency in errant polychaetes for the extrinsic parapodial muscles to cross segmental boundaries. (*a*) *Nephtys* (after Clark and Clark, 1960b), (*b*) *Nereis* (after Defretin, 1949), (*c*) *Aphrodite* (original), in diagrammatic plan views.

podium and are inserted into the opposite wall of the parapodium from that in which they originate. The ventral muscles arise in the sheath investing the ventral nerve cord or on the floor of the segment near it. They run directly to the parapodial walls, as a rule without crossing each other. The ventral extrinsic muscles are generally subdivided and often present a much more complicated appearance than this description might suggest.

In many worms, the insertions of these muscles are distributed over the walls of the parapodium and extend to its tip (*Eunice*, *Nereis*) though this is not always so and in *Nephtys*, the extrinsic parapodial muscles are still inserted near the base of the parapodium. An additional complication, seen in *Nephtys*, *Nereis*, *Aphrodite*, and a number of sedentary worms, is a loss of the strict segmental organization of the parapodial musculature. Muscles arising on the nerve cord sheath in one segment may cross the segmental boundary and be inserted into the parapodium of the adjacent segment (Fig. 75). In *Nephtys* one such muscle exists. It passes posteriorly under the functional septum and is inserted just inside the next parapodium on its anterior face. In *Nereis*, the major parapodial adductor muscle crosses the segmental boundary and runs to the posterior face of the segment behind. In *Aphrodite*, several of the muscles which anchor the intersegmental region cross segmental boundaries. Another important source of variation is the position in the segment of the insertion of these muscles into the ventral nerve cord sheath. If, as in *Nephtys*, those inserted in the anterior face of the parapodium arise in the anterior part of the same segment, their mechanical advantage is slight. If, as in *Nereis*, the same muscles arise in the posterior part of the segment, their mechanical advantage is correspondingly greater, and increased power can be generated during the parapodial stroke.

Most of the intrinsic parapodial muscles are concerned with slight changes of shape of the parapodium; only the acicular muscles may play some part in locomotory movements. As we have already seen, the aciculum and chaetae are generally protracted as the parapodium performs its power-stroke, and are retracted during the recovery stroke. In *Nephtys* and *Aphrodite* the acicular muscles play a more dynamic role and actually produce the power-stroke themselves without the assistance of the extrinsic muscles which are inserted near the base of the parapodium. The acicular protractor muscles are attached to the inner end of the aciculum and are inserted into the parapodial walls near the insertions of the extrinsic parapodial muscles. Contraction of the acicular muscles swings the aciculum forwards, and so moves the distal part of the parapodium backwards. In *Nephtys* the anterior acicular muscles which thus produce the power-stroke are considerably more extensive than the posterior muscles which are responsible for the recovery stroke. The parapodium hinges about the point of insertion of these muscles, instead of about its base as in *Nereis*. The acicular retractor muscles are extensive in all these worms and run from the base of the aciculum to the lateral body wall and sometimes to the outer edges of the ventral longitudinal muscles on the ventral body wall.

With this development in a number of errant polychaetes, we find the first approach to the evolution of a true skeletal musculature for the movement of locomotory appendages. The hydrostatic skeleton still remains essential, however, because the worms lack structurally rigid supports for the locomotory muscles and they can only be provided by the development of a complicated series of muscular braces, balanced against turgor pressure.

THE SEPTA OF POLYCHAETES

The segmental musculature and the method of locomotion of errant polychaetes differ so much from those of earthworms, that the septa cannot serve exactly the same function in the two groups of annelids. In fact, the detailed structure of polychaete septa varies considerably and it is likely that they serve many functions. Some of these functions have been discussed by Clark (1962a).

Although it is true that, in general, polychaetes move in an entirely different way from earthworms, and have a fundamentally different type of body architecture, the families Lumbrinereidae and Capitellidae show remarkable parallels to earthworms. Members of both families have a striking superficial resemblance to earthworms, both burrow through soft sand and mud, and both move by peristaltic locomotory waves in the same manner as earthworms. This entails some modification of the basic organization of the musculature, in particular the development of complete circular and longitudinal layers in the body wall. In lumbrinereids and capitellids, especially the latter, the parapodia are very much reduced; in both families there are well developed and tolerably complete circular muscle coats. The longitudinal muscles are modified in a variety of ways to form an almost complete muscle coat, either by extension of the dorsal and ventral muscles around the sides of the segment, as in *Lumbrinereis*, or by subdivision of the longitudinal muscles and the more or less equal spacing of the resulting blocks around the body wall, as in *Notomastus*. In the thoracic segments of capitellids, i.e. the segments most actively engaged in burrowing, the circular and longitudinal muscles form complete muscle coats. Septation is complete in both lumbrinereids and capitellids, save for the small ventral neural chamber which contains the gonads and runs uninterrupted along the whole length of the body of capitellids. There are radial and circular fibres in the septa of *Lumbrinereis*. In capitellids, the septal muscles are predominantly circular, but they tend to anastomose and form a felt-like mass; in the posterior, abdominal segments, the fibres have a more orderly, grid arrangement. *Capitella* is unusual in that the septa are fenestrated at the sides, but the perforations are bordered by a sphincter muscle and can be closed (Eisig, 1887). The locomotory movements, the arrangement of the body-wall muscles and the structure of the septa are all essentially the same in these worms as in earthworms; presumably the functions of the septa are also similar.

In errant worms in which the circular muscles are reduced, the longitudinal muscles are confined to the dorsal and ventral surfaces of the segments and the extrinsic parapodial muscles are well developed, the septa must serve a different function. Movement is achieved principally by unilateral contractions of the segmental longitudinal muscles, aided by the contractions of the contralateral parapodial flexor muscles. Besides providing forces for forward traction, the longitudinal muscles play an important incidental part in parapodial movements. Contraction of the two longitudinal muscles on one side of a segment extends those of the other side and so increases the effective length of that half of the segment at the time when the parapodial flexor muscles contract. This increases the mechanical advantage of the flexor muscles. In addition, the contraction of one set of longitudinal muscles renders the contralateral parapodium turgid so that contraction of the parapodial flexor muscles moves the appendage as a whole rather than deforming a flaccid appendage. In worms with a hydrostatic skeleton, it is important that pressure changes should be applied where they are most necessary (in this case to the contralateral longitudinal muscles and parapodium) and not be dissipated by dilations of other parts of the body wall. This is best achieved by isolating each segment from its neighbours by watertight septa. As in earthworms, it is unlikely that the septa act as perfectly rigid bulkheads, so that there must be some transmission of pressure changes from one segment to the next by deformation of the septa, but equally, the septa must damp down such pressure differences and permit the establishment of pressure gradients along short lengths of the body in errant polychaetes as they do in earthworms.

In active polychaetes with relatively short segments, and therefore with parapodial muscles with least mechanical advantage, such as *Goniada*, syllids, some eunicids and *Nerine*, the septa are complete and are inserted into the lateral body wall and the inner (coelomic) surfaces of the longitudinal muscles. In a number of other errant worms, the septa are incomplete ventrally, though it appears that when the septal muscle fibres contract, the lower free edge of the septum can be drawn tightly enough across the ventral longitudinal muscles to prevent the transmission of coelomic fluid past it. This is the situation in some polynoids. As we have already observed, the extrinsic parapodial musculature of some worms, e.g. *Nereis*, is more efficiently disposed than in other worms and some of the parapodial muscles may cross segmental boundaries. Parapodial movements are then adequate for locomotion and *Aphrodite* and *Nereis* are able to crawl, albeit slowly, without effective aid from the longitudinal muscles. *Nephtys*, with less efficiently disposed parapodial muscles is not. With this change in the locomotory method, the need for septa is reduced. In *Aphrodite*, which moves solely by parapodial movements and in which the longitudinal muscles are very much reduced, the septa are totally lacking save in a few posterior segments (Fordham, 1925). In *Nereis*, which employs the longitudinal muscles for more rapid crawling, the septa are reduced. Those of the posterior two-thirds of the body are incomplete both dorsally and ventrally (Fig. 76a), but may still be able to seal off the coelom of one segment from that of the next, but in anterior segments (Fig. 76b), these septa are reduced to little more than gut suspensory muscles. In hesionids which are very similar to nereids in their

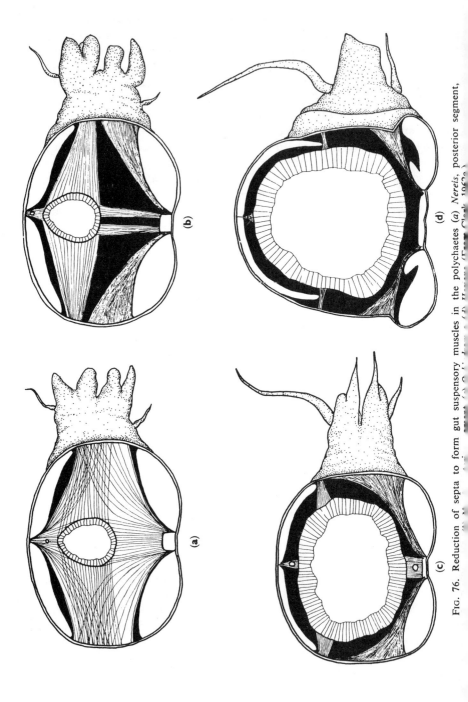

FIG. 76. Reduction of septa to form gut suspensory muscles in the polychaetes (a) *Nereis*, posterior segment, (b) *Nereis* ... (c) ... (d) ... (From Clark, 1962 ...)

construction, all the septa are reduced to gut suspensory muscles (Fig. 76c, d). Although we have no information about the arrangement of the parapodial muscles in these worms, it is likely that they have undergone an improvement in the disposition of the parapodial flexor muscles similar to that in *Nereis*.

Dorso-ventral septal muscles may help to preserve a flattened body-form in some worms. In syllids, the eunicids *Eunice* and *Marphysa* (Fig. 77a) and the spionid *Nerine*, there are particularly conspicuous dorso-ventral fibres in the septa. These are inserted on the inner (coelomic) sheath investing the longitudinal muscles and thick connective-tissue strands run radially through the muscle blocks, linking the insertions of the septal muscles with the connective-tissue stroma of the circular muscle layer and with the sub-epidermal basement membrane. The intestine of these worms is reduced to a narrow dorso-ventral slit in the intersegmental region and, as a result, the dorso-ventral fibres of the septum pass from upper to lower insertions on the body wall without insertions upon the gut. *Eunice*, *Marphysa* and *Nerine* all have dorsal gills and live buried in the sand. We know very little about the habits of these worms, but it is likely that a current of water is drawn over the gills by dorso-ventral undulations of the body, and for this reason a flattened body-form is advantageous. *Aphrodite* also draws a current of water over its respiratory surface by undulations of the dorsal body wall (van Dam, 1940), but in that worm, in the absence of septa, a flattened form is preserved by modified oblique muscles. The body of syllids is not generally flattened and the function of the dorso-ventral fibres in the septa of these worms is unknown.

A comparable function of the septa of some worms is to provide support for the thin intersegmental body wall in addition to that furnished by the oblique muscles. This is perhaps true of the anterior septa of *Nereis* which are inserted into that part of the body wall not supported by the oblique muscles. However, it is unlikely that they, or any septal muscles attached to the gut can withstand substantial lateral forces, for deformation of the intestine wall would nullify any restraining effect that the septal muscles might have on the body wall. In the aphroditids *Leanira* and *Sigalion* (Fig. 77b), the gut occupies a relatively dorsal position in the segment and the greater part of the septum is beneath it. The septal muscles extend from one side of the worm to the other without interruption and, in *Sigalion* particularly, the great development of transverse septal muscles clearly indicates their function in supporting the lateral body wall. The oblique muscles of these two species are correspondingly reduced.

The need for additional support of the lateral body wall, which is not always provided by the septa, is illustrated in a number of ways. Smith (1957) has described a series of circumferential connective-tissue fibres running in the sub-epidermal basement membrane of *Nereis*. They are attached to the collagen tissue of the outer side of the nerve cord sheath, run between the parapodia, over the dorsum of the worm and are inserted in the collagen of the mid-ventral line near their place of origin. They appear to function as restraining bands to which the slender circular muscles are attached. In *Glycera* and *Nephtys*, the oblique muscles are hypertrophied and some fibres run directly

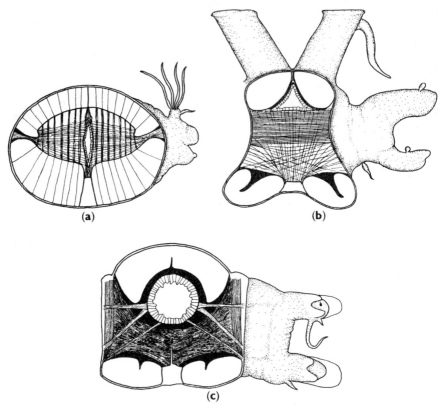

Fig. 77. Some specialized polychaete septa. (*a*) *Marphysa*, anterior region, with strongly developed dorso-ventral fibres, (*b*) *Sigalion*, with strongly developed horizontal fibres, (*c*) *Nephtys*, septa reduced to gut suspensory muscles and the functional septum derived from hypertrophied oblique muscles. (From Clark, 1962a.)

from one side of the worm to the other without making contact with the ventral nerve cord sheath. In *Nephtys* (Fig. 77c) a unique development has occurred (Clark and Clark, 1960b). The nerve cord sheath in the interseg-mental region is drawn up into a long, slender cone which provides a very considerable area for attachment of the oblique muscles. These are inserted over the whole of the intersegmental body wall and form a muscular diaphragm extending to the edges of the dorsal longitudinal muscles. The upper edges of these diaphragms support the gut in a series of slings and, in fact, since the intestine is wedged firmly into the space between the longitudinal muscles, become functional septa and prevent coelomic fluid passing from one segment to the next when the oblique muscles are contracted. It will be recalled that the parapodial muscles of *Nephtys* are not very effectively disposed and therefore isolation of the segments is important, but with the hypertrophy of the oblique muscles, this function has been transferred to them from the true septa which now become redundant and are reduced to gut suspensory muscles comparable to those of *Hesione*. Additional support of the body wall is provided in *Nephtys* by a series of elastic ligaments which extend from the nerve cord sheath to the lateral body wall and also to the parapodia (Clark and Clark, 1960a, b).

The adoption of a sedentary existence in a burrow is generally accompanied by a loss or severe reduction of the septa. This is clear in *Arenicola* and most of the other polychaetes conventionally regarded as sedentary, as well as in *Glycera*. Some of the tubicolous worms, however, retain septa and we must now enquire into their function. *Hyalinoecia*, a tubicolous eunicid, *Sabella*, and the serpulids, all have complete septa which are inserted into the dorsal and ventral longitudinal muscles and into the lateral body wall. The worms extend at the mouths of their tubes for feeding and so are vulnerable to predators at this time. They all have well, sometimes enormously, developed giant axons (Nicol, 1948) and rapid withdrawal reflexes. Such adaptations to a tubicolous life involve also the very great development of the longitudinal muscles, so much so that they occupy much of the volume of the segment and obstruct the extrinsic parapodial muscles. These worms therefore arrive at a condition similar to that of burrowing, sedentary polychaetes in which the development of the body-wall muscles to meet a particular emergency results in the reduction of the parapodia and of the effectiveness of the parapodial musculature. Unlike the burrowing worms in which the circular muscles have become extensive and the longitudinal muscles form a nearly or totally complete muscle coat, these tubicolous species still rely upon the parapodia for crawling, reduced though they be. Furthermore, although the worms can irrigate their tubes by peristalsis of the body wall (Nicol, 1931; Fox, 1938; Wells, 1951), so constricted are they that they are unable to perform more than slight lateral undulatory movements and forward progression depends chiefly upon parapodial movements. As we have already seen, errant worms which have rather inefficiently disposed extrinsic parapodial muscles and yet use them for crawling, have particular need of watertight septa isolating consecutive segments. The need is even greater in these tubicolous species in which the parapodial musculature is cramped and reduced by the enormous development of the longitudinal muscles, and which cannot make use of longitudinal muscle contractions for forward movement because of the limitations imposed by their tubes, and must therefore rely chiefly upon parapodial movements.

STRUCTURE OF THE PRIMITIVE SEPTUM

It is almost impossible to arrive at the primitive structure of annelid septa from an examination of modern worms. Nevertheless, we must make some attempt to do so in order to envisage the evolutionary changes that these structures have undergone. Theoretically, a primitive annelid exhibited complete segmentation of the body-wall musculature, septa and internal organs such as nephridia, blood vessels, etc. Whether any worm ever existed in quite this state is unknown, and certainly there appear to be considerable disadvantages to total segmentation, to judge by the large number of existing worms in which this fundamental body plan is modified.

Presumably the septa were originally complete diaphragms and were inserted directly into the sub-epidermal basement membrane on all sides of the body at the intersegmental discontinuities of the circular and longitudinal body-wall muscles. The distribution of muscle fibres in the septum must have depended on the magnitude and direction of stresses that were normally applied to it. It is clear that the relative development of different sets of muscle fibres in the septum is related to the cross-sectional shape of the segments and to the function of the septum in modern worms, and this must have been so in primitive annelids. The considerable variety in the detailed structure of polychaete septa is obviously related to the great range of variation in segmental organization in these worms and it might therefore be concluded that the primitive septum would have been a good deal less specialized and complicated. This may have been so, but the considerable differences between the three or four oligochaete septa that have been described serve as a warning that even in worms with the same type of body architecture, uncomplicated by parapodia, variation and complexity of structure are still possible.

In the earthworm *Lumbricus*, the radial septal muscles pass through the longitudinal body-wall muscles and are inserted into the circular muscle layer (Newell, 1950). This is as close to the primitive manner of insertion of the septa that has yet been discovered in annelids. In syllids (Malaquin, 1893) and *Eunice rousseaui* (Bütschli, in Hempelmann, 1931), some septal muscles are reported to pass through the longitudinal muscles, but it is more likely that these are radial connective-tissue supports of the septum rather than septal muscle fibres (Clark, 1962a). Generally, in polychaetes, the septa are inserted (if at all) into the inner, coelomic, surface of the longitudinal muscles.

The manner in which the septal insertion became transferred from the sub-epidermal basement membrane to the inner surface of the longitudinal muscles can be readily understood. The muscle fibres of the polychaete and oligochaete body-wall muscles are inserted into a connective-tissue system which divides the muscle into an often irregular series of compartments (Prenant, 1929). The inner surface of the longitudinal muscles may be bounded by a part of this connective-tissue system and, externally, the connective tissue becomes more or less continuous with that of the circular muscle layer and the sub-epidermal basement membrane. The muscle fibres exert their pull on this connective-tissue system rather than on the basement membrane directly. With such an organization it is possible for fibres of the longitudinal muscles to cross segmental boundaries, either passing through the septum or displacing it so that the septum becomes attached to the inner surface of the longitudinal muscle blocks. This is the condition of the musculature in most polychaetes. Although a substantial number of fibres in the longitudinal muscles still may be inserted intersegmentally, the majority are not and the muscle has, to a very great extent, lost all signs of segmental organization.

REDUCTION AND LOSS OF SEPTA

In polychaetes not only is there a pronounced tendency for the segmentation of the longitudinal muscles to be obliterated, but the septa are also often reduced or lost altogether. In the Hirudinea, both processes have been carried to an even more extreme position in which segmental muscles, septa, and even, to a great extent, the coelom have all disappeared. These modifications are correlated with changes in locomotory habit.

In many errant polychaetes—polynoids, phyllodocids, some eunicids and others—the septa, though capable of sealing one segment from the next when the septal muscles contract, are not attached to the longitudinal muscle blocks and so can permit the movement of coelomic fluid from segment to segment in the relaxed animal (Clark, 1962a). A similar result is obtained in the earthworms *Lumbricus* (Newell, 1950) and *Pheretima* (Bahl, 1919), and the polychaete *Capitella* (Eisig, 1887), by the provision of sphinctered perforations in otherwise complete septa.

Sometimes, as in the Nereidae, Hesionidae and Nephtyidae, the septa are reduced to the point of serving only as suspensory muscles of the intestine (Fig. 76). In the Nereidae and possibly the Hesionidae, the loss of functional septa is correlated with an improvement in the efficiency of the parapodial musculature; in the Nephtyidae, with the replacement of the septa by hypertrophied oblique muscles (Fig. 77c).

More commonly, the reduction of septa is associated with a change in locomotory habit to one in which the parapodia no longer play a dynamic role. It should be noted that the worms conventionally included in the group 'Polychaeta Sedentaria' are by no means all sedentary. Such worms as the spionid *Nerine*, which is an active burrower, migrates from the surface to deeper levels in the sand with the advancing tide, and crawls over the surface of the sand, are in many ways comparable to members of 'errant' families. The septa of *Nerine* are not reduced. Active tubicolous worms, such as the

Sabellidae and Serpulidae (Thomas, 1940) among the 'Sedentaria' and the eunicid *Hyalinoecia* among the 'Errantia' also possess complete septa. Septa are lost from the greater part of the body in truly sedentary worms such as terebellids, arenicolids and chloraemids among the 'Sedentaria' and, as a parallel example from the 'Errantia', in *Glycera* also.

In these worms, the body-wall muscles form an almost, if not complete muscle coat about the body, the parapodial musculature is severely reduced, and the parapodia are reduced to ridges or small papillae, almost incapable of movement. Locomotion is by means of peristaltic waves passing along the body-wall musculature. The worms are thus comparable to unsegmented coelomates in their body architecture, locomotory behaviour and in their mode of life.

There can be no question but that the segmental organization of arthropods is derived from that of annelids, although a great many of them move in an entirely different manner from worms and employ locomotory mechanisms which have a different functional and structural basis. Even so, some arthropods, chiefly among myriapods and insect larvae, are partially or completely soft-bodied, burrow into the soil and other substrates, and move in a manner which in many respects resemble that of earthworms. Since arthropods lack septa, they apparently constitute an important exception to the principles we have enunciated.

It is convenient to regard arthropods as animals with a rigid exoskeleton, jointed limbs, and skeletal muscles, but this is, of course, by no means universally true. Hinton (1955) observes that 'in a large proportion of dipterous larvae there is no part of the body that is rigid, and, except for the head, there is no invaginated skeleton or system of phragma'. A similar condition may occur to a greater or lesser degree in the larvae of other endopterygote insects. Forces developed by the body-wall muscles are exerted against turgor pressure (Hinton, 1955) and a hydrostatic skeleton exists in the form of the enlarged haemocoel.

Burrowing, geophilomorph chilopods, like dipteran larvae, are capable of considerable changes of shape (Manton, 1952b). The extended length of *Stigmatogaster subterranea* and *Geophilus longicornis* is 66–68 per cent longer than the contracted length, and there are corresponding changes in the girth of the segments. These chilopods do not have a completely flexible body wall, but each segment possesses tergal and sternal sclerites and intercalary tergal and sternal sclerites, which are separated from one another by extremely flexible joints, so that the sclerites slide over each other when the animal shortens. The pleural region from which the legs arise is flexible and elastic. As the body shortens and the segments telescope, this region bulges laterally.

Unlike dipteran larvae, chilopods retain segmental appendages which are employed when the animals move over the substratum or, presumably, through sufficiently large crevices in the soil. A study of the mechanism of their locomotion forms part of a series of investigations made into myriapod locomotion by Manton (1952a, b, 1954, 1957, 1958a, 1961a). When burrowing through fine, unfissured soil the movements of geophilomorphs become very similar to those of earthworms, though shortening of segments is accompanied by widening of them rather than by a symmetrical, radial dilation as in the earthworms. The principle remains the same, however, the shortened segments form a *point d'appui* while more anterior regions are thrust forwards and more posterior segments are drawn up towards the stationary points.

Unfortunately, we lack a comparative account of the mechanics of locomotion in dipteran larvae when they are burrowing or moving over the substratum. From the slight amount of information available, it appears that the protrusion of prolegs or creeping welts rather than radial dilation of segments provides *points d'appui*, and that longitudinal muscles in one part of the body wall tend to be antagonized by longitudinal muscles elsewhere in the body instead of by circular components as in earthworms. When the larvae of the dipteran *Tabanus* (Hinton and Linn, 1950) crawl, segments shorten by about 37 per cent but dilate proportionately very much less, and are therefore ineffective as *points d'appui*. Fixed points are in fact provided in segments in which the longitudinal muscles are extended and prolegs are protracted by the increased turgor pressure caused by contraction of longitudinal muscles elsewhere in the body. Since the *points d'appui* are formed in the elongated segments, the locomotory waves, unlike those of earthworms, are direct. An essentially similar type of locomotion, save that the animals do not possess prolegs, is employed by blowfly larvae when they crawl over the ground (Fraenkel and Gunn, 1940).

Prolegs and transverse 'creeping welts' are variously developed in dipteran larvae and are used in a variety of locomotory techniques (Hinton, 1955, gives a comprehensive review of the morphology and evolution of these structures). They are generally ventral, and appear then to be an adaptation to crawling on the substratum rather than in it, but they may be present dorsally as well as ventrally, or even completely encircle the segments. Friction between the animal and the substratum may be further increased by the development of setae or spines. Extra traction may be provided by the use of the mandibles which are sometimes hooked into the substratum when the head is extended, and form an anchor while the longitudinal muscles contract and draw the body forward.

Although detailed information to the point is lacking, there is no reason to suppose that dipteran larvae or geophilomorph chilopods are poorer or less efficient burrowers than earthworms. Possibly dipteran larvae devote less of their time to burrowing and hence have longer periods of rest between bouts of locomotory activity. Burrowing chilopods, on the other hand, as carnivores, might be expected to spend a greater fraction of their time burrowing when they are searching for food, though increased intensity of search may be compensated by the greater nutritional content of their diet. These possibilities are at present all imponderables. Earthworms, geophilomorphs and dipteran larvae are all fairly active animals that burrow in somewhat similar manners and for the purposes of the present discussion, may be regarded as roughly comparable. Why, then, do earthworms possess septa, which appear to be essential for continuous burrowing, and the others not?

Localized changes of shape can be produced in soft-bodied worms only if deformation of other parts of the body wall is resisted. In earthworms, changes of shape tend to be localized by the septa which reduce, if they cannot entirely prevent, the transmission of fluid pressures across segmental boundaries. The radial thrust that can be generated by contraction of the body-wall muscles is therefore limited to the force that can be generated within a single segment. Such thrusts are important in widening fissures in the ground through which the body of the worm may pass and, in part, in increasing frictional forces between the worm and the substratum, which, in turn, determine the forward thrust that can be developed. In non-septate worms, however, forces generated by the whole musculature can be brought to bear, if need be, at a single site. We have already observed (p. 124) that *Arenicola* can generate radial pressures $1\frac{1}{2}$–2 times as great as can an earthworm of comparable size, and this difference must be attributed very largely to the absence of septa in the thoracic region of the former. But the geophilomorph chilopod *Orya* can generate radial forces 2–3 times greater than *Arenicola* and can generate its maximum thrust repeatedly (Manton, 1961a). Since *Orya* and *Arenicola* are both non-septate, the greater strength and endurance of the arthropod must be due to factors other than septation.

The muscles which resist deformation of the body wall in non-septate worms, are also the locomotory muscles. In earthworms, locomotory muscles are spared this static role to the extent that the septa succeed in confining pressure changes within segmental boundaries. This must contribute to the greater endurance of the septate worms. Reasons underlying the much greater endurance are less obvious and are certainly complex. Even in arthropods with a cuticle that is wholly or partly soft, changes of shape are much less than those of most worms. The arthropod cuticle is thicker than that of annelids and itself offers some resistance to deformation. Indeed, the tendency for dipteran larvae to rely upon changes of length but not of girth in locomotion, is reminiscent of a similar tendency in nematodes, and in these worms the mechanical properties of the cuticle are a decisive factor in changes of shapes and in movement (Harris and Crofton, 1957).

Despite this restraining influence of the cuticle, changes of shape of the more flexible parts of the body wall must be resisted during changes in internal hydrostatic pressure by muscular forces. Barth (1937) claimed to have distinguished locomotory from turgor muscles in caterpillars, suggesting a division of labour which, in effect, would result in a situation comparable to that in earthworms of relieving the chief locomotory muscles of tonic functions. That there is any precise distinction of function between different muscles is unlikely and certainly remains to be established, but the body-wall musculature of all soft-bodied arthropods is extremely complicated, compared with the simplicity of the annelid body-wall musculature. It is therefore possible, if unproved, that muscles which restrain part of the body wall from deformation during the various phases of the burrowing movements are not those that provide the chief locomotory forces during burrowing.

A more important factor than either the comparative strength of the cuticle or a division of labour among the body-wall muscles is undoubtedly the existence of striated muscle fibres in arthopods. It is difficult to say in what way striated muscles are an improvement over smooth muscles, except that they are capable of faster contraction, but they appear to be characteristic of advanced animals with a skeletal musculature, rather than of the lower Metazoa (Hoyle, 1957).

RETENTION OF SEPTA FOR SPECIALIZED PURPOSES

Although many polychaetes lack septa in the greater part of the body, few, if any, of these worms are completely without them. Septa are often retained in the anterior and posterior parts of the body where they serve special functions not directly related to locomotion. Even the otherwise non-septate *Aphrodite* has two or three diminutive septa in the final segments (Fordham, 1925) although these segments play an entirely negligible role in locomotion.

There may also be an antero-posterior gradient in septal development. In worms such as *Nereis*, in which the septa are present throughout the body but are reduced, those at the posterior end of the series are more nearly complete than those in the middle and anterior parts of the body (Clark, 1962a). Possibly the appearance of a few septa or of more complete septa at the posterior end of the worm is a manifestation of the normal growth processes. Segments are proliferated at the posterior end of the worm from the anterior face of the pygidium; and the youngest segments are also the most posterior. There is evidence, at least in some worms (Clark and Clark, 1962), that the delimitation of new segments as they are formed, follows the formation of the rudimentary septa. As the newly formed segment grows, the muscular sheets develop the definitive features of the adult septum. Whether the initial formation of septa as muscle sheets which are subsequently modified, recapitulates an earlier phylogenetic condition in which all septa were complete, is a debatable and minor point; more likely, it is a consequence of the manner in which the segmental *Anlagen* are formed in the growing worm. Not all annelids undergo this growth process. Leeches, many oligochaetes and some polychaetes have a definitive number of segments and, particularly in the first two groups, rudiments of all the segments are formed during the embryonic stages. The segments are much more nearly of the same age than in worms which continue to proliferate new segments and grow throughout the greater part of their lives. Segmental structures are consequently much more nearly alike throughout the whole body. The leech *Haemopis* thus has a complete series of intersegmental septa at an early stage in its embryonic development when all the segmental rudiments have been formed (Bürger, 1894), but all the septa disappear by the time the young leech hatches.

While the manner in which segments grow and septa are formed is clearly pertinent to the appearance and survival of septa in the posterior part of the body, it is unlikely to afford a sufficient explanation of the existence of septa of the same type throughout a large number of posterior segments. We should expect, rather, to find a graded modification of septa in successive segments. In *Arenicola marina*, the thoracic region bearing the gills is non-septate. It is followed by a narrow, abranchiate, caudal region composed of up to 60 or 70 small segments, all with septa (Ashworth, 1912). The large number of caudal segments and the abrupt transition from non-septate to septate segments makes it unlikely that the existence of the septa is accidental to the growth processes and has no functional justification.

In fact, the function of the caudal septa of *Arenicola* has been considered in some detail by Chapman, Newell and Wells. Coelomic fluid in the non-septate region of *Arenicola* is maintained at a resting hydrostatic pressure of 14–15 cm sea-water by the tonic contraction of the body-wall musculature (Chapman and Newell, 1947). Chapman (1958) observes that the intestine in the caudal portion of the worm serves for the passive accumulation of faeces and concludes that this process is aided by the existence of a low pressure region in the caudal segments, a situation achieved by the presence of a succession of septa separating it from the relatively high pressure region of the thoracic coelom. Furthermore, the pressure difference between the two regions of the body may assist in pushing the anterior gut contents into the caudal region. Wells (1961) has advanced a slightly different interpretation. He concludes that the tail is adapted for the rapid expulsion of faeces and that the septa are retained to anchor the thin-walled rectum when the contraction of the body-wall muscles ejects the faecal matter contained in it. Wells (1953) observed that when the faecal cylinder is shot out, the action is accompanied by a piston-like swelling running backwards along the trunk region. This activity may result in the intestinal contents being pushed into the anterior end of the rectum and in this way assisting defaecation. It is certain that the caudal region is not completely isolated from the anterior part of the body as Chapman (1958) implies, although it may be possible for the two regions to be functionally isolated under certain circumstances. The caudal septa of *Arenicola* are perforated by ventro-lateral clefts (Wells, 1961). Whether or not these are capable of being closed is unknown, and clearly hydrostatic isolation of the caudal region would be dependent upon this ability. Wells' and Chapman's views are not mutually exclusive and there is no reason why the septa should not have multiple functions, including the isolation of the caudal region during the period of faecal accumulation and support for the rectum during defaecation.

A similar interpretation may be placed upon the retention of septa in the caudal region of the other sedentary worms, such as terebellids (Thomas, 1940), though none of these has been studied in detail and no doubt a variety of functions of these septa will be discovered in due course.

Anterior septa appear to be related in some way to a proboscis apparatus, tentacles, or similar eversible and protrusible structures at the anterior end of the body. In *Arenicola* three of the anterior septa are retained, in *Stylarioides* (Chloraemidae) and *Ophelia* there are two, and in *Flabelligera* (Chloraemidae), the Ampharetidae and Terebellidae only one septum survives. It may be that in some worms, the anterior septa isolate the proboscis from the resting pressure maintained in the body cavity and from increases in coelomic fluid pressure when the worm is active, which might otherwise cause proboscis eversion, as Chapman (1958) suggests. But this simple interpretation must be regarded cautiously, for it is clear that many of these septa are highly specialized structures and those few that have been investigated play a much more dynamic role.

In the Arenicolidae, the septum at the anterior border of the first setiger is split to form a retractor sheath and a gular membrane (Wells, 1952b, 1954). These are inserted at the same level in the body wall, but at different positions in the anterior part of the gut. The space between them, the paraoesophageal cavity, represents a cleft within the septum and is in communication with the general body cavity by a dorsal valve. The membranes themselves are imperforate. Despite a general similarity in the structure of the anterior septa of the three species that Wells has studied, *Arenicolides ecaudata*, *Abarenicola claparedii* and *Arenicola marina* (Fig. 78), there are significant variations in each species, and each probably employs a different method of proboscis eversion. Certainly this is so of *A. ecaudata* and *A. marina* which have been studied experimentally and by inference from the anatomy of the anterior region, it is likely to be true of *A. claparedii* also.

The musculature of the gular membrane of *A. ecaudata* is particularly well developed and contraction of it and the septal pouches drives fluid in the paraoesophageal cavity forwards, displacing the retractor sheath and causing eversion of the buccal mass. Protrusion of the pharynx, which completes the process of proboscis eversion, is accomplished by contraction of the longitudinal muscles of the buccal mass and of the body wall of the oral region.

In *A. claparedii*, the musculature of the gular membrane is feebly developed and it is probably the increase in coelomic fluid pressure, produced by contraction of the body-wall musculature, that causes proboscis eversion. The gular membrane and retractor sheath evidently do not prevent the transmission of this pressure to the proboscis because the membrane is displaced forwards as a result of the contraction of the body-wall muscles. There must be a

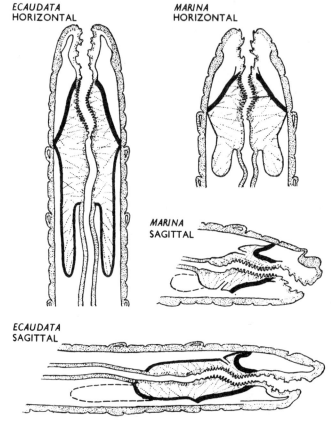

Fig. 78. Horizontal and sagittal sections through the anterior end of *Arenicolides ecaudata*, and *Arenicola marina*. (From Wells, 1954.)

fairly free flow of coelomic fluid into the paraoesophageal cavity through the dorsal valve, which is widely perforated in this species, and there are five wide openings between the body wall and the first septum. As a result, it is likely that pressure is brought to bear directly against the proboscis and not indirectly by way of the retractor sheath, though the movement of the latter may assist eversion of the proboscis.

Arenicola marina is intermediate between these two species. The gular membrane muscles are less massive than those of *A. ecaudata* and in addition to the dorsal valve, there is also a ventral valve leading from the coelom into the paraoesophageal cavity. The body-wall musculature and the gular membrane work together to cause proboscis eversion. The gular membrane alone appears unable to contract sufficiently strongly to cause complete eversion of the buccal mass, and the essential mechanism of proboscis eversion appears to consist of a relaxation of the mouth which permits the general coelomic pressure to extrude the proboscis.

The role of the second and third diaphragms, whether active or passive, has not so far been investigated in any of these arenicolids, although there is some evidence that the third septum may be associated with the growth of the coelomic oocytes and the formation of the egg masses (Wells, 1962). It may be that they partially isolate the proboscidial system from the rest of the body when the proboscis is introverted. Chapman and Newell (1947)

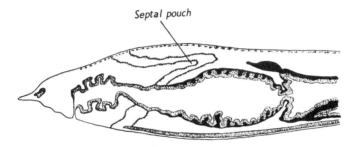

FIG. 79. Sagittal section through the anterior end of *Ophelia*, passing through the septal pouches. (After Brown, 1938.)

observed that during movements of the anterior part of *A. marina*, the proboscis may be everted at irregular intervals, but the movements of the proboscis are not correlated with changes in fluid pressure in the general body cavity, implying the independence of the two fluid systems.

A comparable, though so far as we know at present, somewhat less complicated system occurs in *Ophelia* and the closely related genus *Thoracophelia*. In these worms, all the septa have disappeared except those between segments 2/3 and 3/4. In *Ophelia* (Fig. 79) these form two backwardly directed cones, one within the other, with the apex of each cone forming a caecum or diverticulum dorsal to the oesophagus (Claparède, 1868; Brown, 1938). In *Thoracophelia* the caecum is single and is formed by the fusion of the two septa which remain separate anteriorly (McConnaughey and Fox, 1949). In *Polyophthalmus* there are three such septa, the third between segments 4/5, but in this opheliid they do not form a series of diverticula above the gut (Pruvot, 1885; Saint-Joseph, 1898; Fage and Legendre, 1927). The septa appear to be incomplete at their insertions into the body wall in *Ophelia*, but the apertures can be sealed and coelomic fluid from the thoracic region can be prevented from passing into the head region (Brown, 1938). In

Thoracophelia, there are far fewer coelomic corpuscles in the anterior section of the body and gametes never accumulate in it, suggesting that in this species the septa may be intact (McConnaughey and Fox, 1949). Contraction and relaxation of the septal muscles cause inflation and deflation of the head region. It is generally concluded that the head region is used in burrowing, though in what manner is not altogether clear, and the turgidity of the head is therefore of great consequence in this activity.

The surviving septum of terebellids is a thickened diaphragm inserted at the boundary between segments 4 and 5 (Hessle, 1917; Thomas, 1940). It has four posteriorly directed pouches, the contraction of which is supposed to cause extension of the tentacles (Meyer, 1887; Hessle, 1917). The Ampharetidae and Amphictenidae have approximately the same structures, though they lack diverticula in the anterior septum. The tentacles of these worms are much shorter and less extensible than those of terebellids, so that presumably contraction of the diaphragm alone is sufficient to dilate them (Hessle, 1917).

OLIGOMEROUS ANIMALS

An analogous situation to that found in many sedentary polychaetes occurs in such animals as phoronids and pogonophorans. Those polychaetes which retain only a few septa have secondarily returned to a type of compartmentation of the body which is primitive in the oligomerous phyla. While the functions of this compartmentation may vary, the underlying principles remain the same.

In oligomerous animals, the coelom is divided into five sections during embryological development: an unpaired protocoel, and paired meso- and metacoels. These may suffer considerable modification or even obliteration in the adult, but theoretically, these animals may be regarded as being composed of three segments. It is not to be supposed, nor is it claimed, that this is the same as the metameric segmentation of annelids and vertebrates. As Hyman (1959) points out, the segments of oligomerous animals are not replicates of one another as they are in metameric animals, but contain entirely different organ systems.

In *Phoronis* (Selys-Longchamps, 1907), the coelom of the adult is divided into two separate sections. The first forms the cavity within the epistome, lophophore and tentacles, the second is the coelom of the trunk. The two compartments represent the meso- and metacoels and are separated from one another by a single, complete, transverse septum. The functions of the two coelomic compartments have never been investigated experimentally, but they can be inferred from the anatomy and behaviour of phoronids, and by comparison with other organisms.

Normal movements are restricted to advancing and retreating in the tube (Silén, 1954), but if removed from their tubes, phoronids readily burrow into the substratum again, rear-end first (Selys-Longchamps, 1903, 1907). These movements are accomplished by the familar peristaltic contractions of the circular and longitudinal body-wall musculature. Movements and reactions of the lophophore and tentacles have been discussed by Hilton (1922) and Silén (1954). They are retracted in response to touch, light and other stimulation. The tentacles have little power of independent movement, and they are presumably extended or collapsed chiefly by means of changes of hydrostatic pressure in the common coelomic cavity of the lophophore and tentacles. If a sufficiently strong stimulus is given, the lophophore is collapsed and the entire animal contracts, but it is clear that for many activities, the animal employs two hydraulic systems, that of the trunk and that of the lophophore, independently. The existence of a septum between the meso- and metacoels, assists in providing mechanical isolation of the two fluid systems.

It is instructive to compare phoronids with ectoprocts to which they are allied and which we have already discussed. In the ectoprocts, eversion of the polypide and spreading of the lophophore are parts of the same process. Since the animals are sessile, the coelom is not employed in body movements or locomotion, as it is in phoronids. The ectoprocts therefore have a single hydraulic system, the function of which relates solely to the eversion and inversion of the polypides. The septum between the meso- and metacoels is consequently very incomplete and the two coelomic systems are confluent.

The phoronids and ectoprocts, together with the brachiopods, constitute the lophophorate phyla and are protostomatous; the remaining oligomerous phyla, including the Pogonophora, Hemichordata and Echinodermata, are deuterostomes. Because the latter phyla are three-segmented, they are considered to be related to the lophophorates, but the profound embryological differences between the two groups of phyla suggest that the relationship is not a very close one. Nevertheless, the coelomic compartments appear to serve comparable, if, in detail, different functions and, furthermore, show similar modifications with changing habits and structure in the various phyla.

The coelom of pogonophorans, unlike that of lophophorates, is divided into three distinct and separate sections in the adult. The protocoel is always unpaired and consists of a small chamber at the base of the tentacles, from which a branch enters and runs along the whole length of each tentacle (Fig. 80). In *Spirobrachia*, the tentacles arise from a single cork-screw lophophore and the protocoel extends into this also. The size of the basal tentacular chamber appears to be related to the number of tentacles. In *Siboglinum*, with only one, it is very small; in *Lamellisabella zachsi*, with about 30 tentacles, the basal chamber forms a horse-shoe shaped sac which almost encircles the protosome. Jägersten (1956) claimed that in *Siboglinum ekmani*, an additional branch of the coelomic cavity at the base of the tentacles extends into the cephalic lobe. While Ivanov (1960) admits that the latter cavity may be a derivative of the protocoel, he claims that at least in the (other) species he has examined, the boundaries of the basal tentacular chamber are very distinct and it does not extend into the cephalic region.

The mesocoel is very much more extensive than the protocoel. It is derived during embryological development from paired coelomic pouches and is

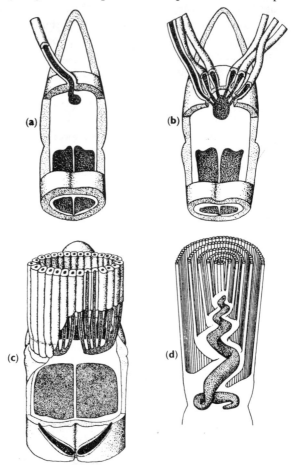

FIG. 80. Protosomal tentacles and the protocoel of pogonophorans. (a) *Siboglinum*, (b) *Oligobrachia*, (c) *Lamellisabella*, (d) *Spirobrachia*. The anterior end of the mesocoel is also indicated in (a)–(c). (After Ivanov, 1960.)

divided in the adult by a longitudinal mesentery which supports the dorsal and ventral blood vessels (there is no gut). Johansson (1939) believed that in *Lamellisabella zachsi*, the protocoel communicates with the mesocoel, but Ivanov (1960), who has studied a considerable number of species, denies this. The protocoel is always separated from the mesocoel by a thick connective-tissue layer and not by a muscular diaphragm.

The whole of the posterior region contains the metacoel. Like the mesocoel, this arises as paired coelomic sacs, initially continuous with the mesocoel, but later separated from it by a constriction which appears between the meso- and metasomes, and ultimately forms the muscular diaphragm between the two coelomic compartments. The metacoel of the adult is divided by a longitudinal mesentery, but this often becomes incomplete, particularly towards the posterior end of the animal. In *Polybrachia* and *Lamellisabella*, the whole middle region of the metasome is filled with spongy parenchymatous tissue which completely occludes the coelom.

When even the anatomy of the Pogonophora is in doubt, it can hardly be expected that it will be possible to discuss the functional morphology of these animals adequately, the more so since very few living animals have ever been seen. However, it is possible to reconstruct something of the normal habits of pogonophorans (Ivanov, 1960). The presence of epibionts on the ends of the tubes of some pogonophorans such as *Lamellisabella* and *Polybrachia*, indicates that they project above the surface of the substratum. The absence of epibionts on tubes of *Siboglinum* and *Zenkevitchiana* suggests that they are completely buried. The greater part of the very long tube, in any case, is buried deeply in the substratum. Because animals may be found at the end of the tube or in its middle, it may be concluded that they are able to move up and down it, and Ivanov (1955a) suggests that once the cavity within the cylindrical tentacular apparatus has become filled with food material, the animal retreats into its tube while external digestion takes place.

The nerve cord includes a single giant axon, and by analogy with other tubicolous worms, it may be assumed that sudden retraction into the tube is a defence mechanism in pogonophorans. The longitudinal muscles by which this would be accomplished are extremely well developed.

It is generally agreed that the numerous papillae, usually crowned with cuticular plates, that are developed on the metasomal body wall and, more especially, the annulus, composed of two or three paired ridges which bear finely denticulate plates, all serve to anchor the worm against the wall of its tube. The analogy with sedentary polychaetes is striking (Caullery, 1944) and, indeed, was the reason for some attempts to link pogonophorans with the Annelida (Uschakov, 1933; Hartman, 1954). The girdle, or frenulum, which is always present on the anterior part of the mesosome, consists of a pair of transverse or oblique ridges of epidermis with a thickened cuticular cap (Fig. 80c). Ivanov (1960) suggests that it serves to wedge the anterior part of the animal in its tube while it is feeding or during movements of the posterior part of the body.

We can now understand something of the likely functions of the coelomic compartments. The protocoel and its branches into the tentacles provide a hydrostatic system comparable to the lophophoral coelomic system (actually a mesocoel) of phoronids, or, in some ways, to the tube-foot/ampulla system of echinoderms (Smith, 1946). Compression of the coelomic chamber at the base of the tentacles, presumably by contraction of the body-wall muscles of the protosome, causes distension of the tentacles and is antagonized by the longitudinal muscles of the tentacle walls. It is unlikely that the tentacles exhibit very great mobility, particularly in the Lamellisabellidae and Spirobrachiidae, in which they are fused together for the greater part of their length, but even so, it is clear that such long structures can be held in an extended position only by turgor pressure. The internal cavity of the cephalic lobe (if it be a part of the protocoel) is unrelated to this function. It is crossed by powerful dorso-ventral muscles which flatten the lobe and, Ivanov (1960) suggests, by rhythmic contractions compress the large cephalic blood vessels and assist pulsations of the heart and dorsal blood vessel in circulating the blood.

The existence of circular and longitudinal muscles in the body wall, bounding the fluid-filled metacoelomic cavity, suggests that, like phoronids, pogonophorans must employ peristaltic waves for locomotion in the tube. Indeed, it can be inferred from the fact that the tube is many times longer than the animal, and that the animals have often been found far from the mouth of their tube, that pogonophorans must move a good deal more than phoronids probably do. This conclusion is also reinforced by the existence of papillae with cuticular crowns, often arranged in a quasi-metameric fashion on the body wall of pogonophorans. These undoubtedly increase the frictional forces between the body and the tube, and must play an important part in locomotion.

Although we may envisage pogonophorans moving up and down their tubes in much the same way as phoronids, and this would satisfactorily account for the existence and form of the metacoel, additional factors must clearly be invoked to explain the great difference between the pogonophoran and phoronid mesocoels. In phoronids, the mesosome bears the lophophore and the mesocoel is functionally equivalent to the protocoel of pogonophorans. The separation of the pogonophoran mesocoel from the metacoel by a muscular diaphragm suggests at once a need for some degree of mechanical isolation of the mesocoel from changes in fluid pressure in the metacoel during peristaltic locomotion. If there is any truth in Ivanov's (1960) interpretation of the function of the frenulum, the mesosome must be wedged against the walls of the tube by dilation of this region of the body effected by contraction of its longitudinal musculature. If, as Ivanov suggests, the mesosome remains stationary while the posterior part of the body is moving in the tube, the need for mechanical isolation between the two hydrostatic systems represented by the mesosome and metasome, is clear. The pogonophoran body therefore includes three independent hydrostatic systems: that in the protosome associated with the tentacular apparatus, that in the mesosome concerned with anchoring the worm in its tube, and that in the metasome, primarily associated with locomotion in the tube. It is obviously important that activity of one hydrostatic system should not interfere with the functioning of the others. In the phoronids, which lack a frenulum or similar anchoring device, the function of the mesosome is entirely different and the body is divided into only two independent hydrostatic units.

In oligomerous deuterostomes, as in lophophorates, the sessile habit has been evolved, and with it, associated changes of structure, particularly of the hydraulic systems, have appeared. The pterobranch hemichordates, which are related to pogonophorans and exhibit various stages on the road leading to a completely sessile habit, are in some ways analogous to the Ectoprocta—the sessile relatives of phoronids. An important difference between ectoprocts and pterobranchs is that the protective outer covering of the colony of the former is provided by a hardening of the body wall, while pterobranchs secrete a rigid coenecium which is completely external to the animals. Only *Atubaria* among pterobranchs lacks a coenecium. It is free-living, entwined around the stems of hydroids (Sato, 1936; Komai, 1949). *Cephalodiscus*, although the members live in a common coenecium, is not truly colonial since the animals are not attached to other individuals. The distal end of the stalk is an adhesive organ and makes only a temporary attachment to the wall of the tube (Andersson, 1907; Gilchrist, 1915). *Rhabdopleura* is truly sessile and colonial. The stalk is attached to a stolon which connects the various members of the colony (Schepotieff, 1907b).

None of the pterobranchs appears to undertake locomotory activity very often. In *Rhabdopleura*, movements seem to be restricted to retraction and emergence of the zooid. Retraction results from contraction of the powerful longitudinal muscles of the stalk, emergence involves the use of the adhesive epistomial lobe (Schepotieff, 1907a). The epistomial lobe of all pterobranchs is very mobile and is used as a locomotory appendage by *Cephalodiscus* when, as they have been observed to do, the zooids emerge completely from the zooecium and crawl around its branches (Andersson, 1907; Gilchrist, 1915). In this type of movement, the stalk is used as a prehensile or adhesive organ, or the animal may adhere by the epistomial lobes of the young zooids attached to the base of the stalk. *Atubaria* also crawls by the use of the prehensile stalk and epistome, but the terminal portions of the two anterior arms of the lophophore are bare and are also used as locomotory organs (Sato, 1936).

The functions of the three sections of the body of pterobranchs differ considerably from those of pogonophorans, and the arrangement of the coelomic compartments is accordingly modified. There are well developed muscular septa between the proto-, meso- and metacoels, and the integrity of the three hydrostatic systems is maintained (Fig. 81).

The mesosome and not the protosome bears the tentacular apparatus in pterobranchs. The coelomic cavity and the body-wall musculature of the mesosome therefore form the hydraulic system by which the tentacles are rendered turgid and are spread.

The protocoel is relatively spacious and occupies the interior of the muscular epistomial lobe (Schepotieff, 1907a, c). This epistomial hydrostatic system has been developed in association with the use of the cephalic lobe as a mobile, powerful locomotory organ, evidently capable of considerable change of shape for which a true fluid skeleton is advantageous if not essential.

The metasome, compared with that of pogonophorans has only minor locomotory functions and has not been reported to display peristaltic movements. Indeed, if reports that the wall of the stalk contains longitudinal but

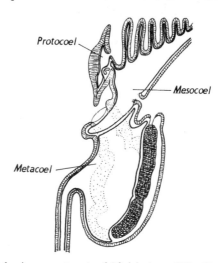

FIG. 81. Coelomic compartments of *Rhabdopleura*. (After Dawydoff, 1948.)

not circular muscles (Harmer, 1905) are correct, it must be incapable of peristalsis without the development of additional specializations such as those found in nematodes. Recovery from a contracted position appears not to be by muscular activity of the stalk, but by the use of the epistome, which drags the zooid forwards towards the mouth of its tube (Andersson, 1907; Gilchrist, 1915). It seems almost inevitable that the complete emergence of the zooid should involve some kind of hydrostatic mechanism, but how this is accomplished is at present unexplained. Consonant with the absence of antagonistic contractions of the body-wall musculature in the metasome, its coelom is reduced. In the body of the zooid, the coelom is almost entirely occupied by the gonads and the digestive system, and in the stalk the coelomic cavity is very largely occluded by muscles and connective-tissue (Harmer, 1905), although it is capable of considerable dilation. The zooids are withdrawn into the shelter of the coenecium by contraction of the stalk muscles and, as in all metazoans in which sudden contraction is a defence mechanism, this can be a rapid process only if there is a true fluid skeleton and changes of shape are unhampered by parenchymatous tissue. Hence, although the pterobranch metacoel is reduced, it remains patent, particularly in the stalk and hence, also, the need for mechanical isolation of the metacoel from the hydrostatic systems of the meso- and protosomes by a muscular septum.

It is at first sight surprising that in the Enteropneusta, which burrow into the substratum, the coelom should be even more reduced than that of pterobranchs. The reduction of the mesocoel in the collar to a mere vestige is readily intelligible since enteropneusts have no tentacular apparatus. The tendency, except in the primitive *Protoglossus* (Caullery and Mesnil, 1904; Burdon-Jones, 1950, 1956), for both proto- and metacoels to be occluded by muscle and connective-tissue is reflected in the changed methods of locomotion in these worms. Burrowing is accomplished in such forms as *Saccoglossus* and *Ptychodera* which have a very long proboscis, by peristaltic

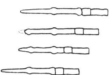

FIG. 82. Peristaltic movements of the proboscis of the enteropneust *Saccoglossus cambrensis*, employed in burrowing. (From Knight-Jones, 1952.)

waves that travel along the proboscis alone and the remainder of the body is dragged along passively (Benham, 1899; Horst, 1940; Knight-Jones, 1952; Rao, 1954) (Fig. 82). In species which have a short proboscis, the peristaltic waves extend onto the collar and anterior trunk regions also (Ritter, 1902). When the animal is made to crawl on the surface, the proboscis again plays a dominant role (Fig. 83). It is extended, attached and the body drawn along by contraction of the proboscis muscles, but peristaltic waves travelling along the body are reported to assist the activities of the proboscis in this type of movement (Crozier, 1915a, 1917; Bullock, 1940; Knight-Jones, 1952). The most active structure of enteropneusts is the proboscis and many authors have observed that it constantly explores the surroundings of the animals, both around the mouth of the burrow and below ground (Ritter, 1902; Assheton, 1908; Horst, 1927–1939; Burdon-Jones, 1952; Knight-Jones, 1952, 1953). Like the pterobranch epistome, it is the most important locomotory organ (Burdon-Jones, 1952). Although there is a rather greater tendency for the protocoel to be filled with connective-tissue and muscle than in pterobranchs, it remains patent in enteropneusts and serves as a fluid skeleton about which the complicated proboscis musculature acts to produce the varied movements of that organ.

The movements of the metasome are relatively slight. It plays very little

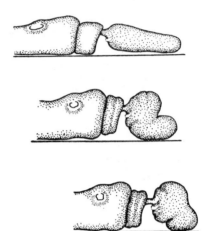

FIG. 83. Use of the proboscis by larval *Saccoglossus horsti*, for creeping along its burrow. (From Burdon-Jones, 1952.)

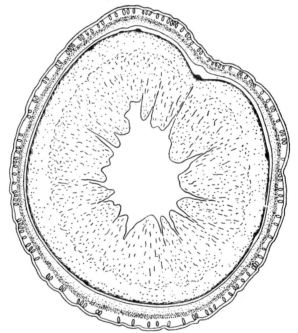

FIG. 84. Transverse section through the proboscis of *Protoglossus koehleri* anterior to the axial complex. (From Burdon-Jones, 1956.)

part in burrowing and although in some species peristaltic locomotory waves appear on the trunk while the worm is moving along its burrow, in several species crawling along the burrow and over the substratum is accomplished by means of cilia (Ritter, 1902; Bullock, 1940, 1945; Knight-Jones, 1952; Rao, 1954).

Enteropneusts are sluggish animals and perform even fewer movements that demand the existence of a true fluid skeleton than pterobranchs, and this must account for the considerable reduction of the coelom in most parts of the body. *Protoglossus* (Figs. 84, 85) is exceptional in that the coelomic cavities of the proboscis, collar and trunk all remain spacious (Caullery and

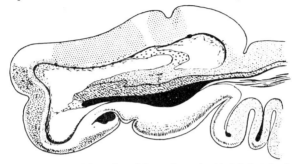

FIG. 85. Coelomic cavity of the collar of *Protoglossus koehleri*, in longitudinal section. (From Burdon-Jones, 1956.)

Mesnil, 1904; Burdon-Jones, 1956). The reason for this is unknown, but Burdon-Jones' observation that it is an extremely active animal provides us with a clue.

THE PSEUDOCOELOMATE PHYLA

Pseudocoelomates lack septa, and they generally lack multiple, independent hydrostatic systems of the kind that exist in segmented worms and in oligomerous animals. The Acanthocephala, however, afford an interesting exception. These pseudocoelomate, intestinal parasites have evolved a hydraulic proboscis apparatus of some complexity which, under certain circumstances, is independent of pressure changes in the general body cavity. The mechanisms of acanthocephalan movements have been considered briefly by such writers as Kilian (1932), Van Cleave and Bullock (1950) and Van Cleave (1952), but their approach has been a rather speculative one, with the result that the acanthocephalan proboscis has been poorly understood until recently. Hammond (1962) has now succeeded in elucidating the mechanics of proboscis eversion in *Acanthocephalus ranae* by direct observation and experimentation. Undoubtedly minor variations both in the structure, and the method of operation of the proboscis apparatus will be discovered when this type of investigation is extended to other acanthocephalans, but it is already evident that these animals have evolved a highly specialized hydrostatic system which, while presenting some unusual features, also reinforces the general conclusions that we have drawn from an examination of coelomate animals.

There are three hydraulic systems in *Acanthocephalus*. The largest of these is the undivided pseudocoel which occupies the greater part of the body and forms the hydrostatic skeleton about which the longitudinal and circular muscles of the body wall act. At the anterior end of the animal, the proboscis cavity is separated from the perivisceral pseudocoel by the proboscis receptacle, a muscular sac into which the proboscis can be withdrawn. The receptacle is anchored to the body wall by a retractor muscle. Finally, flanking the receptacle, there is a pair of sac-like lemnisci, unique structures,

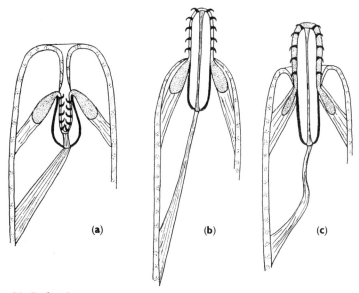

FIG. 86. Proboscis apparatus of *Acanthocephalus*, (a) introverted, (b) everted, (c) contraction of neck muscles, compressing the lemnisci, forcing fluid into the proboscis epidermis, and drawing the anterior part of the trunk over the everted proboscis. (Diagrams based on unpublished observations by R. A. Hammond.)

the function of which has remained mysterious for many years. They are filled with lacunar tissue which communicates with the fluid spaces in the inner layer of the neck epidermis from which the lemnisci originate. Fluid passes freely between the lacunae of the lemnisci and the epidermis of the neck and proboscis, though not into the lacunar system of the trunk epidermis.

The manner in which the three separate systems function can be observed during the cycle of proboscis eversion and retraction (Fig. 86). When the animal is fully contracted, the proboscis lies introverted within the receptacle, and the anterior end of the trunk is somewhat withdrawn into the body of the animal. Contraction of the circular muscles of the trunk and relaxation of the muscle of the neck, extends the body and results in the proboscis apparatus being brought forward to the extreme anterior end of the animal (Fig. 86a). Contraction of the muscular wall of the receptacle now increases fluid pressure in the proboscis cavity and causes proboscis eversion (Fig. 86b). The muscle fibres of the receptacle wall are disposed in a clockwise spiral, and by their contraction reduce simultaneously both the diameter and length, and so the volume, of the receptacle.

The neck muscles run from near the base of the proboscis to the body wall some distance away. These muscles envelop the lemnisci in their course from proboscis to body wall, with the result that when they contract, they compress the lemnisci and force fluid out of the latter into the proboscis epidermis which becomes distended in consequence (Fig. 86c). Since the proboscis hooks are inserted into the connective-tissue underlying the epidermis, distension of the latter forces the proboscis cuticle tightly against the external fangs of the hooks and anchors them firmly into the host tissues. A second effect of contraction of the neck muscles is to draw the trunk forwards over the everted proboscis, so that the whole of the anterior part of the acanthocephalan comes into intimate contact with the mucosa of the host's intestine.

The most advantageous feature of this surprisingly complicated system appears to be that it enables the acanthocephalan to remain attached to the host tissues with extreme economy of muscular activity. So long as the neck muscles remain contracted, the proboscis is 'locked' in an everted position: the lemnisci are compressed, the proboscis epidermis dilated, the hooks retain a tight grip on the intestinal mucosa of the host, and the anterior part of the trunk is drawn over the everted proboscis. The proboscis apparatus is independent of changes of fluid pressure in the main part of the pseudocoel. Such a result is achieved in other animals that we have considered only by separating the proboscis apparatus from the remainder of the body by septa or some other tissue barrier. The chief difference between those animals and the Acanthocephala is that acanthocephalans use the proboscis as an anchor. It therefore remains everted for long periods while the rest of the body performs other movements. This static function of the everted proboscis is quite unlike the functions of the extrovert of other animals and this must account for the very different construction of the organ.

CONCLUSIONS

The performance of movements involving localized changes of shape, as in peristaltic locomotion, is available to all animals possessing circular and longitudinal muscle coats in the body wall. Nevertheless, unsegmented worms are unsuited to this type of movement because a change in the state of contraction of muscles in one part of the body affects all the other body-wall muscles. For this reason, unsegmented coelomates, as we saw in the previous chapter, are slow moving and are almost invariably sedentary. Only worms in which the coelom is subdivided by septa are capable of sustained peristaltic burrowing. Septa impede the transmission to other parts of the body of fluid pressures generated by the local contraction of muscles, and permit pressure gradients to be established over short lengths of the animal. They have therefore some ability to localize the effects of muscle contraction, an ability which is denied to non-septate worms. Ideally, each part of the body between consecutive septa is an independent dynamic unit, and the body-wall musculature is accordingly organized on a metameric basis. In practice, however, this ideal condition may not survive in any existing worms, and segmentation is generally modified to a greater or less extent.

The manner in which septa and the body-wall musculature function in persistaltic locomotion is clearly illustrated by earthworms. Frictional forces between the worm and the ground, and so the magnitude of the forward thrust that can be generated, may be increased by the protracted chaetae, but the fundamental principles underlying peristaltic burrowing are unaffected by refinements of this sort, and are essentially the same as in unsegmented worms. It appears to be impossible to construct a perfectly rigid septum in a biological system such as an earthworm and, in practice, the septa do not prevent some transmission of fluid pressure changes across segmental boundaries. A succession of septa, however, effectively isolates adjacent regions of the body from one another.

Not all septate worms burrow in the manner of earthworms. A number of errant polychaetes have a complete series of septa, but both the arrangement of the musculature and the locomotory methods differ radically from those of oligochaetes. The parapodia are used as appendages and are moved by an extrinsic parapodial musculature that arises on or near the nerve cord in the mid-ventral line, crosses the coelom and is inserted inside the parapodium. The existence of the parapodia and their musculature entails a considerable modification of the body-wall muscles. The circular muscle coat is much reduced and is confined to the region of the intersegmental groove; in some polychaetes it is lacking altogether. The longitudinal muscles are confined to the dorsal and ventral parts of the segments. A secondary restoration of complete circular and longitudinal muscle coats and a return to peristaltic locomotion is possible in these worms only at the expense of the parapodia and of parapodial locomotion. Notwithstanding the extreme differences between oligochaetes and polychaetes, polychaete septa, in general, act in very much the same way as those of earthworms. Their chief function appears to be in helping to provide reasonably rigid insertions for the parapodial muscles by maintaining the turgor of individual segments in which the parapodia are being used. In the same way, they probably permit the parapodium to be moved as a single turgid unit, instead of merely being deformed by contraction of the extrinsic parapodial muscles.

Thus, whether the musculature is organized in circular and longitudinal coats and locomotion is by peristaltic waves, as it is in oligochaetes, or the musculature is very differently organized and locomotion is parapodial, as in polychaetes, septa form an essential part of the hydrostatic system by tending to isolate individual segments or regions of the body from one another.

Although metameric segmentation presents great advantages over the unsegmented condition, at least so far as locomotion is concerned, it is evidently not an unmixed blessing. Almost all organ systems are, of necessity, metamerically segmented; the nervous system in association with the musculature, the gonads, nephridia, blood vessels, etc., as a consequence of the compartmentation of the coelom. Specialization and concentration of function in these organ systems remains very difficult, if not virtually impossible, so long as strict metamerism of the musculature and coelom remains. Segmentation therefore tends to be highly modified and, in particular, the septa are reduced whenever this is possible, that is, whenever septa lose their hydrostatic function.

This is very clearly illustrated in many of the more sedentary polychaetes and in leeches, but a number of errant polychaetes also lack functional septa, although they move in precisely the same way as their septate relatives. This apparent contradiction is explained by the fact that there appears to have been a tendency among the more advanced errant polychaetes for the parapodial musculature to be improved, so that in an extreme case like *Aphrodite*, it approximates to a skeletal musculature and turgor pressure is only of slight importance. The septa are then often reduced to no more than suspensory muscles for the gut and are incapable of sealing off the coelom of one segment from that of its neighbours. With the evolution of a rigid external and endophragmal skeleton and true skeletal muscles in arthropods, both septa and the coelom lose their *raison d'être* and virtually disappear.

A number of burrowing, soft-bodied arthropods, such as dipteran larvae and a few chilopods, present an anomaly. They employ a fluid skeleton and movements of the body wall comparable to the peristaltic movements of earthworms when they are burrowing, yet they lack septa. Unlike unsegmented worms, to which they approximate, they are capable of sustained locomotory activity. Several factors may contribute to the explanation of this exceptional situation, the most important probably being the existence of striated muscles in arthropods.

A few septa are often retained in the posterior and anterior segments of otherwise non-septate polychaetes. It is likely that in many cases these septa isolate the rectum, where faeces collect, or an eversible proboscis or tentacular apparatus from pressure changes in the main part of the body when the animal is moving. Sometimes, as in the proboscis apparatus of the polychaetes *Arenicola* and *Ophelia*, septa may become highly specialized and perform a more dynamic role in proboscis eversion, but this is an exceptional function for these structures. In most polychaetes in which only a few septa are retained they appear to function chiefly in isolating hydrostatic systems which it is important should operate quite independently of one another. This role of septa also appears in the oligomerous coelomates such as phoronids, pogonophorans and pterobranch hemichordates. The proto-, meso-, and metacoels are independent hydrostatic organs and the loss, reduction or occlusion of coelomic compartments in the various phyla in this group of animals, can be traced to changing habits and the loss of function of the coelomic chambers as hydrostatic organs.

Thus compartmentation of the coelom by septa, whether in metamerically segmented animals or oligomerous animals, occurs as an adaptation to providing some degree of hydrostatic isolation between adjacent segments. When, for one reason or another, the need for such hydrostatic isolation disappears, the septa are reduced. The few instances of septa used for other purposes, as part of the proboscis apparatus, as anchors for parts of the body wall or gut, all appear to represent specialized and secondary functions of septa.

5

SWIMMING

AN important method of locomotion that we have still to consider is that type of swimming in which the body is thrown into waves that pass along the animal, generally from head to tail. A great variety of animals swim in this way, and the evolution of metamerism has been seen by a number of authors as an adaptation to this type of locomotion. Snodgrass (1938), Goodrich (1946) and Hyman (1951a) all considered that undulatory swimming movements set up stresses in the body of worms that already displayed pseudometamerism, and that this ultimately led to the metamerism of the musculature and the subdivision of the body into a series of segmental units. Berrill (1955), the most recent advocate of the locomotory theory, has limited his argument to the early chordates in which, it is supposed, the evolution of a segmentally organized musculature provided for more powerful swimming movements than could be generated by a non-segmented musculature.

These theories, like that advanced in this book, have as their basis the mechanical forces that act upon and within animals as they move. They differ from the present theory, at least so far as worms are concerned, in that they look to swimming rather than burrowing for the origin of segmentation. As so many animals swim by undulatory movements, the theory that the evolution of metamerism is related to this type of locomotion is potentially of wide application and for this reason deserves serious consideration. Before accepting or rejecting it, however, a number of questions must be examined and discussed. How does the musculature produce the locomotory waves? Does a segmental organization of the musculature confer any advantages in this type of movement, and, if so, is the segmentation of the musculature peculiarly related to swimming? Is there any fundamental difference between the nature of the swimming movements and the mechanics of swimming in segmented and non-segmented animals? Questions of this sort must be answered before we can attempt to assess the applicability of the swimming theory to the origin of metamerism.

THE INCIDENCE OF UNDULATORY SWIMMING

Almost all macroscopic animals that are long and narrow, and a certain number of microscopic ones, swim by the passage of undulations along the body. As a consequence of this motion, a backthrust is exerted against the water, so providing motive force. Since the thrust is exerted against a fluid which yields as the body moves, the forces available for locomotion are much less than if the animal were able to use solid, immovable objects against which a backthrust could be exerted. On the other hand, if the animal is reasonably stream-lined, frictional forces resisting movement are comparatively slight. For large animals, such as fishes, stresses due to viscosity are negligibly small and the animal relies upon the inertia of the water for the generation of the forward thrust. For smaller animals, viscous forces become of greater importance and Reynold's number:

$$\frac{\text{stress in fluid due to inertia}}{\text{stress due to viscosity}}$$

is reduced from the order of 10^3 or 10^4 for fishes, to about 10^2 for a tadpole (Taylor, 1951). For microscopic animals, Reynold's number is very small

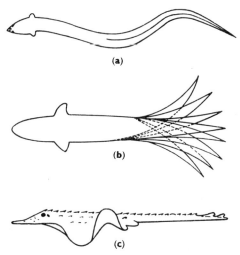

FIG. 87. (a) Undulatory movements of a swimming eel (from Breder, 1926), (b) sculling motion of the tail of a carangid after amputation of the caudal fin (from Gray, 1933c), (c) dorso-ventral undulations of the pectoral fins of a skate (from Breder, 1926).

indeed and stresses due to the viscosity of the medium are many times greater than those due to inertia; the hydrodynamic principles underlying their locomotion are therefore very different from those we shall discuss (Taylor, 1951, 1952a). Nevertheless, many microscopic organisms, including flagellates and spermatozoa, employ undulatory movements for locomotion in a manner broadly comparable to those of larger animals (Gray, 1953a, 1955; Gray and Hancock, 1955).

Undulatory swimming movements are not confined to animals of the same grade of structural organization and their only common feature is their relatively long, ribbon-like shape (Fig. 87a). Excepting only the errant polychaetes, the locomotory waves of all animals pass in the opposite direction to the direction of movement of the animal through the water. In fishes, snakes and polychaetes, the waves are lateral; in nemerteans, leeches and whales they are dorso-ventral. Characteristically, the whole body partakes of the motion, but in some animals such as whales, tadpoles and many teleosts, the anterior part of the body is thick and relatively inflexible, and the locomotory waves tend to be confined to the tail. In extreme examples of this tendency, the tail appears to perform a sculling motion (Fig. 87b) quite unlike the undulations of the whole body that are characteristic of more flexible animals (Breder, 1926; Gray, 1953b), but the two types of movement are fundamentally the same and are functionally equivalent.

Habitual or continuous swimmers generally have a flattened body and thus present a relatively large propulsive surface to the water. The posterior part of the body of sea-snakes is laterally compressed, and pelagic nemerteans have a highly developed dorso-ventral musculature which preserves the flattened form of these worms. Fishes also are often laterally compressed, but with the tendency for the locomotory waves to be confined to the posterior part of the body, even continuous swimmers like the basking shark, *Cetorhinus*, may have an unflattened body if the propulsive forces are generated by movements of a large flattened caudal fin. A comparable modification can also be seen in whales.

Although flattening of the propulsive surface, whether of the whole body or of only the tail fins, is a common adaptation to sinusoidal swimming, it is not essential to this type of locomotion. Pontobdellid leeches lack dorso-ventral muscles and always have an approximately circular cross-section; nevertheless, they do not differ from other leeches in their swimming movements. Even so unlikely an animal as *Sipunculus* has been observed to swim (Fisher, 1954), and a number of cylindrical polychaetes, including *Glycera* and *Arenicola*, can swim, albeit slowly and apparently inefficiently, by modified undulatory movements. Spawning *Glycera dibranchiata*, notwithstanding their circular cross-section, are even able to swim by means of sinusoidal movements comparable to those of *Nereis* and *Nephtys* (Simpson, 1962).

Many animals are only occasional swimmers and need to be sufficiently versatile to meet the conflicting demands of more than one locomotory technique. Multiple adaptations of this sort may be achieved by compromises in body form, or by an ability to alter the shape of the body when the method of locomotion is changed. Thus, while all snakes are able to swim by lateral undulations of the body, they are generally not laterally compressed, since they must present a flattened ventral surface for locomotion over the ground. Swimming efficiency is sacrificed to the demands of terrestrial locomotion in which most snakes engage for the greater part of the time. Sea-snakes, on the other hand, do not face this conflict and do exhibit lateral flattening of the body. The greatest versatility is shown by worms, for they often rely upon changes of body shape to perform a considerable range of activities. Leeches have an approximately circular cross-section when they crawl, but contraction of the dorso-ventral muscles permits them to assume a flattened ribbon-like shape for swimming. The nemertean, *Cerebratulus*, is able to combine adaptations to powerful swimming movements with quite contrary adaptations to burrowing into the substratum in another way. Burrowing is accomplished with the aid of the proboscis and by peristaltic changes of shape of the anterior part of the body. The intestinal region of the worm does not participate in these movements but is dragged along passively. Double bands of dorso-ventral muscles occur between the intestinal diverticula and preserve the flattened form of this part of the body which is used in swimming, but, of course, these muscles preclude its use in burrowing (Coe, 1895).

In some animals, the undulatory swimming movements are confined to the edges of the body (Fig. 87c), and, although the mechanics of this type of swimming are essentially the same as when the whole body performs undulatory movements, it is a manner of swimming with profoundly different evolutionary consequences.

Some relatively broad polyclads, such as *Leptoplana*, are able to swim quite rapidly by muscular ripples which pass along the sides of the worm (Moseley, 1877a; Lang, 1894). The muscular activity involved in this movement is very little different from that in ditaxic muscular creeping that we have already considered in these animals. More elaborate and highly specialized versions of this type of swimming can be found in some tectibranch molluscs in which lateral extensions of the foot form special swimming structures. In *Aplysia* and *Pleurobranchus* (Thompson and Slinn, 1959), the movements of the flexible sides of the foot are recognizably similar to the undulatory waves of other animals (Fig. 88a), but in tectibranchs such as *Akera* (Morton and Holme, 1955) and in the pteropods (Morton, 1954), the structural adaptations become so great and the wave motion so highly modified that they are clearly far removed from locomotory patterns that might be in any way related to the evolution of metamerism (Fig. 88b).

In fact, if the locomotory waves are confined to the sides of the body, the outcome is more likely to be the development of swimming appendages than the evolution of metamerism. Appendages of this sort have been evolved in a variety of animals, including the gastropods and the turbellarian *Convoluta saliens* (Gamble, 1893). In fishes, the use of the lateral fins as swimming appendages is a secondary development and although these structures are clearly metameric in nature, their use for this purpose has no connection with the evolution of metamerism.

The propulsive forces for locomotion, except in certain specialized fishes, are derived from undulatory movements of the whole body or of the tail. Initially, dorsal and lateral fins or fin-folds appear to have served as stabilizers, as they do in modern sharks, and to have controlled pitching and yawing

INCIDENCE OF UNDULATORY SWIMMING 169

perturbations of the movement of the animal (Harris, 1936, 1938) (Fig. 89). A related and important function of the pectoral fins is, and is also likely to have been in some of the earliest fishes, their action as hydrofoils in producing

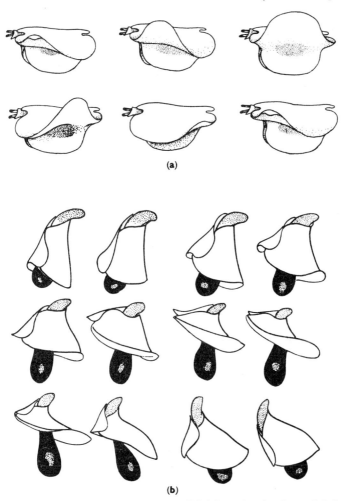

FIG. 88. Successive positions of the epipodial lobes of swimming opisthobranch molluscs. (a) *Pleurobranchus* (from Thompson and Slinn, 1959), (b) *Akera* (from Morton and Holme, 1955).

lift. The Agnatha, sharks, placoderms and acanthodians did not possess lungs or an air bladder, and although buoyant oils and fats may have reduced their density, many of them were heavily armoured and must have been denser than the surrounding water. In these, as in modern sharks, the almost universal presence of the heterocercal tail produced upthrust at the posterior end of the body and the pectoral fins produced upthrust in front of the centre of gravity. As Harris (1936) showed in the dogfish, the equilibrium of such an animal is unstable in the vertical plane and can be controlled by slight adjustments of the inclination of the pectoral fins. The flattened head-shield of the cephalaspids would have served equally well to provide anterior lift, but because of the slight development of the pectoral fins, these fishes would have had poor powers of adjusting the inclination of the body. This, together with the small body musculature, suggests that such fishes lived on the bottom and were only occasional swimmers (Harris, 1953).

A quite different development can be observed in skates and rays. These are

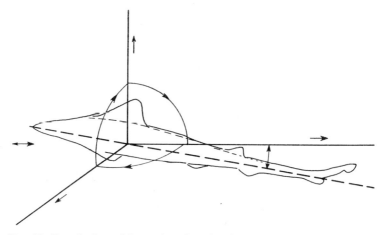

FIG. 89. Perturbations of the motion of a swimming dogfish. (From Harris, 1936.)

exceptional among selachians because the lateral fins, which are hypertrophied and extremely mobile, produce both the propulsive force and lift (Affleck, 1950), and the tail is correspondingly reduced. The locomotory movements of the fins consist of dorso-ventral undulations of a similar nature to the lateral undulations of the whole body of sharks (Fig. 87c). The banjofish, *Rhinobatis*, lies somewhere between these two extremes, since it is able to swim both by movements of the pectoral fins and by lateral undulations of the body.

With the evolution of air sacs in the Actinopterygii, the density of the fishes became close to that of water and the hydrofoil action of the pectoral fin ceased to be necessary. In teleosts, the fins are now much smaller and have narrower insertions into the body than those of sharks (Fig. 90). In consequence, they are more mobile and are used as rudders for steering and braking (Harris, 1937, 1938). Generally the propulsive force is provided by undulatory movements of the body and tail, produced by contractions of the powerful longitudinal musculature, but in some teleosts, the fins have very largely, or even entirely, taken over this function (Fig. 91). The Labridae, Scaridae

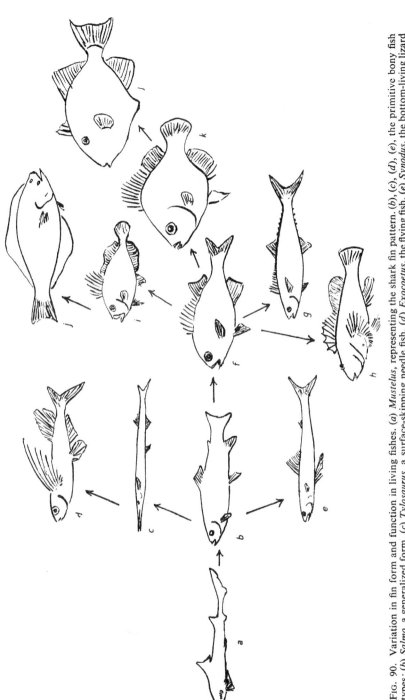

FIG. 90. Variation in fin form and function in living fishes. (*a*) *Mustelus*, representing the shark fin pattern. (*b*), (*c*), (*d*), (*e*), the primitive bony fish types: (*b*) *Salmo*, a generalized form, (*c*) *Tylosaurus*, a surface-skipping needle fish, (*d*) *Exocoetus*, the flying fish, (*e*) *Synodus*, the bottom-living lizard fish. (*f*) to (*l*), percoid fishes: (*f*) *Holocentrus*, the primitive squirrel fish, (*g*) *Euthynnus*, the bonito, an active, continuous swimmer, (*h*) *Scorpaena*, an extreme example of pectoral fin locomotion, (*i*) *Zeus*, the dory, deep-bodied and well-balanced, (*j*) *Hippoglossus*, a typical flat fish, (*k*) *Chaetodon*, another deep-bodied type, swimming with pectoral fins, (*l*) *Balistes*, the plectognath trigger fish. (From Harris, 1953.)

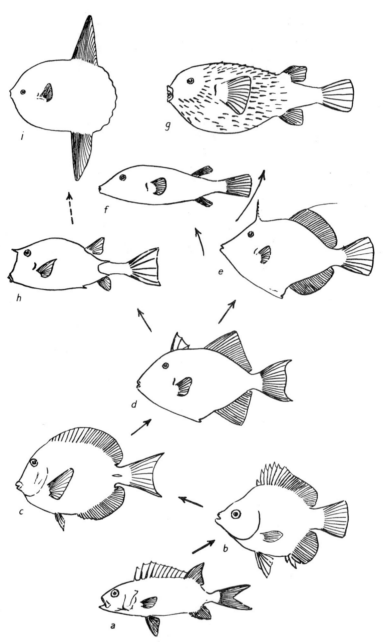

Fig. 91. The evolution of specialized plectognath fishes which swim and balance by undulations of dorsal, anal and pectoral fins. (*a*) *Holocentrus*, representing the ancestral berycoid type, (*b*) *Chaetodon*, a deep-bodied, pectoral fin swimmer, (*c*) *Acanthurus*, the more specialized surgeon fish, (*d*) *Balistes*, the central plectognath type, (*e*) *Monacanthus*, the file fish, (*f*) *Spheroides*, the puffer fish, (*g*) *Diodon*, the porcupine fish, (*h*) *Lactophrys*, the trunk fish, (*i*) *Mola*, the sun fish. (From Harris, 1953.)

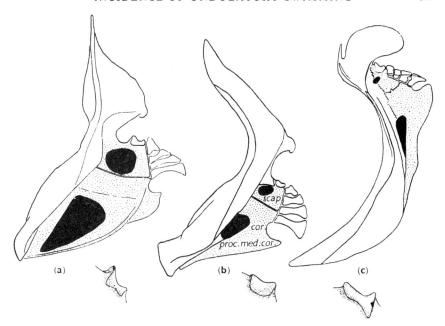

FIG. 92. Development of pectoral girdles and the main areas of attachment of the fin muscles (stippled), in relation to function. (a) *Scarus*, the parrot fish which swims by flapping the pectoral fins, (b) *Epinephelis*, the grouper, which uses the pectoral fins for incidental swimming or braking movements, (c) *Euthynnus*, the bonito, in which the pectoral fins are only thrust out to act as brakes. Below, the articular surface for the first fin ray. (From Harris, 1953.)

(parrot fish) and angel fish rely almost entirely upon movements of the pectoral fins to produce locomotory forces. With this change in locomotory

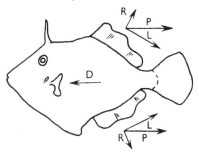

FIG. 93. Forces produced by the dorsal and ventral fins of the balistid *Monacanthus hispidus*. (From Harris, 1937.)

technique, the pectoral fin musculature has become correspondingly enlarged and, with it, the primary scapulo-coracoid girdle from which most of these muscles originate (Fig. 92). In the Balistidae, undulations of the dorsal and anal fins are responsible for locomotion (Fig. 93), while in the Hippocampidae, the dorsal fins alone are used.

In discussing the fin movements of teleosts, Harris (1937) points out that the segmental nature of the fin rays is preserved and the fin is innervated by a regular succession of segmental nerves (Howell, 1933). During forward locomotion, the wave of contraction passes from the top to the bottom of the pectoral fin, that is, from the morphological anterior to posterior, and in the reverse direction when the fish swims backwards. 'The direction of passage of the motor impulses thus corresponds exactly to the direction of passage of the myotome contractions used in normal forward and backward body swimming, suggesting that the whole locomotor system is worked on a similar basic pattern' (Harris, 1937).

The active use of fins for locomotion is clearly exceptional and derivative in fishes. It is not related to the evolution of metamerism and, indeed, clearly postdates the development of a segmental organization of the musculature and nervous system. Similarly in invertebrates, the use of parts of the body for swimming can lead to the development of specialized appendages for this purpose, but can have no influence upon the fundamental body plan which is a basic feature of the evolution of metameric segmentation. In the following discussion, therefore, we shall consider only the mechanics of undulatory swimming movements involving the entire body of the animals, since this alone can be in any way related to the origin of segmentation.

THE MECHANICS OF SWIMMING

The analysis of undulatory swimming presents many difficulties. Attempts to describe and explain how fishes swim have been made by Borelli (1680), Pettigrew (1873), Marey (1894) and Breder (1926), but our present knowledge of the subject derives chiefly from the development over a number of years of a mechanical analysis of swimming by Gray (1933a, b, c, 1946, 1953a), and from Taylor's (1951, 1952a, b) hydrodynamic analysis of some types of undulatory movement in animals of various sizes.

When a fish such as an eel swims, its body is thrown into waves which pass backwards along the body. Each short section of the body performs lateral movements that cross and re-cross the axis of locomotion of the animal (Fig. 94). A fundamental characteristic of the motion of all animals that swim in this way is that the movement of the section has components tangential and normal to its surface (Gray, 1953a). A portion of the body moving tangentially to its surface is like a paddle swept edge-on through the water and exerts no thrust upon the water. Clearly thrust can be developed only if there is a normal component of the motion and the magnitude of the thrust that is available for the motive power is related to the size of the normal, in relation to the tangential, component (Fig. 95). The fact that each section of the body moves at an angle to its surface (i.e. along the resultant of its tangential and normal components) accounts for several features of this type of locomotion

MECHANICS OF SWIMMING

that Breder (1926) and Gray (1933a) discovered from the analysis of cine-films of swimming eels. The undulatory waves pass backwards along the body faster than the fish moves forward, or, in other words, the undulatory waves move backward with respect to the ground. Also, each point on the body traces out a sinusoidal curve in the water of the same wave-length, and this is less than the wave-length of the curve into which the body of the fish is thrown. A consequence of this is that the angle between the body and the transverse

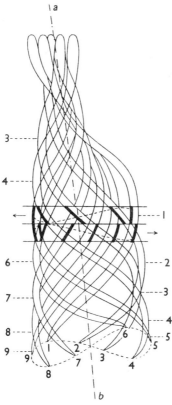

FIG. 94. Successive positions of the body of a swimming eel during the passage of one complete locomotory cycle. (From Gray, 1933a.)

axis (θ_m) is greater than the angle between the path of motion of a section of the body and the transverse axis of motion (θ_b). The difference between these angles ($\theta_m - \theta_b = \alpha$) is termed the 'angle of attack' (Gray, 1933a) and represents the angle between the surface of the section and its direction of motion, to which we have already referred (Figs. 95 and 96). All points on the body do not necessarily suffer the same lateral displacement and generally the tail describes greater lateral movement than the head. Indeed, in many

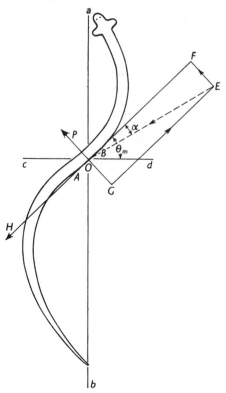

FIG. 95. Components of motion of a short length of a swimming eel, relative to the ground. The animal moves along the axis a–b. (From Gray, 1933a.)

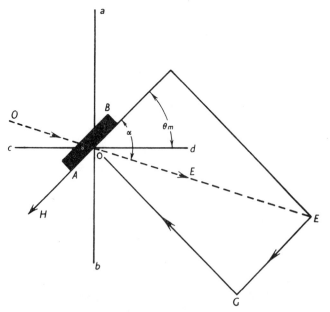

FIG. 96. Components of motion of a short length of a swimming eel, relative to the head. (From Gray, 1933a.)

fishes the lateral movement of the head is imperceptible, but this is a point we shall discuss later.

Each short section of the body describes a figure-of-eight loop relative to the head during the passage of one complete locomotory wave. During each cycle, the left and right sides of the fish alternately form the leading edges and, as they do so, they are inclined and travel obliquely backwards (Fig. 94). The inclination, velocity and curvature of the section all vary. The inclination and velocity are greatest as the section crosses the axis of locomotion and zero at the limits of transverse motion. The curvature is least at the axis, but the section is curved concavely backwards during the first half of the stroke and becomes curved convexly backwards after it crosses the axis of locomotion.

The direction of movement of the section relative to the head (i.e. if the fish is supposed to be stationary) as the section crosses the axis of locomotion is shown in Fig. 96. When the fish is moving, the section also has a forward velocity superimposed on its transverse movement. Its direction of movement and the reaction of the body upon the water as it crosses the axis of locomotion are shown in Fig. 95. Relative to the water, the section moves to the side and forwards, but so long as the angle of attack remains positive, the animal gains momentum. Water is displaced by the movement of the section. Some is deflected above and below the body and contributes nothing towards the thrust, but a proportion of it flows along the body. The flatter the surface presented to the water, the greater this proportion is likely to be and the greater the propulsive force. The initial velocity of the water on encountering the body can be resolved into two components normal and parallel to the surface of the section. The normal component is lost by the water, or, in other words, momentum is gained by the fish. The component of this momentum along the axis of locomotion represents the motive force. As the fish gains speed, the angle of attack diminishes (Fig. 97) and when the fish swims at a constant velocity the angle of attack is ideally zero, though in fact it remains positive, if small, since sufficient force must be provided to overcome the frictional forces acting upon the body. When the fish brakes the angle of attack becomes negative.

It is possible to see from this highly simplified account how lateral movements of sections of the body drive the fish forwards. A more detailed analysis immediately becomes formidably difficult. Some of the complicating factors are:

(1) The inclination and velocity of short sections of the body vary throughout the cycle of lateral movement. Clearly, the propulsive effect of this movement is least as the section nears the limits of its lateral movement and greatest when the section crosses the axis of locomotion, where it travels at highest speed and greatest inclination.

(2) The amplitude of movement, the inclination of the section as it crosses

178 SWIMMING

the axis of locomotion, and the angle of attack are all different at different positions along the body.

(3) A section of the fish does not move through still water, as we have

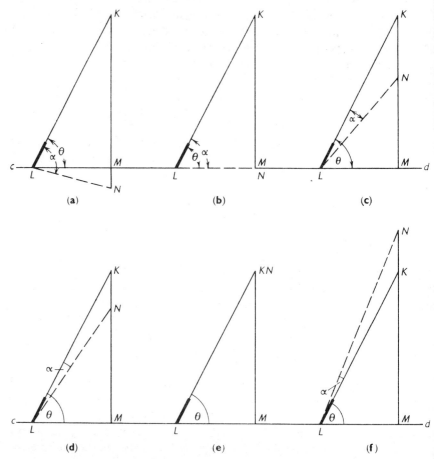

Fig. 97. Effect of an increase in the rate of forward progression of the animal upon the 'angle of attack' (α). (a) Animal stationary, short length of the body moves backwards with respect to the ground, (b) forward velocity of the fish equals the backward velocity of the section, (c), (d) forward velocity of the fish greater than the backward velocity of the section, but 'angle of attack' still positive, (e) zero 'angle of attack', (f) negative 'angle of attack', giving a negative thrust. (From Gray, 1933a.)

assumed, but through water disturbed by the passage of more anterior parts of the animal through it.

(4) As soon as the animal begins to move, frictional forces reduce the velocity of water-flow tangential to the body and retard the fish.

Some of these factors have been taken into account by Taylor (1952b) in his analysis of undulatory swimming. In this theoretical approach to the subject, he has assumed that the body of the animal describes a sine curve of constant amplitude, wave-length and velocity along the whole length of the body, that the cross-section of the animal is circular, and that forces acting upon an element of a flexible cylinder (i.e. the animal) are the same as those acting upon a similar portion of a long straight cylinder. The last assumption implies that the section moves through undisturbed water. Special conditions that exist at the two ends of the animal are neglected. Taylor (1952b) and Gray (1953a) both emphasize the severe limitations of this theoretical analysis and the considerable departure of most animals from the assumptions on

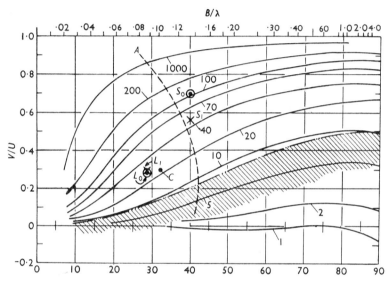

FIG. 98. Relationship between the forward velocity of the animal (V) and backward velocity of the locomotory waves (U), and the wave form of the undulations (B = amplitude, λ = wavelength), for various values of energy output. S, snake; L, leech; C, ceratopogonid larva. ● observed values; X, calculated values. (From Taylor, 1952b.)

which it is based, but despite these necessary simplifications, there is a surprising agreement between the observed and calculated performances of a snake, a leech and a ceratopogonid larva (Fig. 98 and Table 8). Furthermore, several useful deductions can be made from this analysis. Thus, the ratio of the velocity of the animal to the velocity of the waves along its body relative to the head, increases with the amplitude of the waves and decreases with their wave-length; this ratio also increases with the absolute size and speed of the animal.

Undulatory swimming can be compared in a very rough way to sculling with a single oar from the stern of a boat. However, in the eel or snake, the undulatory movement is not confined to the tail, but includes the entire body. This is not invariably the case. The evolution of swimming motions in fishes appears to have been subject to at least two contradictory tendencies. If the waves extend over the whole body as they do in eels, the entire body musculature contributes to locomotion. Further, the muscles of the anterior parts of the eel are capable of developing the greatest thrust. As we have seen, the thrust and therefore the expenditure of muscular energy, is related to the

TABLE 8

Relationship between the Undulations of the Body and the Speed of Movement

(data from Taylor, 1952b)

	V cm/sec	U cm/sec	$\dfrac{V}{U}$	$\dfrac{B}{\lambda}$
Snake	32	46	0·70	0·134
Leech	4·3	15·3	0·28	0·089
Ceratopogonid larva	2·17	7·37	0·3	0·099
Nereis	0·18	−0·76	−0·23	0·18

B, λ, and U are amplitude, wave-length and velocity relative to the ground, respectively, of the undulatory waves. V is the velocity of the animal relative to the ground.

angle of attack of the body. This is greatest at the anterior end of the fish where the musculature is most powerfully developed (Gray, 1933a, b). But although it may be advantageous to employ the whole of the musculature in swimming, there are also corresponding disadvantages. It is likely that a fish swims more efficiently if the locomotory waves are confined to the posterior part of the body like the screw of a ship (Gray, 1933c; Taylor, 1952b). The head and anterior part then form a stream-lined surface offering little resistance to passage through the water. In fact, in both the eel (Gray, 1933a) and the grass-snake (Gray, 1946), the lateral movement of the head is much less than the amplitude of tail movements and this, besides providing some answer to the problem of stream-line flow, must compensate to some extent for the relative weakness of the posterior musculature, since for the same energy output, an increase in the amplitude of the waves increases the speed of the eel through the water.

The exaggeration of undulatory waves in the posterior part of the body is much more marked in the majority of teleosts. The body of many of these fishes is short and thick-set, and much less flexible than that of the eel. Although at first sight the tail of such fishes appears to be waved to and fro in a manner exactly comparable to sculling, as earlier writers claimed, Pettigrew (1873), Breder (1926) and Gray (1933c) have shown conclusively that a fish such as the whiting swims in fundamentally the same manner as the eel. The amplitude of the locomotory waves is small until they reach the tail at which point it increases abruptly. There is some lateral movement of the head and anterior part of the body, but this is very slight. This analysis, although developed originally in relation to the swimming movements of animals such as the eel, therefore applies with few exceptions to all animals that swim by flexions of the body.

The locomotion of snakes, eels and similar animals on land differs from their swimming motions only in that the locomotory waves do not move backward with respect to the ground when the animals are crawling (Gray,

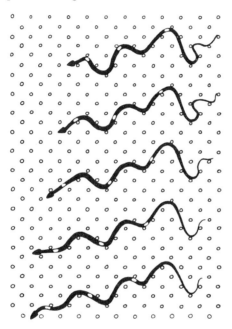

FIG. 99. Movements of the grass-snake, *Tropidonotus*, crawling over a smooth surface with rigid pegs projecting from it and set 1 in. apart. (After Gray, 1946.)

1946; Gray and Lissmann, 1950). The forces acting upon the body of the animal can best be visualized if it is allowed to crawl through an array of vertical pegs (Fig. 99). The mechanics of locomotion are similar to those prevailing when the animal swims, except that, since the thrust is exerted against rigid pegs, there can be no normal component to the movement of a section of the body, though of course there is still a normal force. Each element of the body therefore follows a path identical to the curvature of the body (Gray, 1953a). When the animal swims, each section has a normal component of motion because the water does not offer a rigid *point d'appui* as the pegs do, and the elements trace out a different wave-form from that of the body.

The forces acting upon a short length of an animal moving over land by means of undulatory movements have been discussed by Gray (1953a). As contractile elements on one side of the body shorten, the central axis develops a compression and equal and opposite turning couples are exerted upon neighbouring sections of the body. These forces generate a propulsive thrust providing the neighbouring sections are prevented by external restraints from moving normally to their own axes and if they are sufficiently rigid to resist the shearing forces imposed on them by external forces.

An important variant of undulatory swimming is found in polychaete worms. They differ from all other animals that swim in this way in that the undulatory waves pass forward instead of backwards along the body (Gray, 1939). Swimming in *Nereis* is an extension of rapid crawling which we have already described (p. 125). The parapodia perform their backwardly directed power-strokes as the crest of the locomotory wave reaches them. During this phase of the cycle the whole parapodium is moved in an opposite direction to the undulatory wave and backwards with respect to the ground. This produces a flow of water towards the posterior end of the body which subjects the worm to a forward thrust.

Although it is clear that the movements of the parapodia by cycles of contraction of the underlying longitudinal muscles in combination with the execution of the parapodial power-strokes provides a forward thrust upon the worm, we are still left with the fact that since the undulatory waves pass towards the head faster than the worm moves forwards, the angle of attack of the body must be negative. According to the analysis of undulatory swimming that we have already examined, the movement of the body might be expected to produce a backthrust upon the animal opposed to the forward thrust developed by the movement of the parapodia.

The solution to this paradoxical situation was provided by Taylor (1952b). In his analysis of swimming in smooth animals, he had assumed that the flow of water along the body was laminar, but this assumption ceases to be valid in an animal such as *Nereis* in which the projecting parapodia provide sufficient skin friction to cause turbulent flow of water past the animal. If allowance is made for the great degree of roughness conferred on the animal by the existence of parapodia (assumed in this case to be rigid and immobile), a new family of curves can be derived for the relationship between the ratio of the velocity of the animal to the velocity of the undulatory waves and the wave-form of the undulations (Fig. 100). In this case, unlike that for smooth animals, negative values of the ratio of velocities are physically possible and, indeed, correspond to the situation in polychaetes.

It is clear, particularly from Taylor's analysis, that the power-stroke of the parapodia is not essential for forward progression, though this is not to say that the parapodia are without effect. *Nereis* is a relatively inefficient swimmer, for its speed of movement is slow despite a high ratio of amplitude to wavelength of the undulatory waves (Table 8). *Nephtys* is a better swimmer and the

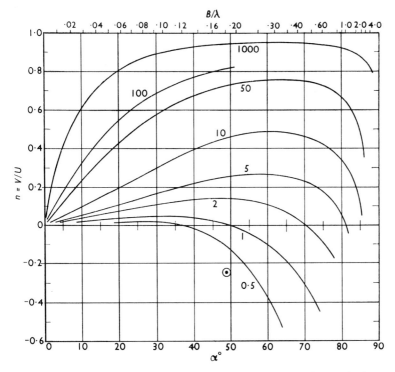

FIG. 100. Relationship between the ratio of forward velocity of animal to backward velocity of locomotory waves, and the wave form of the undulations, for an animal with an extremely rough surface. ● observed value for *Nereis diversicolor*. (From Taylor, 1952b.)

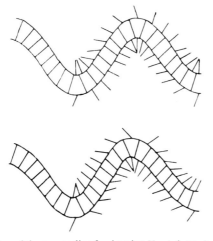

FIG. 101. Inclination of the parapodia of swimming *Nereis* (upper) and *Nephtys* (lower) in relation to the position of segments in the locomotory cycle. Undulations pass from left to right. (From Clark and Clark, 1960b.)

essential difference between these similar polychaetes appears to reside in the manner in which the parapodia function during swimming (Fig. 101).

In *Nereis*, the parapodial power-stroke continues through an appreciable fraction of the locomotory cycle. The parapodia are perpendicular to the surface of the segment when the underlying longitudinal muscles are contracted, i.e. when the parapodium is in the trough of the locomotory wave,

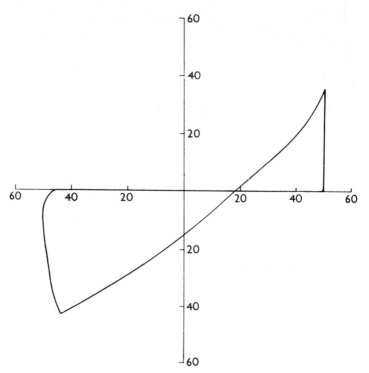

FIG. 102. Relationship between the inclination of the parapodium to the transverse axis of its segment during the passage of one complete locomotory wave (ordinate), and the inclination of the longitudinal axis of the segment to the direction of motion of the worm (abcissa). *Nereis*. (From Clark and Clark, 1960b.)

and as these muscles are relaxed, the parapodium is drawn back in a preparatory movement before executing the power-stroke. The power-stroke begins slowly when the parapodium is about half-way up the leading edge of the locomotory wave, then increases in speed and is not complete until the parapodium is half-way down the trailing edge. A recovery stroke then restores the parapodium to the perpendicular position. This cycle of activity is illustrated in Figs. 101*a* and 102.

Nephtys under comparable conditions has a more concentrated and more rapid power-stroke. It is begun and completed as the parapodium is carried over the crest of the locomotory wave (Fig. 101*b*). Since this is the period when the parapodium is moved by contraction of the longitudinal body-wall muscles backwards with respect to the ground at the greatest rate, its absolute velocity relative to the ground and therefore its absolute rate of movement must be much greater than that of a parapodium of *Nereis* (Fig. 103). Hence, also, the thrust that it exerts upon the water is very much greater.

Although in its fundamentals the undulatory swimming of polychaetes is

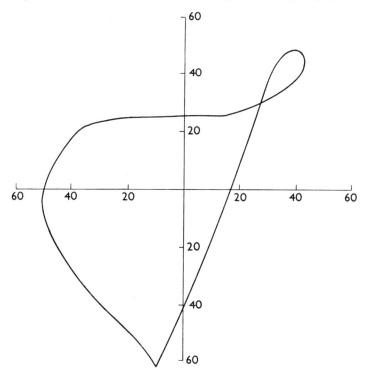

FIG. 103. As Fig. 102, for *Nephtys* swimming slowly. (From Clark and Clark, 1960b.)

similar to that of smooth animals, changes in the rate of swimming involve more complex modifications of the locomotory pattern in the worm than in fishes. When *Nephtys* increases its velocity of swimming, there is, as might be anticipated, an increase in the ratio of amplitude to wave-length of the undulatory waves. In addition, the cycle of parapodial movement is modified (Fig. 104). The total angular movement of the parapodium increases slightly and the power-stroke is completed as the parapodium reaches the crest of the locomotory wave instead of continuing until the parapodium is on the trailing edge of it. The rate of angular movement of the parapodium, and therefore the thrust exerted upon the water, is also greater when the worm swims faster.

A difference between swimming in fishes and polychaetes which probably contributes to the relative inefficiency of the latter, is that in worms the undulatory waves of greatest amplitude are confined to the middle segments of the body. There appears to be no tendency for the region of greatest thrust to be at the posterior end of the animals as we have seen to be the case in teleosts.

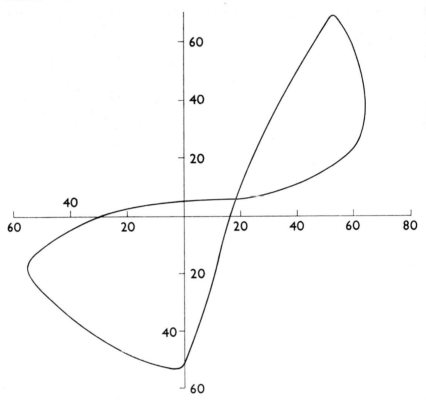

FIG. 104. As Fig. 102, for *Nephtys* swimming rapidly. (From Clark and Clark, 1960b.)

THE BEHAVIOUR OF THE MUSCULATURE

Undulatory locomotion of fundamentally the same type occurs in a wide variety of animals. The essential requirements demanded of the musculature are that it should be capable of producing unilateral contractions of short lengths of the body, and that these contractions should be phased and co-ordinated along the body of the animal.

In all animals that swim by means of undulatory motions of the body, the longitudinal muscles are the only ones to be involved in this type of locomotion, though of course dorso-ventral muscles may contract to produce a flattened body form, as they do in nemerteans and leeches, or the parapodial musculature may play an active part as an adjunct to the longitudinal muscles, as in polychaetes. The contraction of the longitudinal muscles on one side of a short section of the body produces a turning couple which is applied to adjacent sections of the body; it also antagonizes the contralateral longitudinal muscles through the agency of a fluid (in worms) or rigid (in vertebrates) skeleton. A succession of such contractions passing along opposite sides of the body generates the undulations by which the animal swims.

For convenience in the discussion of the mechanics of locomotion, we have considered the behaviour of short sections of the body almost as if the animal were subdivided into units or segments; but in fact, locomotory waves show no such discontinuities. Whether the musculature has a segmental organization, as in annelids and vertebrates, or not, as in nemerteans, turbellarians and nematodes, bears no relation to the ability of the animals to perform undulatory movements, nor, so far as one can tell, to their efficiency in doing

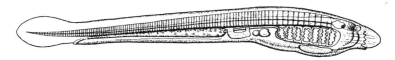

FIG. 105. Ammocoete larva of *Petromyzon*, showing the myotomes as simple, vertical muscle blocks. (After Shipley, 1887.)

so. Individual muscle fibres do not run the whole length of the body in any of these animals and contraction of a small group of them causes a local flexion of the body. The smooth passage of the flexion from one end of the animal to the other is facilitated if there is a degree of overlap of consecutive muscle fibres. This is the situation even in errant polychaetes such as *Nephtys* and *Nereis*, where the fibres of the longitudinal body-wall muscles do not show a strict metamerism, but extend over two or three segments. Many of the fibres are inserted intra- and not intersegmentally.

Only in chordates does the musculature show complete anatomical segmentation, and this does represent an adaptation to swimming. The muscle fibres of fishes are short and are segmentally disposed into myotomes, each separated from its neighbour by an intermuscular septum. The longitudinal muscles are used exclusively for undulatory swimming and the fish does not suffer the changes of cross-sectional shape that are characteristic of worms and which the worms put to use in a variety of locomotory activities.

The myotomes arise ontogenetically as simple, vertical muscle blocks (Fig. 105), but later they become bent into a zig-zag pattern. The mechanical importance of this peculiar shape of the myotomes is fairly obvious (Rockwell, Evans and Pheasant, 1938). The muscle fibres in the myotome are all approximately parallel to the vertebral column; were the intermuscular septa perpendicular to the axis of the fish, contraction of the muscle fibres would produce very little bending of the vertebral column, instead, the pull would be exerted against the adjoining myotomes. However, since the septa are oblique, the pull of contracting muscle fibres is more nearly in the plane of the septum and so is transmitted to the vertebrae. The septa are inserted at the middle of each centrum and the myotomes of fishes are intersegmental; the arrangement is such that contraction of muscle fibres in a myotome exerts a bending movement upon two adjacent vertebrae.

There is a considerable degree of minor variation in the structure of the myotomes, but three main types can be distinguished: those of cephalochordates, those of cyclostomes, with those of the gnathostomatous fishes forming a third group. The differences between these three types reflect an increasing complexity in myotomal structure and have important functional consequences.

The cephalochordate myotome is V-shaped, with the single flexure pointing anteriorly (Fig. 106a). The dorsal and ventral wings are inserted obliquely into

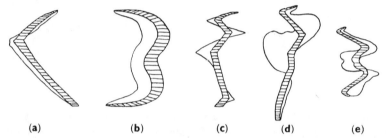

FIG. 106. Myotomes of (a) Amphioxus (after Nursall, 1956), (b) *Myxine* (from Langelaan, 1904), (c) *Squalus* (after Nursall, 1956), (d) the trunk region of *Gadus*, and (e) the caudal region of *Gadus* (from Dietz, 1913).

the mid-sagittal septum, the ventral wing sloping forwards, forming an angle open to the posterior, the dorsal wing slopes backwards and forms an angle open to the anterior end of the animal (Nursall, 1956).

In cyclostomes, the myotome is W-shaped, with the low, rounded middle flexure directed forwards and the two other flexures directed backwards (Fig. 106b). The whole myotome slopes forwards towards the mid-sagittal plane so that it forms an angle open towards the posterior (Langelaan, 1905; Cole, 1907).

The myotomes of the remaining fishes, though W-shaped, are much more complex in their form than those of cyclostomes, and the flexures are much more acute (Fig. 106c, d, e). The musculature of each side of the body is divided by a horizontal septum which coincides with the central, forward flexure. The folding of the myotomes is complex and variable. In general, each myotome is thrown into three cones, corresponding with the three flexures, with the apices of the cones lying close to the mid-sagittal septum. The cones corresponding to the backward flexures are open to the anterior, that of the forward flexure opens towards the posterior end of the fish. In the Chondrostei and Holostei, the forward flexure forms a single cone, but in teleosts there are generally two cones associated with the forward flexure, one above and one below the horizontal septum. In the elasmobranchs (Davidson, 1920), the forward flexure also forms a single, rather asymmetrical cone, but the myotome is complicated by the existence of a second forward flexure in the dorsal part of the body. A second, dorsal forward flexure also appears in some teleosts (Dietz, 1913).

The connective-tissue elements by which the longitudinal muscles exert a lateral torsion upon the vertebral column, are often strengthened or augmented in a variety of ways.

The myoseptum or myocomma which bounds the myotome tends to become attenuated at the apices of the cones and to form connective-tissue extensions into the muscle tissue of the adjacent myotomes. In the Scombridae,

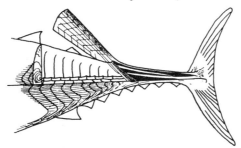

FIG. 107. Tendons in the tail of the mackerel. (After Nursall, 1956.)

which include fast-swimming fish such as the mackerel and tuna, these connective-tissue extensions unite to form tendons running almost the whole length of the body. They are inserted into the vertebral column and pectoral girdle at the anterior end of the fish, and in the caudal region, the tendons insert into lateral processes on the vertebral centra (Fig. 107).

In all gnathostomes, a horizontal septum is formed and extends from the vertebral column to the skin. It divides the lateral musculature into dorsal and ventral, or epaxial and hypaxial moieties, and also provides an additional insertion for the myocommata.

A third longitudinal connective-tissue system may be developed close to the vertebral column. Neighbouring myocommata may become fused between the backward and forward flexures of the myotomes, to form a deep longitudinal ligament at the junction of the myocommata and the mid-sagittal septum within both the epaxial and hypaxial muscles (Nursall, 1956). Longitudinal ligaments of this type are particularly well developed in elasmobranchs, which lack bony supports for the muscles.

The septa, particularly those of the teleosts, are often reinforced by supernumerary bones. These are connected to the vertebral column and not only strengthen the connective-tissue elements, but also transmit tensions in them directly to the vertebrae. The mid-sagittal septum is supported by the neural spine and, in the caudal region, by the haemal spines. In the pre-anal part of the body the ventral ribs lie in the peritoneal fasciae. Dorsal, or epineural ribs may also be developed in the horizontal septum (Fig. 108). They are very variable and may be inserted into the centrum, the transverse process, the neural arch, or even the ventral ribs (Emelianov, 1935). Intermuscular bones are found in the myocommata of many teleosts (Emelianov, 1935; Nursall, 1956), and generally have ligamentous attachments to the neural spines. It has generally been supposed that these intermuscular bones, like the ribs and haemal and neural spines, strengthen the myocommata and transmit tensions

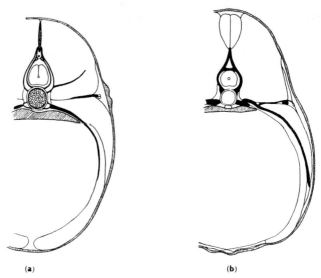

FIG. 108. Ribs and intermuscular bones of (*a*) *Salmo* and (*b*) *Gobius*. (After Emelianov, 1935.)

to the vertebrae, but Jarman (1961) has indicated that there is some difficulty in accepting this interpretation. A detailed investigation of these bones might reveal that they have more complicated functions.

The manner in which the contraction of the longitudinal muscles produces bending of the vertebral column has been considered in some detail by Greene and Greene (1913) and Nursall (1956). Because of the obliquity of the myocommata, contraction of the muscles exerts a tension in the plane of the septum and tends to draw the forward and backward flexures together. As a compensatory movement, one septum tends to slide over the adjacent ones, the greatest movement being at the external margin, the least towards the central axis of the animal. Each myocomma is the boundary between adjacent myotomes and muscle fibres are inserted into both faces of it. It is therefore subject to antagonistic forces which are applied almost simultaneously. Since the myocomma is inserted at different levels along the length of the fish at the forward and backward flexures, a torsional force is applied to the vertebral column when the muscles contract.

The lateral curvature of the body produced by this torsional force is transmitted from head to tail by the serial contraction of the myotomes. Recovery from the flexed position is by contraction of the stretched longitudinal muscles of the opposite side of the body, and also by restoring forces generated in the relatively inflexible spinal column. Rockwell, Evans and Pheasant (1938) point out that the development of bony intervertebral articulations in the higher teleosts may in some cases render the column more flexible, but in extreme examples has the opposite effect of locking the vertebrae together so that the vertebral column is rigid and acts as an axial spring.

It is only in the gnathostomes that both backward and forward flexures are developed in the myotomes. In Amphioxus, the backward slope of the dorsal wing of the myotome acts against the forward flexure and thus provides a functional substitute for a backward flexure (Nursall, 1956). The cyclostome myotome, on the other hand, lacks any such functional equivalent of a backward flexure. The entire myotome slopes forward and is therefore unable to exert a torsional effect upon the spinal column. Nursall concludes that because of this, cyclostomes have not the mechanical apparatus for the generation of a locomotory wave that begins caudally and moves towards the head. He suggests that the extreme flexibility of cyclostomes compensates for this, though in fact they are capable of producing reversed locomotory waves and of swimming backwards (Adam, 1960). The action of cyclostome myotomes has never been studied experimentally and their method of producing reversed locomotory waves remains something of a mystery.

In the majority of fishes, the locomotory waves are most pronounced at the posterior end of the body. The caudal myotomes are much smaller than those more anterior and their effectiveness is increased by a variety of structural modifications:

(1) The obliquity of the flexures progressively increases in more posterior parts of the body. The muscle fibres in the myotomes remain orientated parallel to the axis of the fish and in consequence, the component of the force they generate that is transmitted to the myocomma becomes greater.

(2) Because the myotomes are reduced in size in the caudal region, the mass that they move is less than that moved by more anterior myotomes. This phenomenon can be seen in an extreme form in the carangids and scombrids, particularly the latter, in which there is a caudal peduncle containing bone and ligaments, but little muscle tissue. The myotomes which provide the source of power are more anterior and remain large; they are connected to the tail by tendons.

(3) The vertebrae of the caudal region undergo a variety of modifications depending upon the type of swimming employed by the fish (Goodrich, 1930). In some, the caudal vertebrae are short and numerous, permitting

great flexibility of the tail. In some fast swimmers, the caudal vertebrae are long and few in number, and form an almost rigid rod because of the development of interlocking zygapophyses or the expansion of the neural and haemal spines to form long, flat, overlapping processes as in the marlin, *Makaira mitsukurii* (Kishinouye, 1923) (see Table 9).

(4) The caudal fin (Fig. 109) is similarly subject to a great variety of modifications, the functional significance and evolution of which have been discussed by numerous authors (Ahlborn, 1895; Whitehouse, 1918; White, 1935;

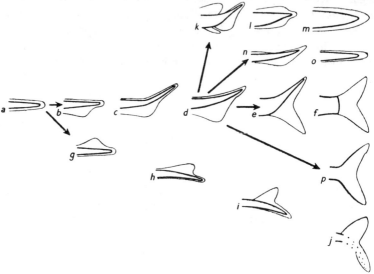

FIG. 109. Tail fins of fishes, showing a possible derivation of the various types. (*a*) Primitive protocercal, (*b*) modified protocercal, (*c*) primitive Gnathostomata, (*d*) Placodermi, Chondrichthyes and early Osteichthyes, (*e*) Palaeopterygii and Holostei, (*f*) Teleostei, (*g*) modified protocercal, (*h*) Pteraspida, (*i*) Anaspida, (*j*) Anaspida, (*k*) Choanichthyes, (*l*) Choanichthyes, (*m*) modern Dipnoi, (*n*) Pleuracanthodii, (*o*) Pleuracanthodii and degenerate forms, (*p*) some Chondrichthyes. Thick lines indicate relatively inflexible regions, thin lines, very flexible regions. Dotted lines indicate regions of unknown flexibility. (From Affleck, 1950.)

Graham-Smith, 1936; Grove and Newell, 1936, 1939; Harris, 1937; Kermack, 1943; Affleck, 1950; Nursall, 1958). Hypocercal and heterocercal tails produce positive or negative pitch, respectively, in addition to forward thrust, and this, in conjunction with the pectoral fins or the anterior end of the body, permits vertical steering by animals that are not particularly flexible in a dorsoventral direction. The teleost homocercal tail is designed primarily to provide a forward thrust, though the unequal development of the hypocaudal and epicaudal lobes may also provide a vertical force component (Fig. 109). In most teleosts the caudal fin is flexible and has a complex intrinsic musculature by which it can be furled or spread, or adjusted in a variety of ways. Fast swimmers, such as the tunnies and marlins, have a rigid fin with a high

TABLE 9
Structural Adaptations to Different Swimming Habits in Teleosts
(from Nursall, 1958)

Group 1 (e.g. eel) undulations of large amplitude and low frequency:
  (*a*) flexible vertebral column of many short vertebrae;
  (*b*) thick muscular body and caudal peduncle;
  (*c*) rounded tail fin with negligible aspect ratio.

Group 2 (e.g. perch and majority of teleosts) undulations of moderate amplitude and fairly high frequency:
  (*a*) flexible vertebral column of short vertebrae, number of which varies from 24 to more than 100, but commonly between 35 and 70;
  (*b*) narrowed, but muscular, caudal peduncle;
  (*c*) flexible, forked tail, aspect ratio 2–4.

Group 3 (e.g. tunny, marlin) undulations confined to tail, of low amplitude and very high frequency:
  (*a*) stiff vertebral column of long vertebrae with well developed zygapophyses; vertebral number 30–45;
  (*b*) slim, non-muscular peduncle;
  (*c*) rigid tail fin, high aspect ratio (5–6).

aspect ratio (Table 10), an adaptive feature to rapid swimming found also in some sharks. In fishes in which the tail executes very rapid oscillations,

TABLE 10
Aspect Ratios of the Caudal Fin of a Number of Fishes
(from Nursall, 1958)

Amia calva (Bowfin)	(extended)	1·0
Salvelinus namaycush (Lake trout)	(furled)	1·0
	(extended)	2·0–2·5
Micropterus dolomieu (Smallmouth bass)	(extended)	2·0
Hoplopagrus guntheri (Striped pargo)	(furled)	1·5
	(extended)	2·9
Perca flavescens (Yellow perch)	(extended)	3·2–3·8
Caranx caballus (Green jack)		3·8
Scomberomorus sierra (Sierra mackerel)		4·1
Euthynnus lineatus (Black skipjack)		5·0
Euthynnus alleteratus (Little tuna)		5·0
Neothunnus macropterus (Yellowfin tuna)		5·2
Makaira albida (White marlin)		6·1

exceptional stream-lining of the caudal peduncle becomes necessary. This is provided by the development of lateral connective-tissue keels in the scombrids and fast swimming sharks like the porbeagle, *Lamna cornubica*, and the vertical flaps at the base of the caudal peduncle of whales. The existence of lateral fins with endoskeletal supports in *Cladoselache* (Harris, 1950) suggests that it, too, was a rapid swimmer and had undergone a similar adaptation.

Most studies of the mechanics of fish locomotion have been concerned primarily with the deep lateral myotomal musculature and have ignored the superficial lateral muscles. These appear to be related chiefly to slow, sustained swimming, while the myotomal musculature is concerned with the development of the more powerful contractions involved in bursts of high-speed swimming. No fish is able to swim both fast and long (Bainbridge, 1960). Some species, such as the pike, swim in short, sharp bursts when hunting or escaping from predators; others, such as the trout, live in swift water currents and require sustained swimming activity to maintain their position in the stream. These two types of fish may be termed 'sprinters' and 'stayers', respectively, to borrow Bösiger's (1950) terminology of the flying abilities of birds. But there remain other fishes which require both staying power for food collection and sprinting ability to escape predators. These varied and sometimes conflicting swimming capabilities of fishes are reflected in the development of the superficial lateral muscles (Fig. 110).

In teleosts, as in more advanced vertebrates, there are two types of muscle fibres. Those of large diameter have little sarcoplasm and a reduced mitochondrial system, they contain little fat and much glycogen, no myoglobin and have a poor vascular supply, whereas the small fibres have an extensive mitochondrial system, contain much fat and, when they are organized into a superficial lateral muscle, a high myoglobin content; they also have an extensive vascular supply (Bösiger, 1950; Braekkan, 1956; George and Naik, 1958a, b; Barets, 1961). These are fast and slow fibres, respectively, and the former appear to be adapted to anaerobic metabolism, the latter to aerobic metabolism.

From the comparative study of the relative abundance and distribution of the two types of muscle fibre in the lateral muscles of fresh-water teleosts by Bodekke, Slijper and Stelt (1959), it is evident that there is a strong correlation between the development of the superficial lateral muscles and the swimming habits of fishes. These authors divided the fresh-water teleosts that they examined into four groups:

(1) 'Sprinters' (e.g. pike, perch) are diurnal carnivores living in still water, which hunt their prey by sudden darts and, if disturbed, rush to the nearest hiding place. The lateral muscles of these fishes are composed exclusively, or nearly exclusively, of large diameter, white fibres.

(2) 'Sneakers' (e.g. eel, burbot) are nocturnal, bottom feeders which have the same slight development of the small diameter red muscle fibres. These are clearly not 'sprinters', and it might have been expected that the superficial lateral muscles would be well developed in them, as they are in the typical 'stayers'. However, the deep lateral muscles, composed of white, large diameter fibres appear to have developed some of the properties of the superficial lateral muscles, in that the sarcomeres are relatively long (suggesting slower contraction of the muscles) and the muscle is exceptionally well vascularized.

(3) 'Crawlers' (e.g. dace, rudd, bream, gudgeon, tench) have a well-

developed superficial muscle and feed upon small and generally sluggish invertebrates; in this respect they are typical 'stayers'. Their escape reactions, however, involve rapid zig-zag swimming in open water, so that the short rapid

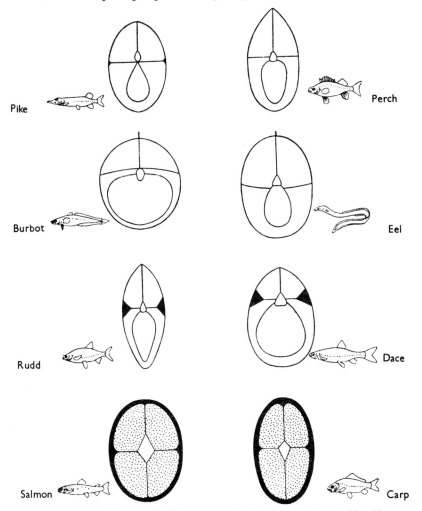

FIG. 110. Extent of the superficial lateral muscles in fishes of various habits. (From Bodekke, Slijper and Stelt, 1959.)

swimming characteristic of 'sprinters' is also within their capabilities. The 'crawlers' are dual purpose fish.

(4) 'Stayers' (salmon, rainbow trout, carp). Both salmon and trout are migratory and live in swift streams. An ability to perform long sustained swimming is therefore essential to them. They have a well-developed superficial lateral muscle and small diameter fibres are also mingled with large

diameter fibres in the deep lateral muscles. The carp, living in stagnant waters, is an unexpected member of this group, but it is suggested that, as it is a large fish and is vegetarian, its food requirements are high and it must therefore spend a considerable proportion of its time foraging.

The same sort of correlation between the development of red superficial lateral muscle and swimming habits may also apply in marine fishes. Braekkan (1956) found that fast swimmers like the tunny have extremely well developed red muscle, while the superficial lateral muscles of fishes like herring and mackerel that swim in shoals, are only slightly less well developed. Bottom feeders like the halibut, on the other hand, have very little red muscle and it is paler in colour than that in herring and mackerel. Braekkan attempted to relate the development of the superficial lateral muscles to the speed with which the fishes swim, but, as we have already observed, it is more likely that the correlation is between sustained swimming and the development of red, superficial lateral muscles. Pelagic fishes like the tunny swim continuously, but benthic fishes such as halibut make sudden darting movements to catch their food and spend much of the time immobile upon the sea-bed. Tunny, herring and mackerel are all 'stayers', while the halibut is a 'sprinter'.

THE CO-ORDINATION OF MUSCULAR ACTIVITY

The co-ordination of locomotory movements may theoretically be achieved by several methods: by a rhythmical discharge in the central nervous system emanating from some such control centre as the brain, or by local reflex arcs in a chain reflex, or by combinations of the two processes. While the maintenance and control of locomotory rhythms has been the subject of many investigations by Gray, Lissmann, von Holst, ten Cate and others (summarized by Healey, 1957), this work has been confined chiefly to vertebrates, and our knowledge of the control of undulatory waves in other animals is still very fragmentary.

At one time it was considered that in fishes the co-ordination of swimming undulations was maintained by local proprioceptive arcs, the stretch on one myotome by the contraction of its neighbours inducing it to contract in its turn. Locomotory co-ordination, in fact, was supposed to be exactly comparable to that in crawling earthworms (Friedländer, 1894; Gray and Lissmann, 1938). This view was strongly supported by ten Cate and ten Cate-Kazeweja (1933) who demonstrated that if the spinal cord and musculature of a dogfish were transected at the level of the pectoral fins, leaving the two parts of the body connected only by the skin and viscera, swimming movements were co-ordinated on both sides of the cut. They suggested that in addition to co-ordination by way of proprioceptive arcs, the overlap of the dorsal spinal nerves permitted the direct nervous transmission of excitation from segment to segment. The chain reflex theory, at least in its application to teleosts, was denied by von Holst (1935) and Gray (1936). They showed that neither clamping a part of the body of eels and other fishes, so that it was incapable of movement and hence of transmitting mechanical stresses to other parts of the musculature (von Holst), nor removing all the skin and musculature from a section of the body (Gray), which has the same effect, prevented co-ordination of swimming movements. Furthermore, Gray and Sand (1936) completely denervated the musculature of a number of segments at the level of the anterior dorsal fin of a dogfish and then clamped this region rigidly; they found that the swimming movements of both parts of the fish remained co-ordinated so long as the spinal cord remained intact. The elasmobranchs thus fall in line with the teleosts, and co-ordination by chain reflexes appears unlikely to occur in any fishes.

This is not to say that peripheral stimulation is irrelevant to undulatory

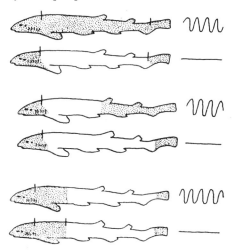

FIG. 111. Effects of deafferentiation upon rhythmical movements of the spinal dogfish. White areas are desensitized regions of the body, the level of spinal transection is indicated by a thick line. In each case, upper figure shows rhythmical movements, lower figure, none. (From Lissmann, 1946.)

swimming movements, clearly it is not, but the extent to which sensory input merely evokes adjustment of the locomotory rhythm, or is actually necessary for its production, is very uncertain, as Gray (1950) has emphasized. The evidence remains conflicting. Von Holst, working chiefly on teleosts, elaborated the view (summarized by Healey, 1957) that the initiation and co-ordination of undulatory movements is entirely a central process, but Lissmann (1946) demonstrated that although a spinal dogfish shows prolonged spontaneous swimming activity, deafferentiation causes a complete loss of rhythmical movements (Fig. 111).

The only non-vertebrate to have been subjected to comparable investigation is the leech *Hirudo* (Gray, Lissmann and Pumphrey, 1938; Kaiser, 1954). As in fishes, the locomotory waves of swimming leeches remain co-ordinated in two parts of the body connected only by the central nervous system, indicating that the chain reflex theory can be applied to swimming leeches no more than it can to fishes. Leeches continue to swim after the supra-oesophageal ganglion has been extirpated and in this respect perhaps resemble the dogfish, though not teleosts, since dogfishes also display prolonged swimming movements after post-medullary section of the spinal cord. However, unlike fishes, the isolated nervous system of the leech shows no rhythmical electrical activity, although if it retains nervous connection with a limited part of the body by only a single segmental nerve, swimming movements are accompanied by a rhythmical activity of the nerve cord of the same frequency as that of swimming.

Gray, Lissmann and Pumphrey (1938) concluded that the only hypothesis to cover all these observations gathered from investigations of fishes and leeches, is that a swimming rhythm emerges only if the general level of excitability of the nerve cord is maintained at an adequate level. In leeches, this level seems to be maintained by proprioceptors in the skin and muscles, but in teleosts this must be supplemented by activity of locomotory centres in the medulla. There appear to be no grounds for supposing the initiation and co-ordination of undulatory movements to be accomplished by a purely central mechanism (Lissmann, 1946). Furthermore, so far as our admittedly limited knowledge of the physiological mechanism underlying sinusoidal swimming goes, there is nothing to suggest a peculiar relationship between metamerism and this type of movement. Indeed, the only theory which might have lent some support to such a view, that of co-ordination by chain reflexes, now appears to be untenable.

THE SEGMENTATION OF VERTEBRATES AND INVERTEBRATES

Having examined the mechanics of swimming and the structure and behaviour of the muscles that produce the locomotory movements, we can now return to the questions posed at the beginning of this chapter. It is clear that, so far as the mechanics of undulatory swimming is concerned, there is no essential difference between vertebrates and invertebrates or between segmented and unsegmented animals; they all swim in the same way provided they are of an appropriate shape. Differences emerge only when we consider how the locomotory waves are produced, and evidently the dichotomy lies, not between segmented and unsegmented animals, but between vertebrates and invertebrates. Among invertebrate worms, the longitudinal muscles are involved in a variety of locomotory activities of which swimming is but one, and whether or not the musculature is segmented is in no way correlated with the ability of the animals to perform undulatory swimming movements. In the more primitive chordates, on the other hand, the longitudinal muscles are devoted exclusively to the generation of swimming movements and there is every indication that the segmental nature of the musculature is fundamental to the performance of this activity. The explanation of this association of metamerism with swimming in the chordates, but not in the invertebrate phyla, lies in the development of a notochord and axial skeleton in the Chordata.

In both worms and fishes, when a locomotory wave is produced, a region of the body is flexed and bending moments are applied to the adjacent sections of the body. In the vertebrates, flexure of the body is produced by the application of a torsional force to the axial skeleton. In invertebrates, possessing as they do a fluid skeleton, it is impossible to produce these movements by applying a torsional force axially; the only force that can be applied to a fluid is compression. Instead, the longitudinal muscles of one side of the body contract, fluid pressure changes are transmitted to other parts of the body wall, causing elongation of the antagonistic longitudinal muscles, and the body is flexed. In worms, the forces of muscle contraction act in a longitudinal direction; in fishes, the forces act transversely.

We have already seen that the myotome muscles of fishes are so arranged as to exert a torsional force upon the axial skeleton. In order to do so, the muscles are inserted into myocommata which, in turn, are inserted into the vertebral column and the median septum. The obliqueness of the myocommata is critical, for were they perpendicular to the muscle fibres, as they are at the time of their first appearance, very little force could be transmitted to the axial system. Flexion of the myocommata is necessary for individual myotomes to exert an axial torsional force, although, in cyclostomes, it is possible for the whole longitudinal musculature to generate torsional forces in the absence of flexures of the myocommata.

The fact that the longitudinal myotome muscles of chordates are inserted axially means that they have little mechanical advantage compared with those of invertebrates, which are inserted peripherally. There are two important consequences of this:

(1) There is a smaller area for direct insertion of muscle fibres upon the axial skeleton and the medial septum than there is at the body wall, despite the fact that because of their smaller mechanical advantage, more, not fewer, fibres are demanded at the axial insertion.

(2) A lateral as well as a longitudinal force is necessary in order to bend an axial skeleton with axial insertions of the muscles. Only a longitudinal force is required to produce bending if the muscles are inserted peripherally into a flexible body wall.

These two problems can be solved most easily by the insertion of the axial muscle fibres into inclined myocommata. Then, both lateral and longitudinal forces are transmitted to the axial skeleton when the muscles contract and there is space for the insertion of a considerable number of muscle fibres. Since the forces are transmitted to the skeleton through the septa, the muscles must be inserted upon them and must perforce be segmentally arranged.

A segmental arrangement of the longitudinal muscles is not necessary for the production of undulatory movements in worms because the tensions produced by muscle contraction are transmitted directly to the body wall and not by way of the septa. Furthermore, smooth transmission of the locomotory waves along the body is, if anything, better assured when there is some overlap of the longitudinal muscle fibres. Smooth transmission of the locomotory waves in chordates is assured by the fact that each myocomma is inserted over a number of vertebral segments, and each segment receives insertions of a number of adjacent myocommata.

THE EVOLUTION OF CHORDATE SEGMENTATION

The strict association of muscle segmentation with swimming in chordates, but not in worms, suggests that, since segmentation subserves fundamentally different functions in the two groups, it is likely to have evolved independently in them. The possibility that metamerism has been evolved but once and has become highly modified in annelids and chordates in response to divergent functional demands is, of course, not excluded, but the former interpretation has the merit of being consistent with most current views of chordate origins.

The view that metamerism was uniquely evolved and is homologous throughout the animal kingdom is implicit in attempts to derive vertebrates from annelids (Dohrn, 1875; Semper, 1876; Eisig, 1887; Minot, 1897; Delsman, 1921-1922) or from arthropods (Gaskell, 1890, 1908; Patten, 1890, 1912). In alternative theories, metamerism is presumed to have arisen independently in vertebrates in the course of their evolution from a nemertean (Hubrecht, 1883, 1887; Jensen, 1960), an echinoderm (Gislén, 1930), or a protochordate stock (Bateson, 1885, 1886; Brooks, 1893; Willey, 1894; Garstang, 1894, 1922, 1928; Berrill, 1955; Carter, 1957; Whitear, 1957; Bone, 1960). Of these, only the annelid and protochordate theories have ever gained wide acceptance, and the latter has gradually superseded the former. Few zoologists today would dispute that vertebrates evolved from protochordates, though the interrelationships between the various protochordate groups and their connection with the vertebrates is still a matter for discussion.

The fundamental similarity of structure between chordates and the tadpole larva of ascidians forms the basis of the majority of the protochordate theories of vertebrate evolution. Willey (1894) and, more recently, Carter (1957) regarded the tadpole as recapitulatory, but Garstang (1928) initiated a fertile line of speculation by indicating the potentialities of larval evolution. If chordates, as he suggested, have been derived by neoteny from the tadpole larva, then a whole range of sessile protochordates must be considered as possible vertebrate ancestors since their specializations to a sedentary existence do not appear in the larva and are not transmitted to its descendants.

Garstang's view has received wide acceptance and discussion now centres chiefly around the nature of the protochordates that invented the tadpole larva. Bone (1960) derives the chordates from the larva of a 'proto-ascidian', but as Tarlo (1960a) and Whitear have pointed out, if one extrapolates the ascidian line backwards, one arrives at an animal very like a pterobranch hemichordate, and the distinction between 'proto-hemichordate' and 'proto-ascidian' is largely one of terminology. Kozlowski (1947, 1948) described a rich and varied sessile hemichordate fauna in the lower Ordovician (Tremadoc), including graptolites, pterobranchs and acanthastids, suggesting that the early Palaeozoic era was a time of considerable radiative evolution of the hemichordates, and it is presumably from among these groups that the chordates emerged (Tarlo, 1960b).

Whatever the precise course of evolution proposed, all the current disputants are agreed upon two things: that chordates evolved from an organism comparable in the broad features of its structure to the ascidian tadpole, and that segmentation of the musculature appeared in the process as an adaptation to more powerful and prolonged swimming.

The appearance of a notochord is generally supposed to have accompanied the development of a lateral musculature in the tail, though its morphological origins remain very obscure. Gislén (1930), one of the few modern opponents of the protochordate theory, attempted to derive the Acrania from carpoid echinoderms and considered the notochord to be homologous with the echinoderm hydrocoel, a homology also favoured by von Ubisch (1929), though this disregards the method of formation of the tadpole notochord from a chain of vacuolated cells in which the central cavity eventually becomes continuous. According to Hubrecht's (1887) nemertean theory, which has recently been revived by Jensen (1960), the notochord is homologous with the nemertean proboscis sheath. Neither interpretation is very convincing and it is more usually assumed that the notochord arose *de novo* in the Chordata. Berrill (1955) draws attention to the fact that the notochord of the ascidian tadpole larva is formed in ontogeny by a group of dorsally situated endodermal cells which become vacuolated and grow out to form a tail. He suggests that 'the localization, in an otherwise previously unspecialized ovum, of a cytoplasmic agent which evokes vacuolation, would result in some form of tail-like outgrowth at a later stage of development'. Despite its important consequences, the development of supporting structures of this type appears neither difficult nor unusual, and Berrill offers the branchial skeleton of sabellid polychaetes and the stomochord of enteropneusts as examples of comparable evolutionary events.

According to Berrill's interpretation, the outgrowth of the notochordal cells stimulates the rapid growth of epidermal and mesenchymal tissues which are carried along with them, so that the development of notochord and lateral muscles are both part of the same process, a view supported by the embryological studies of Tung, Wu and Tung (1958) on Amphioxus. The notochord of the ascidian tadpole consists of a column of 40–42 cells, and it is flanked on each side by a band of muscle cells, the number of which is constant within a species, but varies from one species to another. Neither the notochord nor the musculature of the tail shows any sign of segmentation. Martini (1909) considered that the tail of appendicularians and the ascidian tadpole does show a degenerate metamerism in that each cell corresponds to a myotome, and Gislén (1930) appears to have been prepared to accept some such interpretation also. But there is no correspondence between the cells of the notochord and the muscle cells, and the muscles of each side of the body contract as a unit. A comparable stage exists in the dogfish embryo in which a 'precocious band', composed of myofibrils which do not conform to a segmental pattern, is formed on either side of the notochord (Balfour, 1878; Harris and Whiting, 1954). It is clear that the tail of the ascidian tadpole larva cannot be regarded as segmented from either a structural or a functional point of view. Metamerism must consequently be supposed to have made its appearance at a later stage in chordate evolution with the development of a powerful swimming musculature.

CONCLUSIONS

There is no evidence of any necessary relationship between metameric segmentation and undulatory swimming movements. Many non-segmented animals are able to swim in this way, and both the mechanism of swimming and the behaviour of the musculature that generates the locomotory waves are the same in non-segmented, and most segmented animals. Furthermore, undulatory swimming movements, at least in the metamerically segmented annelids, involve no essential activities that exploit the fact that the animals have a segmental organization. Admittedly in the two polychaete families that have been subjected to detailed investigation, the Nereidae and the Nephtyidae, the segmental parapodia participate in a dynamic sense in the swimming movements, but it is by no means assured that this is true in all other polychaetes, and theoretically, it is not essential to swimming.

Chordates present a different picture. It is by now universally accepted that the metameric organization of the chordate body was evolved independently of that of annelids, and it is very probable that the metamerism of these animals, unlike that of annelids, was evolved as an adaptation to the production of sinusoidal swimming movements. Why swimming should have resulted in the evolution of metamerism in one group of animals, but be quite unrelated to it in another, can be explained only by reference to one profoundly important structural difference between chordates and worms: the presence of an axial notochord in the former.

In worms, contraction of a part of the longitudinal body-wall musculature on one side of the body causes bending of the animal, and the fluid skeleton enclosed by the muscle coat offers no resistance to this movement. In chordates, on the other hand, the notochord, consisting of a column of turgid cells enclosed by a tough envelope, does offer resistance to deformation, and bending can be produced only if torsional forces are applied to it. The contraction of muscles inserted into the sub-epidermal basement membrane, as they are in annelids, would be ineffective in producing this torsional force in the absence of some special means of transmitting it directly from the body wall to the notochord. Forces generated by muscle contraction in chordates are consequently applied medially to the notochord, instead of peripherally to the body wall. The area available for the insertion of muscle fibres into the notochord, or even into a mid-sagittal mesentery, is very limited and yet, because of their medial position, the muscles have little mechanical advantage and must necessarily be massive if they are to produce powerful movements of the body. Furthermore, in order to apply a torsional force to the notochord, these longitudinal muscles must provide a transverse component when they contract. These problems have been resolved by the evolution of a segmented musculature and of septa or myocommata. Myocommata provide a large area of attachment for the massive myotomal muscles, and, since they are inclined to the vertical, permit the forces of muscle contraction to be transmitted directly to the notochord and median septum despite the fact that the myocommata are flexible membranes. In addition, because of their peculiar shape, the myotomes individually exert a torsional force upon the axial skeleton.

From the functioning of the lateral musculature of fishes, it is clear that the myocommata behave in an entirely different manner from the septa of annelids, and they serve a different function. Tensions produced by contraction of the annelid longitudinal musculature are never transmitted to or through the septa. Indeed, the longitudinal muscles of annelids are generally not segmentally organized to any great extent, and the septa are, as a rule, inserted on the inner face of the longitudinal muscles. In those rare annelids in which the septal musculature has a more peripheral insertion, the longitudinal muscle fibres traverse the septum without coming into association with it.

In both annelids and chordates, the segmental organization of the musculature has undergone considerable modification in the course of evolution, and there have been important changes in locomotory techniques with corresponding structural modifications of the animals. In no existing member of either phylum can we see metameric segmentation in its primitive form, although, with a fair degree of probability, we can deduce the functional attributes of the segmented muscles. It is evident that the view that metamerism evolved as an adaptation to sinusoidal swimming movements is correct only for chordate metamerism. As it was discussed by Snodgrass (1938), Goodrich (1946) and Hyman (1951a), the swimming theory included the postulate that segmented animals evolved from those that already displayed pseudometamerism. This, too, is erroneous and pseudometamerism is unlikely ever to have featured in chordate evolution.

The evolution of metamerism in the annelids is a separate matter and remains to be discussed; this, viewed in the wider context of metazoan phylogeny, will occupy us in the next, and final chapter.

6
THE PHYLOGENY OF THE METAZOA

In previous chapters we have examined the structure of organisms from a purely functional point of view, without regard to the phyletic relationships between different groups of animals. From this analysis a number of correlations between locomotory habit and body architecture have emerged. Each type of body movement and each locomotory technique makes certain mechanical demands upon all animals employing it. Generally there appear to be relatively few ways in which these demands can be met, and so widely dissimilar animals may show comparable structural adaptations to the same method of locomotion, or to similar mechanical demands.

This is particularly well exemplified in adaptations to locomotion on a hard, smooth substratum. In order to move, the animal must adhere temporarily to the substrate while it exerts a backthrust. There appear to be only two ways in which animals adhere, by secreting an adhesive mucus or by means of mechanical suckers, and two ways in which locomotory forces can be generated, by cilia or by muscle contraction. Planarians and many gastropod molluscs move in the same way and employ one or other or both methods of adhesion and traction, with the result that the structure of the gastropod foot and of the whole planarian body show the remarkable convergence that we have already observed. Leeches, with the aid of their permanent suckers, show a modified version of the same locomotory technique when moving over solid surfaces, just as some triclads do, and they show similar adaptations to this method of locomotion, with a corresponding reduction of many of their segmental and coelomate features. Asteroids and echinoids, on the other hand, are exceptional in a variety of ways. Those that crawl on rock faces have adhesive suckers on the tips of the podia and the locomotory forces are provided by muscular contraction. However, because of the relative or total rigidity of the body wall, it is impossible for them to use body-wall muscles, and traction is provided instead by the podial muscles. Thus we find totally different structural adaptations to locomotion from those in worms.

Exceptions of this kind are readily intelligible and are fortunately relatively infrequent. Convergence, on the other hand, is quite a common occurrence in the evolution of locomotory structures. We may therefore have some confidence in the correlations we deduce between structure and movement. Furthermore, these correlations must apply equally to extinct as to modern animals since the adaptations are made in response to demands which are, in the main, physical and mechanical and are therefore exactly predictable. With this knowledge at our disposal, we can state the essential structural features an animal must possess if it is to move in a certain way and, conversely, what locomotory powers are possessed by, and hence what environments are available to, animals of any postulated structure.

While these conclusions do not permit us to state the phyletic sequence and evolutionary relationships of the metazoan phyla, they must clearly be taken into account in any attempt to reconstruct the phylogeny of the Metazoa. The history of metazoan animals is largely one of evolution of different types of body organization, represented by the emergence of the main phyla, followed by the exploitation of the potentialities of each during the ensuing periods of radiative evolution. At this level of evolution, mechanical factors appear to be of paramount importance and, because locomotion, particularly of soft-bodied animals, involves activity of the whole body, locomotory adaptations are often related to very profound structural modifications of the type that characterize and separate the major divisions of the Metazoa. It is difficult, in fact, to envisage changes other than locomotory ones that would have such profound effects upon the anatomy of animals.

The final stage of this enquiry must therefore be to consider the evolution of increasing complexity of structure in relation to the penetration of new environments and the evolution of new locomotory techniques, and to see how the existing groups of animals may be related to the scheme. But before we can do this we must decide the structure and habitat of the immediate forerunners of the coelomate and segmented animals. Thence we may discuss the manner in which the coelomate and metameric types of body organization developed and the selective pressures in answer to which they were evolved. Other problems, such as the origin of the Metazoa, or the systematic position of the Cnidaria and Ctenophora, which we must consider briefly, concern us only to the extent that they relate to the structure of the immediate ancestor of the coelomates.

THE ORIGIN OF THE BILATERIA

With few exceptions, all coelomate and segmented animals can be shown to be bilaterally symmetrical or to have been derived from bilaterally symmetrical animals. This fact is reflected in the common use of the term 'Bilateria' which was first used in Hatschek's (1888–1891) classificatory system to designate all Eumetazoa above the level of coelenterates, and despite considerable modifications to the system, the term and concept are still widely accepted (Hyman, 1940; Marcus, 1958). The essential structure of the most primitive bilaterally symmetrical metazoans which must form the starting point of the present inquiry is disputed and depends upon the view taken of the mode of evolution of the Metazoa. We must therefore begin by considering the claims of the rival theories in this field.

Theories of metazoan origins, including those canvassed in recent years, fall into two main categories: those in which multicellular animals are held to have arisen by the integration of protistan colonies, and those in which the Metazoa are supposed to have evolved by the internal division, or cellularization, of multinucleate or syncytial protozoans. The former view, that metazoans originated from colonial protistans, was first proposed by Haeckel (1874a, b, 1875; see also Lankester, 1876) and after many years of rather uncritical acceptance by the majority of zoologists, has been revived in slightly modified versions by Ulrich (1950), Remane (1950, 1952, 1954), Jägersten (1955), Sachwatkin (1956), Marcus (1958) and others. Theories that metazoans evolved by the cellularization of multinucleate protistans descend from speculations by Jhering (1877), Saville-Kent (1880–1882) and Sedgwick (1894), though these authors appear to have been anticipated by Schmidt (1849) and Diesing (1865) who classified the ciliated Infusoria with the lower worms as Prothelmintha. These views have been revived and extended by Steinböck (1937), Hadži (1944, 1953) and their supporters (De Beer, 1954; Pax, 1954; Hanson, 1958; etc.).

The form taken by the earliest postulated multicellular animals differs substantially in the two kinds of theory. According to the Haeckelian view, they were hollow balls of cells; in the opposing theories of Steinböck and Hadži, they were solid animals. A third group of theories, originating in the speculations of Lankester (1877), Metschnikoff (1883) and Ziegler (1898) and supported in more recent years by Snodgrass (1938) and Hyman (1940), bridge the gap between the two. According to these writers, metazoans originated in colonial protistans forming a solid mass of cells and later came to resemble the coelenterate planula larva.

Although these theories differ even in their interpretation of metazoan origins, it is in the logical development of them, when the course of evolution of the major phyla is considered, that the divergences become pronounced and irreconcilable.

The Syncytial Protistan Theory

It is claimed that the metazoans are derived from a multinucleate, ciliate-like protistan stock and that the most primitive existing metazoans are therefore the partially cellularized Acoela (Fig. 112). The evolution of greater structural complexity than that found in Turbellaria is not generally discussed by the proponents of this theory and controversy centres chiefly around two questions: whether or not the lack of complete cellularization, particularly of the gut, of the Acoela is a primitive or secondary feature, and whether the structural organization of the Cnidaria and Ctenophora is to be regarded as primitive or not.

As to the latter question, in the syncytial protistan theory the structure of cnidarians and ctenophores is regarded as derivative and these two phyla are therefore considered to be reduced triploblastic animals, having evolved probably from rhabdocoels. Following Hadži (1953), this necessitates a reversal of the conventional view of the Cnidaria, so that the Anthozoa, which show traces of bilateral symmetry, are the most primitive and the Hydrozoa, showing none, are the most advanced (i.e. reduced) cnidarians. This view is strongly opposed by such authorities on the coelenterates as Hyman (1959)

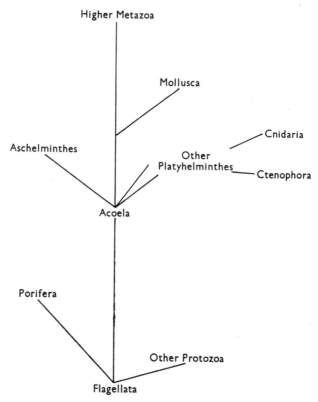

FIG. 112. Phylogeny of the lower Metazoa according to Hadži (1953).

and Pantin (1960), either because it involves a reversal of the usual sequence of the cnidarian classes (Hyman) or because of the great and fundamental differences between the turbellarian and cnidarian genital, nervous and muscular systems (Pantin). Steinböck (1958), while remaining a supporter of the acoelan theory, has offered a modified phylogeny of the lower Metazoa (Fig. 113) in which the Cnidaria are derived independently from pre-ciliate ancestors, thus avoiding these criticisms.

This modification of the theory does not, however, answer the criticism that the syncytial nature of the acoelan gut is a primitive feature, a view which is contested by a number of helminthologists (Böhmig, 1895; Karling, 1940;

Ax, 1956). This is clearly one of the cornerstones of the syncytial protistan theory and if this is not accepted, the theory, while remaining possible on purely theoretical grounds, lacks any substantial evidence in its favour.

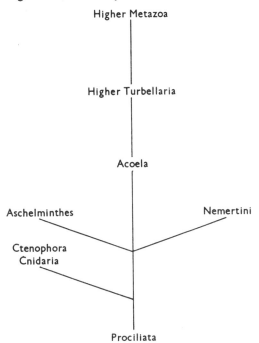

FIG. 113. Phylogeny of the lower Metazoa according to Steinböck (1958).

The gastraea theory

Haeckel's gastraea theory states that the Metazoa originated from colonial protistans which formed first a solid ball (morula), then a hollow ball of cells (blastaea) rather in the manner of *Volvox*. The critical evolutionary step was then the separation of locomotor and nutritive functions of the anterior and posterior cells, respectively, in the colony, and the invagination of the one into the other, so that the cavity within the animal was bounded by a double wall. This is the gastraea stage which was ancestral to all existing Metazoa and which is still represented in the embryology of many animals by the gastrula. The Porifera, Cnidaria and Ctenophora are still at the gastraea level of organization (Fig. 114). Jägersten (1955, 1959) departs slightly from Haeckel's theory in supposing the gastraea to have been bilaterally symmetrical rather than radially symmetrical, but other proponents of the theory (Ulrich, 1950; Remane, 1950, 1954; Marcus, 1958) do not introduce this additional postulate. All these modern writers couple the gastraea theory with an enterocoelous origin of the coelom. Thus four coelenteric pockets are formed at the gastraea stage, so giving rise to a basic, tetraradiate cnidarian stock (Fig. 12a, p. 20). In Jägersten's version of the theory it is supposed that six pockets were formed and that the Octocorallia are the most primitive coelenterates (Fig. 7, p. 13). Subsequent evolution involves the separation of the coelenteric pockets from the digestive cavity to form an unpaired anterior coelomic and two succeeding pairs of coelomic sacs, the final pair having been produced by a splitting of the terminal gastric pouch. The most primitive triploblastic animals are therefore oligomerous (i.e. with the body divided into three 'segments') and all acoelomate or unsegmented animals are secondarily reduced. It is this feature of the modern version of the

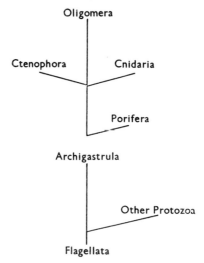

FIG. 114. Phylogeny of the lower Metazoa according to Marcus (1958).

gastraea theory that proves to be the chief obstacle to its general acceptance. Not only is it difficult to conceive of the phoronids, pogonophorans and pterobranchs as ancestral to all other Bilateria, and of all unsegmented animals as degenerate, but the theory is also heir to all the disadvantages of the enterocoel theory that we have discussed previously.

The planuloid theory

It has always been supposed that in the early multicellular animals, there would be considerable advantage to a separation of locomotory from trophic functions, and that the adaptive changes along these lines accounted for the appearance of the endoderm. Lankester (1877) suggested that the food would be passed to the inner ends of the cells in the morula and that these would then become cellularized to form the endoderm. Metschnikoff's (1883) discovery that intracellular digestion is characteristic of the lower Metazoa gave added weight to this suggestion, particularly when it was pointed out that this fact rendered the formation of a mouth and digestive sac superfluous at the stage

ORIGIN OF THE BILATERIA

postulated by Haeckel. Thus, according to these authors, the most primitive Metazoa were solid planulae from which both the coelenterates and the turbellarians evolved (Fig. 115). Recent investigations of turbellarian digestion by Jennings (1957, 1959) reveal more variation than was originally supposed, but Metschnikoff's generalization still appears a tenable one.

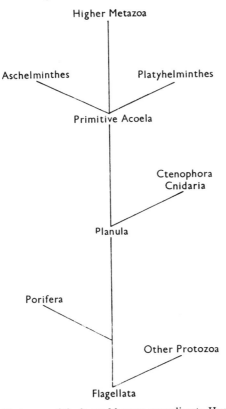

FIG. 115. Phylogeny of the lower Metazoa according to Hyman (1940).

In the Cnidaria, the most primitive living metazoans, endoderm is generally formed by ectodermal cells wandering into the centre of the blastula, rather than by invagination, and the larva is a planula and not a gastrula, both facts being interpreted as recapitulatory and therefore as offering additional support to the theory. This theory has never received the volume of support accorded to Haeckel's theory, but it has its modern advocates in Snodgrass (1938), Hyman (1940, 1959) and Hand (1959).

Although these theories differ substantially from one another in the types of organism from which the Metazoa are held to have been evolved, and in the relationship of existing phyla to the earliest metazoans, they have some common features. It is invariably supposed, for example, that metazoans evolved as pelagic animals swimming by means of cilia and subsequently became benthic. The various interpretations of the structural organization of the early multicellular animals also differ less profoundly than might appear at first sight and to a great extent such differences as exist can be resolved into questions of the time of appearance of the main structural features of triploblastic animals.

(1) Bilaterality appears rather later in the gastraea theory of Remane and Marcus than in the bilaterogastraea theory of Jägersten or in the acoeloid and planuloid theories.

(2) A patent digestive cavity is an important feature of the gastraea, but is regarded as a relatively late development among the Metazoa in the planuloid and acoeloid theories.

(3) Coelomic compartments are held to have evolved at an earlier stage in the gastraea and bilaterogastraea theories than in the alternative theories.

Both the development of bilateral symmetry and the evolution of a mouth and gut must have occurred early in the history of the Metazoa, though we are unable to say at precisely which stage in the evolution of these animals these events took place. A resolution of these problems is clearly of fundamental importance to any discussion of metazoan origins, but they are of relatively minor importance in discussions of the subsequent course of evolution of the triploblastic animals. The question of the time of origin of the coelomic compartments cannot be so easily dismissed. If, as I have attempted to show in previous chapters, the coelom serves primarily as a hydrostatic organ, it is difficult to conceive of any function that it might have subserved in small animals moving by ciliary activity; rather, we should expect the coelom to make its first appearance in relatively large animals and not during the early stages of metazoan evolution. This view of the function of the coelom implies a rejection of the gastraea theory in its modern form. Whatever other criticisms may be advanced against the gastraea theory, its most disturbing feature is the absence of any functional justification of the evolution of the coelomic compartments at so early a stage in metazoan phylogeny, or of any indication of the selective advantage of three over any other number of pairs of pouches.

If the gastraea theory is rejected, we are left with a planuloid or an organism comparable in its fundamental structure to the modern Acoela as the most primitive bilaterian from which the remaining Metazoa evolved. Our next task is therefore to consider the properties and evolutionary potentialities of such an animal.

THE PROTOBILATERIA

Whether the Protobilateria evolved from colonial or solitary protistans, they are likely to have been ciliated externally. They are likely to have had, or to have evolved modest powers, comparable to those of ciliates, of changing their shape by means of peripheral contractile elements. Composed of relatively few cells, they are likely to have fallen in the size range of the larger ciliates and smaller Acoela, that is of the order of 1 mm in length. The size of the animals is important for it helps to determine the habitats that were available to them. Modern organisms that answer to this description are able to swim by ciliary activity and generally do so. It is likely that the Protobilateria also led a free-swimming existence. Had they been smaller, they might have been able to seek shelter as an interstitial fauna (Wieser, 1959), but animals of this size, possessing a weak musculature, are not small enough to creep between the sand grains nor powerful enough to thrust them aside. This environment is therefore precluded and these animals must have lived on the surface of the substratum or swum above it. Possibly they did both.

A further assumption may be that, as in many other phyletic series, the way to evolutionary advancement lay by way of increase in size and the exploitation of new environments (Rensch, 1959). Indeed, for the Protobilateria, increase in size demands the penetration of a new environment, for the modern Acoela are themselves near the limiting size for animals that swim by ciliary action and the larger members of this order as well as the Rhabdocoela crawl over the substratum with the aid of cilia and swim sluggishly and infrequently. With very little increase in size, therefore, the Protobilateria, too, would have been forced to abandon ciliary swimming for ciliary creeping.

The substratum, receiving as it does a constant rain of detritus, presents an environment with an abundant food supply. There is therefore considerable advantage in the adoption of a benthic existence. There is an element of inevitability in the transition from a pelagic to a benthic life with increasing size: increase in size extends the range of food organisms available to the animal, but at the same time entails increased nutritional demands; the increase in size makes both swimming and food capture more difficult, but there is an abundant supply of food on the substratum on which the larger animals are forced to live. It is likely, too, that at about this stage in the evolution of the Bilateria, a mouth and gut were evolved. Larger animals tend to be less selective in their food intake unless they possess special organs for capturing and sorting organic particles. Detritus lying on the substratum is mixed with a considerable fraction of inorganic particulate matter so that unselective microphagous animals inevitably suffer a reduction of the gross nutritional value of their food. More ingested material must therefore be exposed to the digestive processes, and this demands bulk feeding which presupposes a mouth and gut.

Locomotion in this new environment was ciliary, as it is in many modern planarians, and a new development in the Bilateria must have been the appearance of an extensive mucus gland system in the epidermis for the secretion of a mucus trail which is necessary for ciliary creeping over soft substrata. In order to preserve an adequate locomotory surface in contact with the substratum, some degree of dorso-ventral flattening is necessary, but so long as a flattened shape is preserved, a considerable further increase in size is possible without demanding a more powerful locomotory apparatus. To judge from the Tricladida and Polycladida, however, there is likely to have been a tendency for the use of muscular locomotory waves to supplement and even supplant the ciliary beat in the larger animals. This must have involved a considerable strengthening and development of the body-wall musculature which had hitherto been responsible only for slight changes in shape.

Modern Platyhelminthes may well be the end-product of radiative evolution of organisms which were structurally not greatly different from the Protobilateria. The flattened form has become fixed by the development of dorso-ventral and other muscle fibres which traverse the mesenchyme, and other modifications of the musculature and also of the nervous system may be related to the variety of locomotory specializations in these animals. In the polyclads which rely chiefly upon muscular pedal waves, the nervous system remains a diffuse plexus, but triclads are capable of both ciliary and muscular locomotion and furthermore, many of them perform movements, as in leech-like crawling, in which the body-wall musculature acts as a concerted whole. In these worms the nervous system is more concentrated. But the most dramatic changes within the phylum are due to the adoption of a parasitic habit. This has had profound consequences for the structure of the reproductive system and for reproductive biology.

THE EVOLUTION OF THE COELOM

In early stages in the evolution of the Bilateria, the body-wall musculature can have consisted of little more than a few contractile elements sufficient to produce slight and feeble changes of shape. A subsequent enlargement of the musculature, which would have permitted greater and more powerful changes of shape, is likely to have happened only after the animals had abandoned a free swimming existence for one on the substratum where a more effective musculature would have become increasingly necessary as the animals become larger and ciliary locomotion became less efficient. From the analysis given in Chapter 2 of the movements of worms of this grade of organization (i.e. with a solid body) it is evident that the only practicable arrangement of muscles that would permit reversible changes of shape is in circular and longitudinal layers. With this apparatus the animals can perform a great variety of body movements including leech-like creeping and the production of peristaltic waves of contraction in the body wall. The limitations imposed upon these movements by a solid body are illustrated by the platyhelminths which can perform only relatively weak and slow muscular movements. In particular, they are too weak to burrow, and this must also have been the situation among the primitive Bilateria.

In spite of this disability caused by the lack of a true fluid skeleton, these animals were pre-adapted to a burrowing existence by virtue of possessing circular and longitudinal muscles in the body wall. So long as longitudinal muscles alone provide the main locomotory forces, the body may be dorso-ventrally flattened or, indeed, adopt any shape. But when the circular musculature is employed, as in peristaltic movements, the cross-sectional shape of the animal must perforce be circular. Animals, when they assume a circular cross-section, are ill-adapted to locomotion over the substratum (except by highly specialized means, as in leeches), because only a small surface area can be presented to the ground. But this body form is ideally suited to burrowing, since the body wall is then in contact with the substratum on all sides and consequently, the entire body-wall musculature is able to contribute to the total locomotory force.

This represents an explosive evolutionary situation. Because of the existence of circular and longitudinal muscles, the worms have great potentialities, but they are prevented from realizing them for lack of a true fluid skeleton. Once this is acquired, however, the animals become capable of performing powerful muscular movements, the most important consequence of which is that they can now form a burrow in the substratum.

Although the surface of the substratum presented a rich new environment to the pelagic Protobilateria, with an increasing population and variety of organisms exploiting it considerable advantage attached to a burrowing existence. Modern planarians are rather unselective carnivores and scavengers, and many of them are cannibalistic. It is likely that the rhabdocoel, triclad and polyclad specializations, including the evolution of a prehensile proboscis and the carnivorous habit, appeared at an early stage as soon as the population of macroscopic animals living on the sea-bed became sufficiently large to support carnivores. The ability of smaller microphagous and detritus feeders to burrow into the substratum would have afforded an escape from predation while not excluding them from the rich food supply on the surface.

Just as the adoption of a benthic existence was an inevitable step in the evolution of the swimming Protobilateria as they increased in size, so the habit of burrowing into the substratum rather than living on it must certainly have been developed at some stage among the bottom living forms. For this to have been possible, it was essential that a true fluid skeleton be evolved, and any fluid-filled cavity between the gut and the body wall serves this function equally well. There is in fact, every reason to suppose that the necessary conditions for a period of radiative evolution existed at the time of the first appearance of the coelom. In view of the great advantages of this advanced type of body structure, it is likely that this result would be attained by whatever means were most readily available to the animals. Indeed, the embryological and morphological evidence gained from modern animals, to say nothing of the number of totally irreconcilable theories of the evolution of the coelom, suggest that a fluid skeleton was evolved independently several times, and unless there is evidence to the contrary, there is no reason to assume that the secondary body cavity is homologous throughout the animal kingdom. In some cases, as for example, the pseudocoel and coelom, there is indisputable embryological and morphological evidence that the cavities have a different origin and are of different natures, but even a 'true' coelom may be polyphyletic. The tacit assumption that it is a homologous structure in all coelomates underlies the great controversies about the origin of the coelom. It is preferable to avoid making assumptions of this sort until they become essential for the further development of a theory, and to be guided instead by such factual evidence as exists.

In some nemerteans, technically a group of solid bodied worms, an approach to the acquisition of a fluid skeleton has been made by the severe reduction in the number of parenchyma cells and an increase in the volume of the interstitial fluid. Nemerteans show considerably greater powers of muscular movement than planarians and undergo much greater changes of shape; furthermore, some of them are able to burrow. In addition to this quasifluid skeleton, nemerteans possess a true one in the rhynchocoel. While this permits rapid and powerful eversion of the proboscis, it cannot provide an effective alternative to a fluid skeleton about which the body-wall muscles might operate.

An increase in the volume of interstitial fluid in the parenchymatous tissue appears to have been a relatively unprofitable solution to the problem of providing a fluid skeleton, but the alternative, the evolution of a coelom, has been exploited with considerable success by a wide variety of animals. The coelomic cavity may have been derived from enlarged and cavitated gonads in some animals, as postulated in the gonocoel theory. A gastric pouch or pouches may have become separated from the main digestive cavity in others, as claimed by supporters of the enterocoel theory; this would imply that the merits of a fluid skeleton outweighed the advantage of a large absorptive area in the digestive system, but this might well have been so if the evolution of the coelom had coincided with the beginnings of extracellular digestion. It is conceivable, too, that in some animals a cavity appeared *de novo* within the mesoderm and that this is reflected in modern animals by the schizocoelous method of coelom formation. There is evidence, chiefly embryological, that might be quoted in support of all these possibilities; none of it is conclusive and it is most unlikely that more reliable evidence will ever be forthcoming, so that the precise method of evolution of the coelomic cavities in modern phyla must always remain in doubt. We are left, however, with a strong belief that conditions favouring radiative evolution existed at the time of appearance of the coelom, a variety of theories to account for its evolution in different groups of animals, and conflicting evidence as to its origin. The conclusion that it is polyphyletic seems inescapable.

The mechanical properties of a fluid skeleton were exploited in other ways than simply by using peristaltic contractions of the body-wall muscles for burrowing. The existence of a large fluid-filled cavity within the body of the animal made the evolution of eversible structures a simple matter and the long, prehensile proboscis of sipunculids is but one example of such an additional development in coelomate worms. It is likely that with the evolution of a secondary body cavity came a variety of adaptations of this sort, associated with the exploration of new habitats by these animals.

Although the early coelomates, by virtue of their efficient fluid skeleton, would have been able to excavate a burrow in the substratum, they are unlikely to have been able to engage in sustained burrowing. Probably the majority of them were relatively sedentary animals, living in much the same way as do modern unsegmented coelomates like sipunculids and echiuroids. A variety of food collecting devices—proboscis, mucus net, ciliary filter-feeding structures, and so on—must have been evolved. These are adaptations that can be observed in all modern sedentary animals whatever their evolutionary status. If modern sedentary and tubicolous animals are a reliable guide, the early coelomates are likely to have pumped a respiratory current of water through their burrow by peristaltic contractions passing along the body wall; the existence of a well developed circular muscle layer makes this activity readily available to the worms. Co-ordinated activity of the body-wall musculature is no longer associated solely with locomotion, but now becomes a feature of normal maintenance activities when the worms are stationary. This increased use of the muscular system and co-ordination of muscular activity, particularly along the length of the worm, must have been accompanied by a corresponding improvement of the nervous system. The tendency for a concentration of the nervous system from a diffuse plexus into longitudinal trunks which occurs in triclads and may well already have been a feature of the more advanced Protobilateria, must clearly have been continued in the early coelomates.

In addition to these immediate adaptations to a relatively sedentary life, which follow upon the evolution of a coelom, the existence of a body cavity has a number of secondary, but ultimately extremely important advantages which are not directly related to the colonization of new environments.

Whether the coelom originated as a gonocoel or not, the gonads have become intimately associated with it, with the result that it serves as a repository for maturing gametes. While solid-bodied animals are not incapable of storing eggs, obviously an important effect of the existence of a large cavity would have been to increase the reproductive potential of the animals by permitting the accumulation of a large number of gametes over a long period, without impairing locomotory ability by the development of discrete, enlarged and turgid gonads. Synchronization of breeding activity, giving a short but intense breeding season, can then be coupled with high productivity. Synchronization is advantageous in two ways: if all members of a population breed at the same time, the chance of successful fertilization of the eggs is improved, and if the time of reproduction is geared to external events, young may be produced and released at a time when conditions are optimal for them (Barnes, 1957). All this is possible only if there is some means of storing gametes and of delaying maturation of the gametocytes.

Gonadial development and maturation is delayed with respect to somatic growth in nearly all eumetazoans by the action of inhibitory hormones produced in the central nervous system (Clark, 1961). It is now becoming accepted that neurosecretory cells are the most primitive source of hormones, and inhibition of gonadial maturation may well be one of the most fundamental functions of the hormones they produce (Clark, 1962b); it is certainly one of the most widespread. While it may seem a far cry from the evolution of the coelom as a fluid skeleton to the first appearance of endocrine mechanisms, the increased reproductive flexibility conferred on animals by their possession of a secondary body cavity would have demanded more precise hormonal control and co-ordination, if it did not actually provide the *raison d'être* for the evolution of an endocrine system.

With increasing size, a blood vascular system must have made its appearance. The precise point at which it evolved is unimportant and, in any case, appears to be unrelated to the development of a secondary body cavity. But since all organs are bathed by coelomic fluid, the coelom also has an important function as a fluid transport system. This would have been more particularly true of the transport of metabolites about the body than with the transport of oxygen, since in general, coelomic fluid, unlike blood, is separated from the epidermis by the muscle coats of the body wall and is therefore less able to take up oxygen from the environment. As a consequence of this additional function of the coelomic fluid, instead of excretory organs necessarily being multiplied and situated within a short distance of all tissues in the body, it is now possible for excretory products to diffuse into the coelomic fluid from which they can be removed by a much smaller number of more efficient nephridia.

It seems likely, then, that the evolution of a secondary body cavity was accompanied by a host of other modifications and changes in the structure and biology of the Bilateria. Not only did the existence of a true fluid skeleton enable the animals to make powerful muscular movements and so to burrow into the substratum, but the mere existence of a fluid-filled cavity within the body had important and far-reaching physiological consequences also. If the Bilateria appear to have gained little by being able to burrow, being at the same time condemned to a relatively sedentary existence, the secondary consequences of the evolution of the coelom make for much greater biological success in these animals than in their predecessors and tip the scales in its favour.

THE PSEUDOCOELOMATE PHYLA

The pseudocoel is comparable to the coelom in that it is a fluid-filled cavity between the body wall and the gut, but it is of a different morphological nature and has an entirely different origin from the coelom. Nevertheless, in many respects, the radiation of the pseudocoelomate animals parallels that of the coelomates.

The pseudocoelomate groups include the Nematoda, Nematomorpha, Kinorhyncha, Gastrotricha, Rotifera, Endoprocta, Acanthocephala and possibly also the Priapulida. Some, possibly all, of these animals are included in the Aschelminthes (Grobben, 1908), and the recent tendency has been to include all but the nearly related Endoprocta in a single phylum (Lang, 1953; Hyman, 1959). The interrelationships between these animals are obscure and uncertain, and there is still disagreement about the groups which should properly be regarded as pseudocoelomate. Their common feature is the possession of a body cavity which is derived from a persistent blastocoel (Hyman, 1951b). This definition of a pseudocoelomate, although it appears precise, is difficult to apply in practice because the embryology of some of the more obscure groups is very poorly known. The lack of an epithelial lining of the body cavity of the adult may be indicative of a pseudocoel, but it is not a reliable guide. The embryological evidence for the coelomate nature of the Chaetognatha, for example, is unequivocal, but the body cavity of the adult is a secondary formation and is not lined with a coelomic epithelium (Burfield, 1927). The peculiar difficulties of deciding the affinities of some putative pseudocoelomates are also shown by the Priapulida. The lining of the spacious body cavity of these worms contains no nuclei; it is therefore not usually regarded as a true peritoneum (Hyman, 1951b; Lang, 1953) and this suggests that the animals are pseudocoelomate. On the other hand, the presence of longitudinal and circular muscle coats in both the gut and body wall suggests that the body cavity may be a coelom (Hyman, 1951b; Boettger, 1952). The crucial embryological evidence which might answer this question is lacking.

As our knowledge of the structure and embryology of these relatively poorly known animals grows, it may be necessary to revise periodically the composition of the pseudocoelomate phyla. Nevertheless, the existence of a definable group, the Aschelminthes, containing the majority of these animals seems to be generally accepted. But acceptance of the existence of this group does not imply unanimity about the significance that is to be attached to it. Marcus (1958) and other adherents of the gastraea and enterocoel theories agree in general with Hyman's (1951b) conclusions about the composition of the Aschelminthes, but whereas to Hyman the distinction between a coelom and a pseudocoel is fundamental, Marcus attaches very little phyletic significance to it. Thus, alone among modern authorities on bryozoans, he continues to unite the Ectoprocta and Endoprocta in a single phylum, while acknowledging the coelomate nature of the former and the pseudocoelomate nature of the latter, since he regards the endoprocts to have evolved from ectoprocts by neoteny (Marcus, 1939, 1958). It is implicit in this view that the pseudocoelomate condition has been evolved independently more than once, and hence, that the mere existence of a pseudocoel is no guarantee that an organism is therefore allied to the other pseudocoelomate phyla.

If the pseudocoel is indeed a persistent blastocoel, it follows that all pseudocoelomates either originated as neotenous animals or are derived from ancestors which themselves originated by paedogenesis, for the blastocoel is an embryonic cavity and it is difficult to account for its survival into an adult organism except as a result of neoteny. In the larger of the modern Aschelminthes, the pseudocoel serves as a fluid skeleton comparable in most respects to a coelom, but this is unlikely to have been so in any larval stage. Because of the small size of most larvae, locomotion is generally by means of cilia and changes of shape are slight, mechanically weak, and unimportant. It is therefore difficult to see any functional advantage of a fluid skeleton in organisms the size of most larvae. The blastocoel, when it persists into the larval phase, is an accident of the mechanics of gastrulation and almost always disappears in the late larva or at metamorphosis. If the stem group or groups of the Aschelminthes originated as neotenous larvae, they would have acquired a body cavity which, although functionless in the larva, is a potential fluid skeleton in the subsequent radiation of the group.

At first sight, these theoretical speculations appear to accord unusually well with deductions that might be made from the structure and interrelationships of existing pseudocoelomates. The Rotifera are often regarded as a stem group from which other aschelminths evolved, and they have long been supposed to be neotenous trochophores. This suggestion was first made by Huxley (1853) and received the support of many of his contemporaries and successors, including Lankester (1877), Hatschek (1878, 1888–1891), Balfour (1880) and Kleinenburg (1886), and this interpretation is still subscribed to by Marcus (1958). The trochophore from which rotifers evolved is usually taken to be that of polychaetes, but since this type of larva is common among the lower worms, this evolutionary step might have taken place at any one of a number of stages in the evolution of the Eumetazoa. However, Beauchamp (1909), Remane (1929–1933) and Pejler (1957) among students of the Rotifera, have produced a considerable body of evidence that benthic and periphytic rotifers are the more primitive members of this phylum and that the pelagic habit has been evolved independently several times. This view has gained support in recent years and quite clearly destroys the force of the argument, for which there is now little evidence, that these animals are derived from neotenous trochophores.

Alternatives to the trochophore theory are that the rotifers evolved from benthic, creeping forms which moved by means of cilia on the ventral surface rather in the manner of planarians (Hartog, 1896; Beauchamp, 1909; Remane, 1929–1933; Hyman, 1951b; Pejler, 1957). While this theory has general, though not unanimous support, it fails, in its present form, to account for the persistence of the blastocoel in the adult animals. Despite this deficiency, it remains the interpretation that is most in agreement with our present knowledge of the evolutionary trends within the phylum.

If the Rotifera can no longer be considered as neotenous trochophores, there is no shortage of other pseudocoelomates for which similar claims have been made. Remane (1929) has suggested that the Gastrotricha have arisen in this way and that this group, in turn, has given rise to nematodes and kinorhynchs. Brien (1959) claims that the Endoprocta also evolved from trochophores which settled by the apical region and underwent a 180° rotation with a corresponding reorganization of the larval structures. There is, however, considerable disagreement. Hyman (1951b) suggests that endoprocts are derived from the Rotifera; Marcus (1939, 1958), that they are closely related to the Ectoprocta from which they evolved by the settlement and neoteny of the larva of the latter.

In the face of this bewildering array of conflicting opinions about the interrelationships of the aschelminth phyla, it is impossible to form a coherent picture of the evolution of the animals. Nevertheless, it seems to be generally implied, if not always explicitly stated, that the pseudocoelomates are polyphyletic and that gastrotrichs, endoprocts and probably also rotifers had an independent origin among the early metazoans by paedogenesis. The trochophore larvae from which they are supposed to have stemmed is characteristic in one form or another of several coelomate phyla, and it is from the early coelomates that we may envisage the pseudocoelomate condition to have originated.

The pseudocoelomates are often small animals and appear originally to have been benthic, either sessile, or creeping on the surface of the substratum, or living as an interstitial fauna. It cannot be argued that they evolved in response to the selective advantages of prolonging the pelagic larval phase, as is the case in the evolution of the chordates from a neotenous protochordate tadpole larva (Berrill, 1955; Whitear, 1957; Bone, 1960). The fact that paedogenesis occurred several times in the history of the early Metazoa suggests, rather, that the timing of sexual maturation with respect to somatic development was not very precisely controlled in these animals. Since at this stage in the evolution of the Bilateria, hormonal mechanisms, particularly those controlling growth and reproduction, are likely to have made their appearance, it is to be expected that they would be relatively imperfect.

At this period more than any other, neoteny is likely to have been a relatively common occurrence and it is surprising, not that it happened at all, but that it did not happen more frequently.

THE OLIGOMEROUS PHYLA

Another important development among the primitive coelomates or, at an even earlier stage, among acoelomates, was the appearance of oligomerous animals with their body divided into three regions or 'segments', the proto-, meso-, and metasomes. The oligomerous phyla include the lophophorate protostomes (Phoronidea, Ectoprocta and Brachiopoda) and the deuterostomes (Hemichordata, Echinodermata and Pogonophora, and, by descent, protochordates and chordates). The Chaetognatha might possibly be added to this assemblage; they are deuterostomatous and, like other deuterostomes, show the enterocoelous method of coelom formation. However, only two, instead of three, body regions are evident during their embryological development and in the adult, and for this reason some doubt exists about their phyletic affinities.

Although these phyla generally receive little attention in most theories of metazoan phylogeny, they occupy a crucial position in the views of Ulrich (1949), Remane (1950, 1952), Jägersten (1955), Marcus (1958) and other modern supporters of the gastraea-enterocoel theory.

The phylogenies proposed by these authors (Fig. 116) involve the following postulates, some of which are generally acceptable, others not. The higher Metazoa (Eumetazoa) can be separated into two main lines of descent on the criterion of the fate of the blastopore (mouth or anus), and with this there is usually also associated determinate or indeterminate cleavage, and a characteristic manner of coelom formation (schizocoelous or enterocoelous). Less reliably, a characteristic larval type (trochophore or dipleurula) is associated with these embryological features. Various names have been proposed for these two divisions of the Eumetazoa (Hyman, 1940; Marcus, 1958); all of them are open to some criticism and none have replaced Grobben's (1908) Protostomia and Deuterostomia, which now have almost universal usage. Oligomerous animals occur in both branches of the Eumetazoa. The lophophorates are protostomatous and all deuterostomes are either oligomerous (Echinodermata, Hemichordata, Pogonophora) or are related to these unequivocally oligomerous forms. Since the three-segmented body plan is supposed to have been uniquely evolved, oligomerous animals are held to lie near the stems of both protostomatous and deuterostomatous lines. Indeed, Ulrich (1949) unites all oligomerous phyla into a single group, the Archicoelomata, from which all the Bilateria evolved. Enterocoely, a feature of all the deuterostomes and of the articulate brachiopods among the protostomatous oligomerous animals, is regarded as the most primitive method of coelom formation, schizocoely as a modified and derivative method. The most primitive eumetazoans are envisaged as possessing unpaired anterior and posterior coelomic pouches, and a pair of medial pouches (or two pairs of medial pouches according to Jägersten, 1955), and such organisms evolved from animals at a level of structural organization comparable to that of cnidarians, with four (or six) gastric pockets which form the forerunners of the coelomic pouches.

The features and consequences of this reasoning which make the theory

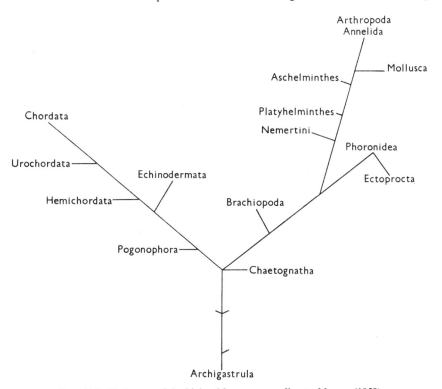

FIG. 116. Phylogeny of the higher Metazoa according to Marcus (1958).

unacceptable to many zoologists are: (1) enterocoely is regarded as primitive and other methods of coelom formation are derivative, (2) acoelomate and unsegmented phyla are regarded as secondarily reduced and (3) the evolution of the oligomerous conditions is not considered in relation to its functional significance. The several versions of the gastraea-enterocoel theory can be criticized on matters of detail, but many of these are not crucial to the theory and, were the criticisms accepted, it is probable that the theory could in most cases be modified to accommodate them without serious disturbance.

In alternative theories (Figs. 117, 118), the division of the coelom into protocoel, mesocoel and metacoel, and the enterocoelous manner of coelom formation are accepted as fundamental characteristics of the deuterostomes (Hyman, 1959), but neither is regarded as a primitive feature of the Eumetazoa. The lophophorates have clearly protostomatous affinities in that the blastopore becomes the mouth of the adult, but the cleavage pattern and the manner of coelom formation are variable and may be taken as indicating some relationship between the lophophorates and the deuterostomes. The oligomerous condition is therefore assumed to have arisen in the protostomatous coelomates, and animals with this type of organization to have given

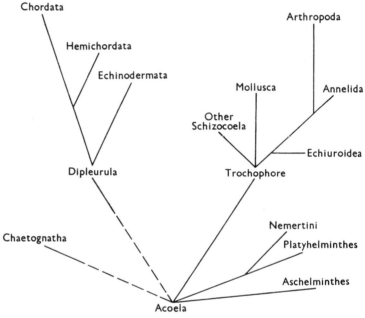

FIG. 117. Phylogeny of the higher Metazoa according to Hyman (1940).

rise to the deuterostomes, with their very different embryological and larval features, several of which are foreshadowed in the lophophorates.

This, or a very similar analysis is inoffensive and is probably acceptable to the majority of zoologists, but it can hardly be regarded as satisfactory because of its vagueness and also on the grounds that it, too, offers no functional explanation of this major evolutionary step. Yet it is clear that it is only by considering its functions, that the nature and origin of a structure can be made intelligible. The evolution of both the coelom and metameric segmentation is clearly related to mechanical considerations, and the same is true of the oligomerous type of morphological organization.

Some oligomerous animals, such as the echinoderms, are so highly modified that little of phyletic significance can be concluded from their structure. Of the others, all, with the exception of the problematical Chaetognatha, are sedentary and either permanently attached to the substratum, as the Ectoprocta and most pterobranchs and brachiopods, or living in burrows and tubes, as the lingulid brachiopods (François, 1891; Morse, 1902; Yatsu, 1902a), the pterobranch *Atubaria* (Sato, 1936; Komai, 1949), the Phoronidea, Pogonophora and Enteropneusta.

Although the interrelationships of the lophophorate phyla are very obscure

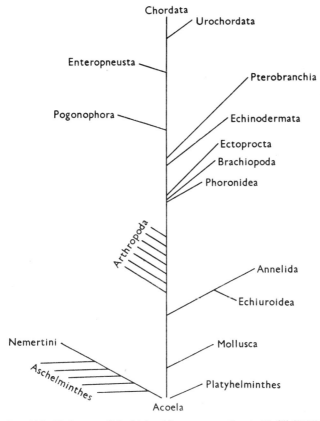

FIG. 118. Phylogeny of the higher Metazoa according to Hadži (1953).

(Ulrich, 1950; Marcus, 1958; Hyman, 1959), it is generally supposed that the attached condition of the ectoprocts and brachiopods is secondary, although this does not prevent both phyla from retaining a number of primitive features. The tubicolous phoronids are therefore the least modified of the lophophorates, at least so far as gross structure and manner of life are concerned. In the hemichordates, on the other hand, it is the burrowing forms, the Enteropneusta, that are generally regarded as having been derived from the sessile forms, the pterobranchs, having lost their tentacles in the process (Schepotieff, 1908; Burdon-Jones, 1952; Hyman, 1959), although it need not be supposed that the sessile habit and colonial structure of most pterobranchs were necessarily features of the earliest hemichordates. Rather, the Pogonophora, which have undoubted hemichordate affinities (Dawydoff, 1948; Ivanov, 1952, 1955b, 1960; Jägersten, 1956), may give a closer impression of the earlier deuterostomes, notwithstanding such obvious specializations as the disappearance of the digestive system and the peculiarities of their embryological development, which rule them out as a stem group (Manton, 1958d). The superficial resemblance between the pogonophorans and phoronids, both apparently preserving something of the general body form of the earliest deuterostomes and lophophorates, respectively, is significant. There is little doubt that this resemblance is indeed superficial and that the two phyla are not closely related (Johansson, 1937, 1939; Ivanov, 1960), but, as in other coelomates, much can be learned by considering the function of the coelomic compartments of animals with the same type of structure.

In adult phoronids, the lophophore is carried on the mesosome and the trunk constitutes the metasome. It has sometimes been claimed that the epistomial lobe represents a reduced protosome (Masterman, 1896; Schneider, 1902; Schultz, 1903), but since the entire larval pre-oral region is shed at metamorphosis (Roule, 1900) and the epistome appears as a small fold above the mouth, generally late in development, there is considerable doubt whether more than two body regions exist in the adult (Masterman, 1897; Selys-Longchamps, 1907; Hyman, 1959). Indeed, there has long been controversy over the existence of a protocoel in the larva and many students of phoronid embryology have regarded the most anterior compartment as a persistent blastocoel rather than a true coelom (Ikeda, 1901; Goodrich, 1903; Selys-Longchamps, 1904, 1907) although the subsequent discovery of cilia on the protonephridia in the pre-septal space of the actinotroch larva may be regarded as evidence in favour of the latter since ciliated epithelia do not normally occur in blastocoelic cavities (Goodrich, 1909, 1946). There is no doubt, however, that the perivisceral cavity of the entire trunk constitutes the metacoel and that a mesocoel exists in the lophophore.

It is clear that except for the presence of a separate tentacular hydraulic system, phoronids are comparable to the unsegmented coelomates in the existence of a body wall including circular and longitudinal muscle layers, bounding the coelomic perivisceral cavity which serves as a hydrostatic skeleton. The closest parallel to the organization of the coelomic system of the phoronids can be found in some sedentary polychaetes, such as terebellids, in which only a single septum survives, isolating the tentacular region from the remainder of the uncompartmented body.

But the present adaptations of phoronids are inadequate to account for the origin of these oligomerous animals, for the embryological evidence suggests that either the phoronids (and presumably the other lophophorates) never possessed a protosome, or else it has been severely reduced in consequence of their sedentary life (Hyman, 1959). To understand more of the significance of the tripartite body we must therefore turn to the deuterostomes. Among these, the pterobranchs offer something of a parallel to the Phoronidea, in that the lophophore is again born on the mesosome, but the epistome of pterobranchs is well developed and is used as a prehensile organ when the animals are moving up and down their tube, or during locomotion outside it (Andersson, 1907; Schepotieff, 1907a; Gilchrist, 1915). The division of the coelom into proto-, meso-, and metacoels as independent hydrostatic systems of the epistome, lophophore, and body, respectively, is intelligible in the pterobranchs, but it is not certain that this represents the primitive function of the three body regions. In the Pogonophora, the tentacles arise from the protosome and, in a highly specialized way, the mesosomal and metasomal hydrostatic systems act in concert as locomotory devices (Ivanov, 1960). However, the Pogonophora present so many aberrant features, both in embryology and in adult structure, that it cannot be concluded that the use of the three body regions in these animals is any more or less primitive than their use in pterobranchs.

The original function of the proto- and mesocoels remains for the present an insoluble problem. At most we can conclude that in the primitive oligomerous animals, the metacoel was the main perivisceral cavity and that one of the two anterior body regions bore a tentacular feeding organ and contained a coelomic compartment which served as a hydrostatic organ for the distension and erection of the tentacles by turgor pressure. The other anterior body region seems likely to have formed some kind of accessory locomotory organ with its own, independent coelomic hydrostatic system.

The oligomerous animals very probably followed a similar evolutionary derivation to the other early coelomates. We have suggested that the uncompartmented coelom evolved as a hydrostatic skeleton among those early Bilateria that adopted a burrowing existence and exploited the rich detrital deposit on the sea-bed while gaining the advantage of protection from predation by living buried in the substratum. The oligomerous animals similarly gained the protection of the substratum, but exploited a different food source by the evolution of a tentacular apparatus with which they collected the settling detritus rather than feeding on that which already formed the surface layers of the substratum. For this to have been possible, a minimum of two coelomic compartments was essential. As in the burrowing, unsegmented coelomates, a true fluid skeleton is a necessary adjunct to both peristaltic movements in the burrow and activity during its construction. A second coelomic compartment was associated with the tentacular apparatus. But since these oligomerous animals necessarily exposed part of the body above the surface of the substratum, movement up and down the burrow was even more necessary than in the unsegmented coelomates if the animals were to escape predators by retracting into the burrow and to emerge again for feeding.

As an adaptation to this frequent movement within the burrow or secreted tube, an additional hydrostatic organ was developed at the anterior end of the body to provide an accessory locomotory organ comparable to the pterobranch epistomial lobe or the pogonophoran mesosome with its frenulum.

Because of the uncertainty about the existence of a protosome in phoronids and the variable position of the tentacular organ, carried on the protosome of pogonophorans and the mesosome of pterobranchs, it is impossible to decide whether all these phyla had a common oligomerous ancestor. It is quite likely that they are polyphyletic, just as the unsegmented coelomates might well be. If this interpretation is correct, the oligomerous condition becomes comparable with the coelomate and pseudocoelomate conditions, and represents a grade of construction rather than a characteristic of a natural group of animals.

THE EVOLUTION OF METAMERISM

It is almost axiomatic that major evolutionary advances occur only among relatively unspecialized animals. Specialization involves a commitment to evolution along set lines and leads only to further specialization, though it is possible for new evolutionary trends to be initiated in groups of animals that acquire unspecialized structures secondarily by means of neoteny. The radiation of the Bilateria and the appearance of the coelomates must consequently have taken place shortly after the Protobilateria had forsaken the plankton for the sea-bed, among animals that had not evolved dorso-ventral muscles or other structures which preserved a flattened body form as an adaptation to creeping on the substratum. A dorso-ventral musculature would, in fact, have hindered the generation of the peristaltic locomotory waves necessary for burrowing. In the same way, the evolution of metamerism must have been an early development among coelomates which had not yet become specialized for a sedentary life.

It seems certain that the body-wall musculature can have been strengthened and enlarged only as an accompaniment to the evolution of a secondary body cavity, for rapid and powerful changes of shape which can be produced by such a musculature are possible only in animals with a true fluid skeleton. These associated developments, the appearance of the coelom and the enlargement of the musculature, were necessary if the animals were to burrow into the substratum, and both appear to have evolved as an adaptation to burrowing. However, the appearance of a single (or even a paired) coelomic compartment, as we have seen, resulted in muscular activity anywhere in the body affecting the entire musculo-hydrostatic system and condemned the animal to a relatively sedentary existence. Only metamerically segmented worms have the ability to perform strictly localized muscular activity economically and are capable of sustained burrowing, even though worms that lack segments are able to perform precisely the same locomotory movements. It is clear that the compartmentation of the coelom by septa and the evolution of a segmentally organized musculature and nervous system must have been early developments among the coelomate, but unsegmented worms.

The alternative view, proposed by modern supporters of the gastraea-enterocoel theory, is that the oligomerous animals are ancestral to the metamerically segmented worms. The advantages and consequences of a subdivision of the metasome of oligomerous animals into segments are exactly the same as those of a subdivision of the whole body of unsegmented coelomates. From a purely mechanical point of view, therefore, there is no reason to prefer unsegmented coelomates to oligomerous animals as the immediate forerunners of the annelids. However, the evolution of the oligomerous condition appears to have been in direct association with the development of a tentacular apparatus and a specialized mode of life. On the grounds of specialization alone, then, the oligomerous phyla are the less likely to have given rise to metameric animals. Even supporters of the gastraea-enterocoel theory suggest that the evolution of metameric from oligomerous forms involved the suppression of the proto- and mesocoels and the incorporation of the proto- and mesosomes into the head of the derivative animals. In other words, it is supposed that the oligomerous animals underwent a period of regressive evolution, returning secondarily to a condition comparable to that of unsegmented coelomates, before metamerism was evolved. There is thus no disagreement about the gross morphology of the immediate ancestors of the annelids, but since there are good grounds for supposing unsegmented coelomates to have evolved independently of the oligomerous animals and to have co-existed with them, to claim that the latter gave rise to metamerically segmented worms is to introduce unnecessary complication.

The crucial step in the evolution of metamerism was the evolution of septa. Unless the fluid skeleton is divided into compartments, there is neither any necessity for, nor selective advantage to a segmented body-wall musculature, for segmentation of the musculature alone would not have increased the locomotory repertoire or efficiency of these animals. Judging by the structure of the septa of modern annelids, these muscular diaphragms were primitively inserted into the sub-epidermal basement membrane. This entailed a subdivision of the muscle layers in the body wall into segmental units, each, necessarily, with its own nervous supply. Just as it is impossible to separate the evolution of a powerful body-wall musculature from the evolution of a fluid skeleton, so it is impossible to consider the evolution of a segmental nervous system independently of the evolution of a segmented musculature. With the new type of body architecture, each segment acts as a quasi-autonomous unit but, at the same time, its activities must be co-ordinated with those of the rest of the body. The intrasegmental nervous system provides the means of co-ordinating activities within the confines of a single segment and meets the longitudinal elements of the nervous system, by which overall co-ordination of activities is intermediated, in the segmental ganglia.

Other organ systems must also have acquired a segmental organization at the same time as the coelom became subdivided by septa. Transport of oxygen and other metabolites in the coelomic fluid which may have supplemented the blood vascular system in this function in the early, unsegmented coelomates, is now precluded except for intrasegmental flow, and an effective vascular supply in each segment becomes essential. In the same way, the excretory system must necessarily have acquired a segmental organization. In unsegmented coelomates, a relatively small proliferative zone could have produced a large supply of gametocytes which then matured in the spacious coelomic cavity. This is not possible in septate worms and, unless the protoannelids were to have suffered a marked reduction in their reproductive capacity, gonadial tissue must have been contained in nearly every segment. Each segment must obviously also have contained genital ducts.

The modern worms most nearly resembling the protoannelids are oligochaetes. In these worms, the segmental organization of the body-wall musculature and compartmentation of the coelom by intersegmental septa and dorsal and ventral longitudinal mesenteries can be seen in an almost completely unmodified form. Additional adaptations to burrowing that are exhibited by oligochaetes, include the evolution of chaetae which, when protracted, increase the frictional forces between the body wall and the substratum, and so increase the propulsive forces that can be generated.

Although oligochaetes may resemble the protoannelids in their structure and locomotory habit, it is still debatable whether they or the polychaetes should be regarded as the most primitive living annelids from which the others evolved. At first sight the polychaetes appear to be the stronger contenders for this position. They are almost exclusively a marine group, they are dioecious, many have small eggs, and their development includes a pelagic larval phase. Their reproduction is often an uncomplicated process involving shedding gametes into the sea without elaborate preliminaries. Furthermore, the group includes species with the most primitive type of excretory organ found in coelomates, the protonephridium. The reproductive system is unspecialized and the gonads are frequently diffuse and widespread throughout the body. Most important, the worms, particularly errant polychaetes, may exhibit metameric segmentation that is almost idealized in its lack of modification from the postulated archetype, with a non-segmental prostomium, a large number of identical segments, followed by a non-segmental pygidium.

The fact that oligochaetes, especially in the structure of their reproductive system, show a loss of metamerism and acute specialization, whereas the most primitive polychaetes do not, is not necessarily evidence that the Polychaeta is the most primitive group if it can be shown that these specializations are related to ecological necessity and what metamerism survives in oligochaetes is of a more primitive type than that found in polychaetes. When we strip oligochaetes of features which are clearly adaptive to the non-marine habitats of these worms, we are left with a metamerism as unmodified as that of errant polychaetes. The only advanced feature of oligochaetes is their colonization of fresh-water and terrestrial habitats, and from this, most of their peculiar anatomical and physiological features spring.

Oligochaetes have been able to penetrate and survive in these environments only by virtue of an efficient osmoregulatory and excretory system. They are not all equally successful in this respect, of course; among earthworms, the Lumbricidae are relatively poorly adapted (Ramsay, 1949), and the tropical and very successful Megascolecidae are highly adapted to life in dry environments (Bahl, 1947). Primitive and presumably inefficient excretory organs, the protonephridia, exist only in a few exclusively marine polychaetes (Goodrich, 1946) and their occurrence in terrestrial planarians and nemerteans—not noticeably successful invaders of dry land—is due to the impossibility of developing metanephridia with open funnels in solid-bodied worms.

The genital system and reproductive biology of oligochaetes is also extremely specialized and relates directly to their life in fresh-water and in the soil. In common with the majority of fresh-water invertebrates, limicoline oligochaetes have no pelagic larval phase. Development is direct and the eggs are consequently large and yolky, and only a few are produced. The small number of eggs poses problems during the process of fertilization, for if there are only few eggs, special provision has to be made to ensure that they are all fertilized. In oligochaetes, this consists of copulation and transferring of sperm from one animal to another so that the eggs are fertilized as they are laid. The fact that the worms have to meet one another leads to further complications. The nature of the environment prevents them coming in contact very frequently, but the reproductive capacity of the animals can be much increased if every such meeting results in eggs being fertilized. By the evolution of hermaphroditism, this is assured. Indeed, two worms are fertilized at each meeting. This combination of copulation and hermaphroditism entails specialization and concentration of the genital system and a reduction in the number of genital ducts, and in many worms, the development of accessory genital glands. An unmodified metamerism of the genital system is therefore incompatible with the environments in which oligochaetes live.

If the lack of primitive characteristics in the reproductive and excretory systems of oligochaetes is set aside, the only remaining morphological justification for the view that the worms are not primitive but are derived from polychaetes, is in the disposition of the chaetae. It is generally agreed that the lumbricine arrangement of the chaetae, with four pairs in each segment, is more primitive than the perichaetine arrangement in which the chaetae form a complete or nearly complete ring around the segments. The perichaetine state can be shown to have evolved independently several times and by differing methods in various families in the class (Michaelsen, 1914). The two pairs of chaetae on each side of a segment in lumbricine worms are held to represent the noto- and neuropodial chaetal sacs of a polychaete parapodium (Stephenson, 1930) and this, apart from the specialization of the oligochaete reproductive system, is the chief reason for supposing polychaetes to be the more primitive group. In the absence of other evidence that oligochaetes are reduced polychaetes, this seems a slender thread on which to hang so important a conclusion. It might be deduced with equal probability that parapodia were derived in worms which already possessed chaetae disposed in the lumbricine fashion and that this determined the form of the parapodia. It is, of course, impossible to dispose conclusively of the possibility that oligochaetes are derived from a polychaete stock, but if this is so, the oligochaetes show a remarkably complete return to the ancestral type of body architecture that polychaetes such as the Capitellidae, which have adopted a similar method of locomotion, do not.

We may conclude then that although the reproductive and excretory organs of oligochaetes are specialized and modified, in the structure of their body-wall musculature and the partition of the coelom, oligochaetes are much closer to the protoannelids than the errant polychaetes are. It is but a slight step from the unsegmented coelomate to the segmented condition as we see it in oligochaetes. No fundamental change in locomotory behaviour is involved, and the only anatomical changes are the development of septa and, with them, the metamerism of the muscle coats of the body wall. In errant polychaetes, on the other hand, the method of locomotion is entirely different from that of unsegmented worms and the musculature is profoundly modified. The longitudinal muscles are divided into dorsal and ventral blocks and the circular muscles are restricted to the intersegmental region; extrinsic parapodial muscles are developed. If the Polychaeta represent the group of annelids which evolved from unsegmented coelomates, we must envisage the simultaneous occurrence of a radical change in the method of locomotion, the evolution of metamerism, an extensive modification of the musculature, and the evolution of lateral appendages. Fewer problems are posed if it is assumed that polychaetes evolved from worms that were already segmented and had a structure which was basically that of oligochaetes.

The evolution of the polychaete architectural plan, as in other important evolutionary steps that we have discussed, was related to the exploitation of new habitats by the evolution of new locomotory techniques. The early annelids were burrowing animals and for the success of their particular method of locomotion, depended upon a reasonably consolidated substratum. Without this, dilations of segments produced by contraction of the longitudinal muscles would not have wedged the worm against the walls of its burrow and would not have prevented slipping as a thrust was exerted against the substratum by the anterior end of the animal. In order to live and move in less firm substrates and, in particular, in the surface detrital ooze, modifications of locomotion and structure were necessary.

A change of habitat of this sort must also have entailed changes in the feeding habits of the worms and additional evidence that the most primitive polychaetes lived on or in the surface layers of the substratum can be adduced from the evolutionary history of the polychaete stomodeum. Eisig (1914), Jägersten (1947) and Dales (1962) have drawn attention to the similarity between the muscle bulb in the floor of the mouth of most archiannelids, and the buccal bulb of larval ariciids and adult cirratulids and the comparable muscular bulb in the Ampharetidae, Terebellidae and Oweniidae, and have suggested that these are all homologous structures. Dales regards these sedentary worms, together with the Eunicidae, all of which have a ventrally directed proboscis, as the polychaete families most closely related to the archiannelids. He also suggests that the buccal bulb evolved in an archiannelid stock as an adaptation to surface deposit feeding. This morphological development in the proboscis apparatus, which relates to a change in feeding habits and of habitat of the forerunners of the polychaetes, is matched by structural changes of even greater significance associated with the new locomotory habits of the worms.

Lateral segmental extensions, or parapodia, provide considerable resistance to movement and when protracted reduce the tendency for segments forming the *points d'appui* to slip. The parapodia must, of course, either be retracted, or be deflected backwards against the body while the segment of which they form part moves forwards. But although the parapodia of modern errant polychaetes play a dynamic role in locomotion and themselves exert a backthrust against the substratum, they are unlikely to have done so at their inception.

If the musculature of a segment is completely symmetrical, as it is in earthworms, the extent of radial dilation is the same in all directions and depends upon the extent to which the longitudinal muscles can contract. This is limited by the physiological properties of the muscle fibres and to a smaller extent by the resistance to deformation offered by the organs contained within the segment. If, now, part of the body wall is weakened by a reduction in the thickness of its muscle coat, contraction of the longitudinal muscles results in the weakened area being dilated to a greater extent than the rest of the segmental body wall. Assuming the same contractions of the longitudinal muscles, it is obvious that there is much greater dilation of a localized region of the body wall than if the dilation is equally distributed over the whole circumference of the segment. If the longitudinal muscles are confined to the dorsal and ventral surfaces of the segments, the reversible changes of shape now take place in the longitudinal and lateral directions instead of longitudinal and radial. Such a modification in the distribution of the longitudinal muscles alone, would not have rendered the worms incapable of burrowing through the substratum but it would have reduced their efficiency, and it would not have assisted greatly in burrowing through less well consolidated substrates. It is therefore likely that the circular muscle coat became concentrated at the anterior and posterior ends of the segments at the same time as changes occurred in the distribution of the longitudinal muscles. The effect of these modifications would have been to have left a window in each side of the segment with only a thin layer of muscles, while the rest of the segmental walls remained highly muscular. Under these circumstances, contraction of either the circular or longitudinal muscles would have caused dilation of the thin part of the body wall, or, in other words, a permanently dilated blister was formed on each side of the segment. This, strengthened by the development of skeletal rods in the form of acicula, and provided with chaetae to increase frictional resistance to slipping through the substratum, formed the basis of the polychaete parapodium.

With the development of the parapodia, the region of application of propulsive forces to the substratum tends to shift from the entire body wall to the parapodia. This fact results in important changes in both the method of locomotion and the morphology of the segment.

It must be assumed that elements of the body-wall musculature remained associated with the parapodium and became converted into the intrinsic and extrinsic parapodial musculature. Its function initially may have been no more than to position the parapodium so that it was directed laterally when the segment was stationary, and reflected back along the body in moving segments. As locomotory forces tended to be applied more and more by way of the parapodia, these muscles must have been increasingly developed to resist the torsional forces on the parapodium. They ultimately become sufficiently powerful to enable the parapodium to perform a power-stroke during the locomotory cycle instead of remaining passive.

The septa acquired a new function during the initial stages of this evolutionary trend. It became important that changes in fluid pressure in one part of the body should not cause dilation of the parapodial lobes in other segments. The parapodia, being thin-walled, are particularly liable to dilation by relatively small pressure changes and it would obviously be a severe impediment to locomotion if parapodia were protracted in moving segments or retracted in stationary ones. Septa have therefore been retained in some errant polychaetes, to help localize pressure variations, but with the increasing efficiency of the parapodial musculature, they have tended to become very much reduced or to acquire new functions unrelated to the hydrostatic system.

A third modification of considerable importance was the change from bilaterally symmetrical locomotory movements to unilateral movements. In cylindrical worms which lack parapodia, and in which burrowing is performed by antagonistic contractions of complete circular and longitudinal muscle coats, it is inevitable that the segment should contract symmetrically. This is not so in errant polychaetes in which propulsive forces are generated in each segment unilaterally and alternately. Unilateral contraction of the musculature is, of course, not a new development; all worm-like animals swim by this means. If the polychaetes lived in or on the surface layers of the substratum, it is likely that they swam much more frequently than the primitive, burrowing annelids which, in fact, may never have left the deeper layers of the substratum at all except, possibly, for breeding. Polychaetes may also have performed swimming rather than burrowing movements through the semifluid surface layers of the substratum. Since the circular muscles remained completely passive during this type of locomotion, their importance declined, and bilaterally symmetrical locomotory movements, already rendered inefficient by the reduction in the circular muscle layer with the development of parapodial lobes, were superseded by unilateral locomotory movements in which the longitudinal muscles and the extrinsic parapodial muscles provided the motive force.

ADVANCE AND REGRESSION

We have so far considered the evolution of only a selected group of phyla which illustrate the development of increasingly complex types of structural organization. The analysis has been based on the premise that, particularly in the evolution of major groups of animals, important changes in structure are generally associated with changes in locomotory techniques and the colonization of new environments, and that mechanical factors have been of the first importance in determining the course of evolution of the Metazoa. If this thesis is correct, it should apply equally to animals other than those we have considered in the development of the theory. It is claimed that the coelom was evolved as a hydrostatic skeleton and that metamerism appeared as an adaptation to burrowing through the substratum. What of coelomates that no longer make use of a hydrostatic skeleton, and segmented animals that do not burrow? It might be expected that changes in structure or habit which result in a loss of their primary function would be correlated with a substantial modification of the coelom and of segmentation. Changes of this sort, both progressive and regressive, can be found in both lophophorate and deuterostome oligomerous animals and also in metamerically segmented worms, and these modified forms provide a testing ground for the theory.

Lophophorates

The evolutionary trend among lophophorates has been from a tubicolous existence, in animals with a structure similar to that of *Phoronis*, towards the adoption of a completely sessile habit in the Ectoprocta and Brachiopoda. The metasomal body-wall muscles of phoronids bound the hydrostatic skeleton formed by the metacoel and are responsible for producing the changes of shape associated with extension and contraction of the worm and its movements up and down the tube in which it lives. Although ectoprocts are sessile and the hydrostatic system is not employed in body movements or locomotion, the coelom remains patent and, at least in phylactolaemes and a few gymnolaemes, the metasomal body-wall musculature is not significantly modified. The metasomal hydrostatic system is used in the eversion and retraction of the polypide and since, for this, the kind of deformation imposed on the body wall is unimportant so long as the necessary hydrostatic pressure is generated, the animals are free to develop a rigid, protective exoskeleton with the associated modification of the body-wall musculature that we have already observed in the cheilostomatous and cyclostomatous gymnolaemes. Both eversion of the polypide and spreading of the lophophore are parts of the same process in ectoprocts, unlike the situation in phoronids in which the lophophoral system is independent of pressure changes in the metasome. The ectoprocts therefore have a single hydrostatic system which relates solely to the eversion of the polypide, and the septum between the meso- and metacoels is consequently very incomplete and the two coelomic systems are confluent.

Brachiopods have suffered an even greater modification of the coelomic system than ectoprocts. They are sessile and have no eversible parts, and, furthermore, the lophophore is supported by a 'cartilagenous' or collagenous brachial skeleton (Prenant, 1928) and the tentacular filaments often contain supporting spicules. The loss of hydrostatic mechanisms appears to be almost complete but there is still a good deal of uncertainty about this.

As in other lophophorates, the protosome is only doubtfully present. Possibly it is represented by the median part of the brachial fold over the mouth, but in articulates, this is filled with parenchyma and in inarticulates, the coelomic cavity in it is continuous with the mesocoel (Blochmann, 1892, 1900), and there appears to be no true protocoel. Other divisions of the mesocoel extend into the lophophore and filaments where, it was suggested by Hancock (1858), they provide turgor pressure by which the brachia are extended when the shells gape and the animals feed (Orton, 1914; Atkins, 1959, 1960). However, at least in *Lingula* this is not so, and the animals are able to spread the lophophore after the distal ends of the brachia have been cut and the coelomic fluid leaks out (Chuang, 1956). The extension of the brachia of this animal is brought about by the relaxation of the brachial retractor muscles and the release of tension on the brachial frame.

The main part of the mesocoel surrounds the oesophagus and, except in *Crania* in which an intact septum survives, it is in open communication with the metacoelic perivisceral cavity. Special conditions obtain in *Crania*. In addition to the retention of the primitive septum between the meso- and metacoels, the anus occupies a median posterior position and the rectum is surrounded by an undivided coelomic cavity, the anal chamber, which is formed by the dorsal and ventral mesenteries of the intestine dividing and passing to the lateral body walls, leaving this isolated median coelomic cavity. Unfortunately, too little is known of the functional morphology of brachiopods to account for the survival of the septum, but the anal chamber appears to be related to the protrusion of the rectum. Whether it serves as a hydrostatic organ in this function is unknown. Since anal protrusor muscles exist (Blochmann, 1892), it is unlikely that it does so. Chiefly, it seems to permit movement of the rectum, which would be impossible were the dorsal and ventral intestinal mesenteries continued to the posterior end of the animal.

Apart from the very doubtful cases of the coelomic canals in the lophophoral filaments and the anal chamber, the only part of the coelomic system that may still serve a hydraulic function is that in the pedicel, and even here it is generally very much modified. The pedicel is an extension of the posterior part of the ventral mantle lobe and contains an extension of the mantle coelomic canal system, but it remains patent only in lingulids (Yatsu, 1902b; Ashworth, 1915). This family of brachiopods is the only one in which the adult is not attached to the substratum; instead, it lives in a burrow and is able to move up and down it by extension and contraction of the stalk (François, 1891; Morse, 1902; Yatsu, 1902a; Chuang, 1956). The terminal bulb at the distal end of the stalk is covered by an adhesive secretion to which sand grains stick, and this forms an anchor in the substratum. It may be concluded that the stalk coelom provides a fluid skeleton about which the muscles in the walls of the pedicel act, but since these are described as consisting either of exclusively longitudinal elements (Schaeffer, 1926) or of crossed spiral fibres (Blochmann, 1900), it is evident that for reversible changes in length of the stalk to take place, a highly specialized and peculiar mechanism must have been evolved in the lingulids for this activity. Other brachiopods have much less extensible stalks. In discinids, the coelomic cavity is almost completely filled by large muscles which adjust the position of the animal on its stalk, while in articulates, the movements of the stalk are even more restricted and it is filled with connective tissue and lacks muscle elements altogether (Ekman, 1896).

For most purposes and in the great majority of brachiopods, no use is made of the coelomic chambers as hydrostatic skeletons, and yet the coelom, except that in the stalk, shows no signs of reduction or occlusion. The likely explanation of this apparent exception to the thesis presented here, is that the coelom has been retained as a fluid transport system, supplementing the blood vascular system. The coelomic fluid is in constant two-way circulation, driven by cilia in the mantle canals, the coelomic canals of the tentacles and lophophore, and, in *Lingula*, in the stalk. Numerous authors have discussed the large number of free cells of a variety of types that occur in the coelomic fluid (Blochmann, 1900; Morse, 1902; Yatsu, 1902c; Prenant, 1928; Ohuye, 1936, 1937); some of them contain echinochrome (Ohuye, 1936) and also haemerythrin (Kawaguti, 1941). The former was claimed as a respiratory pigment, although it probably is not (Cannan, 1927), but the latter pigment is. The blood contains no free cells and, although it is probably similar in composition to the coelomic fluid, it seems not to contain respiratory pigment, since only haemerythrin has been identified in brachiopods and it invariably occurs in corpuscles, not in solution in the plasma. Although the physiology of respiration has yet to be investigated in these animals, it seems inevitable that the coelomic system should be chiefly responsible for gas transport in the body. A considerable area of the coelomic canals is exposed to water currents flowing in the mantle cavity and the coelomic fluid apparently circulates much more vigorously than the blood. Indeed, whether the coelomic fluid contained a respiratory pigment or not, it is difficult to see how it could avoid serving as an oxygen transport system.

Deuterostomes

The deuterostome, oligomerous animals have undergone striking radiation which culminates in the evolution of the chordates, but as in the lophophorates, the general tendency in this branch of the animal kingdom has been towards the development of more or less sessile forms, with a resulting great modification of the coelomic compartments. Of the two groups that might be considered to have retained something of the body form of the most primitive deuterostomes, pogonophorans and pterobranchs, there can be no doubt that the latter have been the more productive of new forms. From animals akin to them, we can trace the origin of enteropneusts, echinoderms and ascidians. However, because of the evidently complicated history of all these groups, elucidating their phylogeny and, in particular, the modifications that the coelomic compartments have suffered, is extremely difficult and makes the radiation of the deuterostomes a rich source of confusion and controversy.

The enteropneusts are very closely related to pterobranchs as, for example, the presence of a post-anal adhesive tail in the young enteropneusts indicates (Burdon-Jones, 1952). Enteropneusts have adopted a burrowing existence, have acquired a worm-like body, and have lost the lophophore in the process. With the loss of the lophophore, the mesocoel, which served as its hydrostatic organ, has accordingly been reduced to ill-defined cavities in the collar. The locomotory techniques of enteropneusts are rather different from those of the majority of other coelomate worms, since they tend to rely upon the use of the proboscis or upon cilia and the greater part of the body is dragged along passively. Thus the metacoel, like the mesocoel, has only minor mechanical functions, and it is very largely occluded by muscle and connective tissue.

As in the pterobranchs, the pre-oral lobe is the most important locomotory organ and although in both classes of hemichordates there is some tendency for the protocoel to be filled with connective-tissue and muscle, it always remains patent and serves as a hydrostatic skeleton for the complicated movements of the pre-oral epistome or proboscis.

Despite a rich fossil record, even the interrelationships of existing classes of echinoderms cannot satisfactorily be resolved, and the origins of the phylum are quite uncertain. The peculiarities of structure of adult echinoderms, and of the later stages of their embryology, indicate that the history of the phylum has been a complicated one and that it includes a period when the animals were attached to the substratum. It is assumed that the animals were originally bilaterally symmetrical, but that upon attachment to the substratum by the anterior region, they have suffered a reduction of coelomic compartments and organs of the right-hand side, which gives them a profoundly asymmetrical structure. The subsequent acquisition of pentaradiate, and sometimes bilateral, symmetry has introduced additional modifications of structure.

Notwithstanding the obvious complexity of echinoderm history, a number of authors have concluded that these are essentially animals of simple structure, little more differentiated than coelenterates; so much so, that MacBride (1896) regarded them as one of the earliest products of the 'Protocoelomata'. Because of the close similarities between the enteropneust tornaria larva, the bipinnaria larva of asteroids and the auricularia of holothurians, there can be no doubt that hemichordates and echinoderms are closely related. Since the former have a more complex and more highly differentiated tissue structure, it has been concluded that they are more advanced derivatives of the Protocoelomata.

This view is still very generally and unquestioningly accepted, but it obviously cannot be reconciled with the interpretation, which I offered on a previous page, of the earliest oligomerous deuterostomes as vermiform, tubicolous animals. Usually, the common ancestor of deuterostomes is regarded as a free-swimming animal with three pairs of coelomic compartments, the hypothetical dipleurula, which is really little more than the lowest common denominator of echinoderm and hemichordate larvae. Waterhouse's (1941) interpretation of the Cambrian *Peridionites* as having the essential structure of a dipleurula and representing an extremely primitive group of echinoderms from which the other classes were derived, hardly bears examination. This animal was unattached, but heavily armoured and 9 or 10 mm long; it must therefore have possessed quite different locomotory devices from the dipleurula or any other larva and, accordingly, a substantially different structure. Clearly, little of phyletic value can be concluded from it.

The dipleurula theory, with its strong overtones of Haeckel's theory of recapitulation, was first proposed by Sémon (1888) and almost immediately won the nearly universal acceptance that it has enjoyed ever since. In the form in which it was summarized by Bather (1900) (Fig. 119), the free-swimming dipleurula, presumably ciliated, is supposed to have adopted a creeping mode of life on the substratum and then to have become attached by its preoral lobe. In order to account for the subsequent fate of the coelomic cavities, it is necessary to assume that the animal became attached towards the right side and that in consequence structures on the left-hand side of the animal increased at the expense of those on the opposite side. The mouth, together with the left hydrocoel, migrated to the posterior end of the dipleurula, which was now directed upwards. The right hydrocoel and hydropore

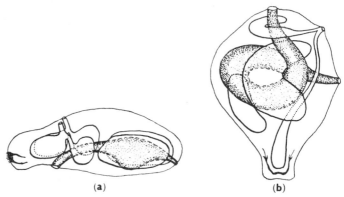

FIG. 119. Dipleurula theory of the origin of echinoderms. (*a*) The dipleurula, (*b*) postulated early pelmatozoan after settlement and torsion. (After Bather, 1900.)

disappeared. The effect of these changes was to throw the gut into a coil, and the pressure of the oesophagus not only pulled the hydrocoel upwards, but pressed it into a horseshoe curve. The somatocoels (metacoels) were also involved in this torsion of the gut, so that the right somatocoel became aboral and the left was carried upwards towards the mouth of the attached animal.

Bather envisaged the later development of the sessile animal to include the growth of three tentaculiferous grooves from the mouth, two of which later subdivided to produce a total of five. These grooves were accompanied by outgrowths from the hydrocoel. Hyman (1955, 1959), in a critical discussion of the dipleurula theory, complains that this is hardly a satisfactory explanation of the origin of the ambulacral system and points out that the present structure of the hydrocoel is much more intelligible if it is viewed as the hydrostatic system of a tentacular apparatus; furthermore, embryological evidence suggests that the water vascular system arose as tentacles. This appears to have been in the mind of some of the earlier authors and, in fact, for some years before the appearance of Bather's (1900) volume, it was being suggested by several writers, including Bury (1895), MacBride (1896) and even

Sémon (1888) himself, that the ambulacral system took its origin in the development of hollow circum-oral tentacles containing extensions of the hydrocoel, at a stage known as the pentactula, preceding the attachment of the organism to the substratum (Fig. 120).

While the pentactula theory may account for the association of the hydrocoel with the ambulacral system that subsequently developed from the

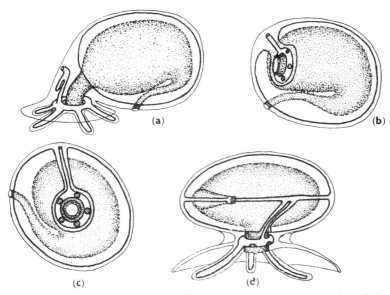

FIG. 120. Pentactula theory of the origin of echinoderms. (*a*) Lateral view of the bilaterally symmetrical pentactula, (*b*) and (*c*) two stages during the torsion of the pentactula, viewed from the dorsal and original left sides, respectively, atrium and circum-oral tentacles omitted, (*d*) lateral view of the radial form showing complete torsion, half the atrium and two tentacles omitted. (After Bury, 1895)

tentacles, it leaves unanswered the question of how the hydrocoel came to form a circum-oral horseshoe or ring. For it is assumed already to have been unpaired and disposed in this manner in the bilaterally symmetrical, and unattached, pentactula (Bury, 1895). MacBride (1896, 1914) achieved a compromise between the pentactula and dipleurula theories, which avoids this difficulty. He placed the first appearance of tentacular structures at an even earlier stage than Bury had, and envisaged the dipleurula as possessing five tentacular outgrowths arising from the hydrocoel on each side of the body (Fig. 121). What function, if any, this tentacular system was supposed to have is obscure in the extreme, but leaving aside this problem, it was supposed that the tentacular system of the right-hand side was subsequently lost and the other formed a circum-oral ring of tentacles of the type postulated in the pentactula, at a later stage, when the animal settled on the substratum and the gut underwent torsion.

It is possible, by a series of manoeuvres of this sort, to account for the evolution of the ambulacral system in some version of the dipleurula theory, but its chief defect remains: it is impossible to account for the existence of coelomic cavities in the dipleurula. If, as it is assumed, the dipleurula was a small, free-swimming, ciliated organism, it is difficult to conceive of any activity for which it would require coelomic cavities, except possibly to provide turgor pressure for a tentacular apparatus. This might explain the existence of the hydrocoels, but not of the somatocoels, nor yet of the axocoels, if these are regarded as separate entities in echinoderms. And yet it

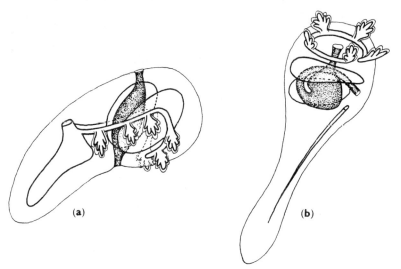

FIG. 121. Dipleurula theory of the origin of echinoderms. (a) Hypothetical dipleurula ancestor of asteroids and crinoids with paired hydrocoel, (b) suppression of right hydrocoel, coupled with torsion after settlement. (After MacBride, 1896.)

seems necessary to suppose that echinoderms originated from coelomate ancestors. The alternative, that the echinoderm coelom originated independently, is unacceptable because of the similarities of coelom formation in echinoderm and hemichordate embryology. These similarities are too great to be regarded as the product of convergence, hence the common ancestor of both groups must have possessed coelomic compartments. The fact that echinoderms are undoubtedly highly modified animals, far removed from any common ancestor of them and the hemichordates, implies that the hemichordates bear the closer resemblance to that ancestor. Since the enteropneusts may fairly be regarded as derived from pterobranchs, we return, once more, to animals with the structure we have postulated for the earliest deuterostomes.

A pterobranch origin of the echinoderms was first proposed by Grobben (1923), and this theory of echinoderm origins differs from the dipleurula theory, as developed by MacBride (1896, 1914), less than might have been expected. The dipleurula with paired tentacular structures that MacBride envisaged (Fig. 121), is essentially the same as a pterobranch with paired lophophores bearing five tentacles on each side (Fig. 122), with the difference

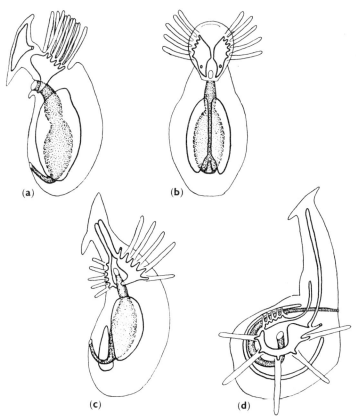

Fig. 122. Pterobranch origin of echinoderms. (a) Lateral, and (b) oral views of bilaterally symmetrical pterobranch ancestor, (c) and (d) suppression of right hydrocoel and torsion after settlement. The animal is presumed to have settled by its pre-oral lobe in (c) and (d), but is shown in an inverted position to correspond with (a) and (b). (From Hyman, 1959, after Grobben, 1923.)

that the animal must be regarded as relatively sedentary instead of ciliated and free-swimming. The difference is not one of structure, but of size and habit. In both the dipleurula and the pterobranch, the tentacular apparatus arises from the mesosome, and it must be assumed that the protocoel became connected with the mesocoel in the echinoderm line of evolution. The pterobranch epistome, used as a locomotory organ, is equivalent to the attachment pit of crinoid and asteroid larvae. The torsion of the gut and the reduction of the organs and coelomic compartments of the right-hand side of the animals was associated with the adoption of a completely sessile, attached existence by the echinoderm ancestor, and follows the same course as that proposed in the dipleurula and pentactula theories.

Even were it not evident from their structure in echinoderms, it might have been deduced that the adoption of a completely sessile existence must have resulted in a loss of the original functions of the proto- and metacoels. Only the surviving mesocoel, now transformed into the water vascular system, still served as a hydrostatic organ, maintaining turgor pressure in the podia. This function remains and has been exploited in holothurians, echinoids and asteroids, in which the podia are extensible and retractable and have a sufficiently powerful musculature for them to be used as locomotory organs (Smith, 1946, 1947). The other coelomic compartments have either become reduced or have acquired new functions.

The left protocoel, or axocoel, is always reduced and becomes, to all intents and purposes, a part of the water vascular system to which it contributes the madreporite, madreporic ampulla and part of the stone canal. Sometimes, as in most holothurians, its reduction is even greater, and it is vestigial. The right protocoel does not appear at all in holothurian and crinoid embryos, and in those of the remaining classes it is very small. Even so, it is of considerable morphological importance since it produces a small muscular structure which becomes the madreporic vesicle, generally considered to be homologous with the cardio-pericardial vesicle of pterobranchs and enteropneusts. During the larval development of enteropneusts, this organ originates as an ectodermal invagination near the hydropore, or as a mesenchymal space (Bateson, 1884; Morgan, 1894; Rao, 1953) and so appears to have little to do with the coelomic system. But what is regarded as its true nature is revealed during regenerative growth, when, in *Glossobalanus minutus*, it is produced as an outgrowth of the protocoel (Dawydoff, 1909). The cardio-pericardial vesicle of pterobranchs has a similar origin in asexually reproducing individuals (Schepotieff, 1907b); the embryonic development of these animals is insufficiently known to permit any conclusion about its origin in the larva.

The metacoels, or somatocoels, are represented in adult echinoderms by the perivisceral cavity, the genital sinuses and the perineural and hyponeural canal systems. In crinoids and ophiuroids, the perivisceral cavity is reduced to such an extent that it forms no more than a series of sinuses. In crinoids, particularly, it tends to be occluded by connective-tissue often impregnated with calcareous deposits. Numerous free coelomocytes are found in the coelom of members of both classes and it must be assumed that the chief surviving function of the metacoel is as an accessory circulatory system; certainly, with the evolution of brachial ossicles and a true skeletal musculature, the hydrostatic function of these coelomic compartments has vanished.

It is only in holothurians, asteroids and echinoids that the metacoel is spacious. It is likely that this is a secondary condition and the factors underlying the evolution of a large perivisceral cavity undoubtedly vary from class to class and, because of a sad lack of information, they are incompletely understood. Some contributory factors are clear, however.

From our previous discussion of holothurian locomotion, it is evident that these echinoderms have acquired a functional organization which approaches that of unsegmented, coelomate worms with a body-wall musculature arranged in circular and longitudinal layers and in which the perivisceral coelom serves as a fluid skeleton. The burrowing, apodous holothurians, lacking locomotory tube-feet, present the closest parallel to worms. In echinoids, with their completely rigid test, the coelom has an entirely different function. The existence of a large, fluid-filled cavity permits the enlargement of the gonads and changes in the volume of the gut, but in addition, it is involved in short term volume changes when the podia are retracted. Such relatively transient changes in volume are compensated by dilation of the peribuccal membrane and eversion of the branchial papillae. A further consideration which must apply, particularly in the regular echinoids, is that by virtue of their hollow body, a considerable saving of weight is achieved in these heavily armoured animals. Although the body wall of most asteroids has some degree of flexibility, it is doubtful if the perivisceral coelom ever serves as a hydrostatic skeleton and these animals appear to approach the echinoids much more closely than holothurians in the functional attributes of the metacoel.

The structure of tunicates, like that of echinoderms, has been profoundly modified as a consequence of their adoption of a completely sessile existence. As a result, the derivation of the phylum from preceding forms is beset with difficulties and is largely speculative. The relationship between tunicates and the more advanced forms, the Chordata, is supported by overwhelming evidence of the close and fundamental similarities between the ascidian tadpole larva and chordates (Berrill, 1955), but deciding the relationship between ascidians and more primitive animals depends almost entirely upon the fact that the manner in which the pharyngeal gill slits are formed is identical in enteropneusts, ascidians and cephalochordates, and that these 'protochordate' phyla are the only animals, other than chordates, in which the pharynx is perforated in this way.

If, as it is generally concluded, pharyngotremy was uniquely evolved, it follows that this event must have taken place among the common ancestors of hemichordates, tunicates and chordates; that is, among the early deuterostomes with a structure similar to that of modern pterobranchs. It is rather unexpected that animals with a lophophore, which has both a respiratory and a food-collecting function (Burdon-Jones, 1960), should have evolved pharyngeal gill slits which often serve precisely the same functions, and we can only guess at the probable course of evolution of these structures. The primitive oligomerous animals, which were sedentary and tubicolous, must have possessed a food-gathering organ and this, as we have seen, took the form of a lophophore or similar tentacular structure. Respiration is likely to have been of secondary importance; many sedentary worms lack specialized respiratory surfaces. The tentacular apparatus, of course, presents a large surface area and is ideally suited to the respiratory exchange of gases as well as food collection, and in most filter- and ciliary-feeding animals these functions are combined in a single organ. The only obstacle to the use of the lophophore in this way was the compartmentation of the coelom. There must inevitably have been free gaseous exchange between the coelomic cavity of the tentacular apparatus and the surrounding water, but this coelomic cavity was separated from that of the rest of the body for mechanical reasons. It is only with the evolution of a blood vascular system which penetrates into the lophophore that a dual function of this organ is possible. Before this stage was reached, however, other respiratory organs must have become increasingly necessary, particularly if the animals increased in size, and these appear to have taken the form of pharyngeal gill slits.

Whatever the origin of pharyngotremy the adoption of a sessile life by tunicates has been accompanied by an enormous hypertrophy of the pharyngeal chamber and a multiplication and subdivision of the gill slits. In other respects, the animals have suffered a reduction, more commonly a total loss, of structures we suppose to have been present in sedentary, but unattached ancestral forms. For many years it was supposed that the coelomic cavities were included among the structures that had vanished, a regression that might be considered to accord very well with the almost immobile habit and lack of hydrostatic organs in tunicates. But in recent years, it has become more widely accepted that the epicardia (also known as the perivisceral sacs in *Ciona*) are coelomic (Selys-Longchamps, 1938; Drach, 1948; Brien, 1948; Berrill, 1955). These paired, but unequally developed compartments appear late as outgrowths from the pharyngeal region. They clearly serve no hydrostatic function, but have instead gained new functions. Garstang (1928) considered the epicardia to form a budding organ, but this view has been vigorously contested by Berrill (1935, 1951, 1955) who regards them as possessing primarily an excretory function. The extreme regenerative capacity of the coelomic epithelium is then merely a reflection of the unspecialized nature of this tissue. The epicardia must also have the purely mechanical function in common with the pericardial cavities of other animals, of providing a fluid-filled space within which the heart can beat, an activity which would otherwise present some difficulty.

Because of the interpolation of a larval phase of such different structure from that of the lower deuterostomes, and the late appearance of the coelomic rudiments, developmental studies provide no clue whether the ascidian coelom represents proto-, meso- or metacoel. From the fact that the coelom is perivisceral, we may hazard a guess that it is metacoelic.

The two major phyletic lines descending from the early deuterostomes have both led to the development of animals with a completely sessile habit, with an accompanying regressive evolution that has affected many of the organ systems and not least the coelomic compartments. It is little short of astonishing that from this unpromising beginning, one of the most successful types of structural organization, that of vertebrates, should have emerged. It is likely that the acute specialization occasioned by the sessile habit has been circumvented by paedomorphosis, and that at some stage in the evolutionary sequence from hemichordates to ascidians, a tadpole larva was evolved which gave rise to the early chordates (Garstang, 1928; Berrill, 1955; Whitear, 1957; Bone, 1960). The three coelomic divisions of the early deuterostomes are still represented in chordates, although the proto- and mesocoels have no function and have virtually disappeared, making only a vestigial and transitory appearance in the embryological development of some chordates (Goodrich, 1917; De Beer, 1955). Only the metacoel survives, and even so it has lost its mechanical function, is secondarily subdivided, and has acquired a number of new functions. The pericardial chamber, for example, like the ascidian epicardia, provides a fluid-filled space permitting pulsations of the heart (Goodrich, 1930). The most significant developments in the Chordata have been the evolution of an internal skeletal system with true skeletal muscles, and a segmented musculature which relates primarily to the generation of swimming movements. Neither makes any demands upon a fluid skeleton.

All in all, we can conclude that deuterostomes reveal the same fundamental principle as the lophophorates: the coelomic compartments appeared originally in soft-bodied forms in which they served a hydrostatic function. Subsequent evolutionary changes have produced animals of profoundly different habit and structure from the primitive tubicolous worms, and with these changes the coelomic compartments have undergone regression or change of function.

Metamerically segmented animals

The interrelationships of the oligomerous phyla are not clearly understood, and conclusions about the fate of the coelom during the radiation of these groups must necessarily be based largely on speculation. When we turn to the metamerically segmented worms, we are on surer ground because much relevant variation of habit and structure occurs within a single phylum—the Annelida. Annelids and their undisputed relatives, the Arthropoda and Mollusca, reveal almost the entire gamut of variation on a single type of structural organization, although, as we shall see, the interpretation of the last two phyla proves to be exceptionally difficult.

Even in the Polychaeta, which retain a patent coelom and are metamerically segmented, we have already observed substantial modifications of the primitive annelid structure that is exemplified in the oligochaetes. With the evolution of parapodia, peristaltic burrowing movements were superseded by others in which bilaterally symmetrical changes of shape of the segments were no longer appropriate. In consequence, the circular muscle coat has been progressively reduced in errant polychaetes, and in some it is lacking altogether. The locomotory movements of the most primitive polychaetes are much more akin to sinusoidal swimming than burrowing and, for this, metamerism of the longitudinal muscles is unnecessary. The longitudinal muscles are therefore massive but have lost almost all signs of their originally segmental organization. Despite these changes in locomotory habit, errant polychaetes have retained a spacious coelom and in many of the errant worms there are also complete septa. The functions of the coelom and the septa are no longer precisely the same as in oligochaetes, but a hydrostatic skeleton and segmental compartmentation of the coelom are still essential for the functioning of the parapodia of the most primitive polychaetes. In some worms with a more advanced type of parapodial musculature, an approach is made to the development of skeletal muscles, and the need for the hydrostatic isolation of individual segments is reduced. Septa then disappear, or are reduced to gut suspensory muscles and the like, as in the Nereidae, Hesionidae, and some of the Aphroditidae. The problem of providing suitable insertions for the skeletal muscles in these soft-bodied animals appears to have been met by bracing certain points on the body wall with other muscles, and keeping them rigid while they are in use by turgor pressure. Thus, although the need for septa and a segmented body-wall musculature may vanish, the coelom remains essential as a hydrostatic organ and is always spacious in polychaetes.

The evolution of a quasi-skeletal musculature is an advanced feature that appears in only the errant worms, and the Polychaeta are much better known for their regressive changes. The sedentary worms, including some, such as the Glyceridae, which are conventionally regarded as 'errant', have tended to revert in both habit and structure to a condition very like that of their unsegmented, coelomate ancestors. The worms live in permanent, or near-permanent burrows in the substratum and undertake little locomotory activity. The septa almost entirely vanish, leaving a spacious, undivided coelomic cavity, and the body-wall muscles lose their segmented character and are arranged, rather, in uniform circular and longitudinal layers. The parapodia and parapodial musculature are reduced, of course, but very rarely disappear altogether. The parapodia, bearing special hook-like chaetae, form low ridges which can be slightly protracted to wedge the worms against the walls of their burrows when, for example, they are pumping water or, less frequently, when they are crawling along their burrows.

If many sedentary polychaetes show a reversion to a condition similar to that of unsegmented coelomates, leeches show even greater regressive evolution and have achieved a state which is functionally not greatly different from that of the acoelomate worms. There is no doubt that leeches are derived from oligochaetes and that they represent worms that have abandoned a burrowing existence in the substratum, for a carnivorous and ectoparasitic life in water, swimming and creeping over the substratum. With the very slight resistance to movement that has to be overcome in their environment, and with the evolution of oral and posterior suckers, an entirely different method of locomotion from peristaltic burrowing is available to the leeches. Temporary adhesion between the worms and substratum is provided by the suckers; the chaetae which perform this service in other annelids are therefore redundant and disappear. The performance of looping movements entails alternating contractions of the entire longitudinal and circular muscles, and there are no localized antagonistic contractions of the circular and longitudinal musculature. The whole *raison d'être* of segmentation has therefore gone, and with it go septa and segmentally organized body-wall muscles. Finally, but equally far-reaching in its effect, looping movements make slight demands upon the hydrostatic skeleton and, since the external resistance to movement is minimal, powerful locomotory forces do not need to be generated. The need for a true fluid skeleton is removed and the coelom is almost completely obliterated by the invasion of botryoidal tissue. The result of these changes is the development of an animal which in its habitat, movements and mechanical attributes, does not differ greatly from planarians, and consequently its essential structure also resembles that of planarians. Some organ systems—the gonads, nephridia, and nervous system—remain unaffected by these changes and are still segmentally organized. The fact that they are unchanged despite the changes in other structures, together with the fact that it is precisely these organ systems which show pseudometamerism in planarians and nemerteans, reinforces the conclusion that they are not peculiarly related to segmentation or its evolution.

The great similarity between the early stages of molluscan and polychaete embryology suggests a close affinity between these animals, although with few exceptions, molluscs show no signs of segmentation. If the existing classes of molluscs are extrapolated backwards, one arrives at the hypothetical 'archimollusc' which has something like the structure of a gastropod but without the torsion of the visceral mass on the foot that is characteristic of modern gastropods. For many years the archimollusc was considered to represent the condition of the common ancestor of all molluscs and this convenient fiction could continue to be usefully employed until the discovery of *Neopilina* (Lemche, 1957; Clarke and Menzies, 1959). Now that the structure of one of the archaic Monoplacophora is known in detail (Lemche and

Wingstrand, 1959), it is necessary to reconsider the origin of the Mollusca and the structure of the most primitive molluscs.

It seems reasonably clear that the early molluscs possessed a large, flattened foot on which they crept over the substratum by cilia on a mucus trail, as *Neopilina* probably does (Lemche and Wingstrand, 1959, 1960; Clarke and Menzies, 1959; Menzies, Ewing, Worzel and Clarke, 1959) and some gastropods certainly do, or by the generation of muscular pedal locomotory waves as in the Polyplacophora and many gastropods. Both methods of locomotion are also characteristic of planarians and make virtually no use of a hydrostatic skeleton. Molluscs have an open blood vascular system and when hydraulic mechanisms are employed, as, for example, in the protrusion of the foot or the extension of tentacles, the haemocoel and not a coelomic system provides the necessary fluid skeleton.

In most molluscs, the only cavity that can be regarded as coelomic is the pericardial sac. This structure is generally interpreted as a dilation of the genital ducts that permits the expansion and contraction of the heart, and its small size is held to be a primitive feature. This interpretation and its phylogenetic consequences have recently been re-stated by Fretter and Graham (1962) who suggest that the Mollusca evolved from a turbellarian-nemertean stock by an increase in volume of the digestive system, causing a departure from the flattened body form and the evolution of a visceral mass which later became protected by the secretion of a shell. With this reduction of exposed surface area and increase in body volume, came the need for efficient respiratory organs and a circulatory system. It was at this point that the pericardial coelom was evolved by a dilation of the genital ducts, which permitted the development of a contractile heart to this end. The animal might then be supposed to have possessed all the essential molluscan attributes and to have corresponded to the familiar 'archimollusc'.

It is implicit in this interpretation that the gonocoel theory is unquestioningly accepted, otherwise the pericardial space cannot be regarded as coelomic, and even in terms of the gonocoel theory it is no more than an incipient coelom. This interpretation of the morphological status of the pericardium has therefore been attacked by opponents of the gonocoel theory, and Sarvaas (1933) refuses to regard the molluscs as coelomate, claiming that the pericardium is no more than it appears to be—a dilation of the genital ducts. This difficulty might be avoided if the molluscan coelom was once very much larger and the pericardium represented a vestigial coelom. Such a view was expressed by Naef (1924), who considered the extensive perivisceral cavity of cephalopods to represent a more primitive condition than that obtaining in gastropods and lamellibranchs, but this interpretation has not generally found favour with malacologists.

A second implication of the more orthodox view is that the molluscan coelom evolved independently of that of other coelomates, because it first appeared in conjunction with the evolution of the shell and other typically molluscan features, and not in a common ancestor of the Mollusca and Annelida. And not only did it evolve independently but it had an entirely different functional significance from the outset. On these grounds also, it seems doubtful if the pericardium can properly be regarded as a coelom.

Finally, it is very difficult to accommodate *Neopilina* within this scheme, since it possesses additional coelomic compartments (Fig. 123): a pair of extensive dorsal coelomic sacs and two pairs of ventral gonads, the cavities in the ducts of which might be regarded as coelomic if the gonocoel theory

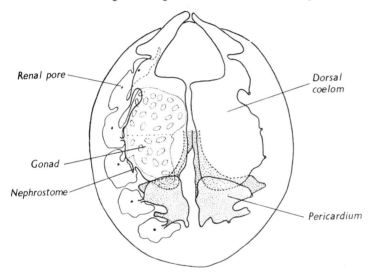

FIG. 123. Coelomic compartments of *Neopilina*. (After Lemche and Wingstrand, 1959.)

is accepted. Furthermore, *Neopilina* shows an apparent metamerism of the gills, nephridia, gonads, blood vascular system, pedal muscles and nervous system. If, as Lemche and Wingstrand (1960) suggest, *Neopilina* bears a very close resemblance to the hypothetical ancestral mollusc, much of the subsequent evolution of the Mollusca must have been regressive and many structures in most existing molluscs are in a secondarily reduced, and not a primitive condition.

The repetition of so many organ systems in *Neopilina* has prompted Lemche (1959) to revive the views of Pelseneer (1898–1899), Heider (1914), Naef (1924) and Söderström (1925b) that molluscs are segmented animals and have a close affinity with the Arthropoda. The small coelomic sacs of both phyla, contrasted with the expanded coelomic sacs of annelids, are regarded as primitively so, but, as we shall see, this raises considerable difficulty when we examine arthropodan origins. It is a corollary of this theory that most molluscs show a loss of metamerism, but signs of it can still be found in

Neopilina, in the renal ducts, nervous system and musculature of the Polyplacophora, and in the heart, gills and kidneys of tetrabranch cephalopods.

The view that molluscs were descended from metamerically segmented animals did not receive wide support before the discovery of *Neopilina*, and it is not at all certain that the serial repetition of various organ systems of this animal represents anything more than pseudometamerism such as occurs in some nemerteans and turbellarians. Portmann (1960) comments that the only coelomic structures that show seriation in *Neopilina* are the two pairs of gonads and these cannot be homologized with the gonads of any other molluscs. The gonad of polyplacophorans, for example, appears to correspond with the pair of dorsal coelomic sacs in *Neopilina*. Yonge (1957) has also argued against the view that molluscs are or ever were metamerically segmented, chiefly on the grounds that the evolution of the ctenidia and structures related to them is intelligible only on the assumption that molluscs are descended from animals with a single pair (Yonge, 1947), although a secondary multiplication of ctenidia, as in the Polyplacophora (Spengel, 1881; Yonge, 1939) and presumably the Monoplacophora, is possible. This argument may be countered by supposing molluscs originally to have had several pairs of ctenidia, as in *Neopilina*, but that they were subsequently reduced to a single pair in the stock from which all other molluscs evolved (Fretter and Graham, 1962).

If the molluscs are not regarded as metamerically segmented animals that have been subjected to a secondary suppression of their metamerism, nor as more primitive animals directly descended from a platyhelminth stock, it might be considered that they are evolved from unsegmented coelomates that possessed a spacious coelom. Lemche and Wingstrand (1959) suggest that the division of the coelom of *Neopilina* into anterior dorsal sacs and pericardial chambers may represent a more advanced condition than the undivided and relatively large coelom of *Nautilus* and some other cephalopods. Nothing is known of the function of the large coelomic cavity of *Nautilus*, still less that of the dorsal coelomic sacs of *Neopilina*, and it is impossible to assess the probability that a large coelom evolved independently in molluscs. If, however, the molluscan secondary body cavity is a legacy from an ancestral, unsegmented, coelomate worm, its reduction in the great majority of molluscs must clearly be correlated with the change in locomotory habit from peristaltic movements to pedal locomotion. This would have resulted in an emphasis upon the longitudinal musculature of the ventral body wall and a reduction in the now unused dorsal body-wall musculature. The reduction of the coelom, the enlargement of the digestive system and development of the visceral mass, and the evolution of a protective shell followed. But against any theory that implies the reduction of a large coelom by mere contraction of the coelomic compartments, must be set the fact that in leeches the reversion of an unequivocal coelomate to a planarian type of locomotion which makes the coelom redundant, is followed by the occlusion of the coelom by parenchymatous tissue. There is both embryological evidence and the evidence of the Acanthobdellidae and Branchiobdellidae, which show characters intermediate between oligochaetes and leeches, that the reduction of the leech coelom has been achieved in this way. There is neither embryological, nor convincing comparative anatomical evidence that the mollusc coelom has been reduced.

The almost identical embryology of the more primitive molluscs and polychaetes, up to the stage of the development of the early trochophore larva, makes it inconceivable that the Mollusca and Annelida are not very closely related. Indeed, on the strength of the embryological evidence, it could be claimed that the two phyla must be more closely related to one another than either is to any third group of animals. Despite this, the origin of molluscs remains obscure. It is possible that when the embryology of *Neopilina* becomes known, the relationship between molluscs and other phyla will become clearer, but, judging by the complication that the discovery of adult *Neopilina* has introduced, the outlook is not promising.

The phylogeny of the Arthropoda presents very similar problems to that of molluscs. Arthropods are clearly related to annelids; as in molluscs, the coelom is reduced and, in a general way, the reduction of the coelom can be related to the locomotory habits of the animals. But when we attempt to consider in more detail how the present condition of arthropods might have arisen and what, if any, relation it bears to the functional morphology of the segmented coelomate worms, we encounter the greatest difficulties.

From a purely functional point of view, the structure of arthropods represents the culmination of a tendency we have noticed in errant polychaetes, and with the development of a rigid exoskeleton and endophragmal skeleton, the hydrostatic system of the body-wall muscles and coelom becomes replaced by true skeletal muscles. Having lost its function, the coelom, although it may make a transitory appearance in embryology, is reduced to such vestiges as the end organ of excretory organs, the cavities within gonads and genital ducts, or becomes incorporated into the haemocoel. The body-wall musculature is similarly highly modified. The great advance shown by the arthropodan system appears to be that control of body shape is transferred from the musculature to the skeleton, making more muscle available for locomotory purposes, a principle that has been beautifully demonstrated in a series of studies of myriapod anatomy and locomotion by Manton (1952a, b, 1954, 1957, 1958a, b).

Although the rigid exoskeleton is characteristic of arthropods and is probably the most important single factor determining their form, it does not preclude the use of a hydrostatic system. The Onychophora and many insect larvae have a soft cuticle and locomotion involves to varying degrees the use of a fluid skeleton comparable to that of annelids. The extension of the body and cirri of the barnacle *Lithotrya* (Cannon, 1947) and of the legs of spiders (Parry and Brown, 1959a, b), as well as other systems in arthropods (Manton, 1958c), also involve hydraulic mechanisms. But in all these cases, it is the blood in the haemocoel and not coelomic fluid that serves this function.

A variety of postulates has been advanced to account for the insignificance of the arthropod coelom. Several authors have attempted to relate molluscs to arthropods and have argued that the coelom of these animals is primitively small (Heider, 1914; Naef, 1924; Lemche, 1959). But here we encounter in an even more acute form, a difficulty we have already met when considering the systematic position of the Polychaeta. It is extremely difficult to envisage the evolution of a complex intrinsic musculature of appendages in animals that did not already possess a segmented body-wall musculature and a subdivided coelom. There is unequivocal embryological evidence that the arthropodan coelom is segmented in its organization, whatever its subsequent fate, and this suggests that it was once very much larger but has suffered a reduction, as most students of arthropods affirm. Lankester (1904) was certainly of this view and suggested that the haemocoel evolved by dilation of the peripheral parts of the blood vascular system at the expense of the coelom, a process which he called phleboedesis. This implies that expansion of the vascular system and reduction of the coelom happened at the same time. An alternative view has been proposed by Sarvaas (1933) who suggested that reduction of the coelom followed upon the development of a rigid exoskeleton and skeletal muscles, and the loss of hydrostatic mechanisms. Presumably then, in arthropods with a soft cuticle, when the need for a hydrostatic skeleton reappeared, it was supplied by an expanded blood vascular system, the coelom no longer being available for this purpose.

The stage at which an enlarged haemocoel appeared is unknown and it is uncertain whether it has been evolved more than once in the Arthropoda (Tiegs and Manton, 1958). Tiegs and Manton are strong supporters of the view that the modern Onychophora are representatives of the primitive arthropodan stock from which myriapods and insects evolved. This interpretation of the position of *Peripatus* is not universally accepted, but whatever the view taken of arthropod evolution, it is clear that if *Peripatus* is in any way representative of a pre-arthropodan or primitive arthropodan stock, then the evolution of a spacious haemocoel predates the evolution of a rigid exoskeleton, and this raises many difficulties.

Manton (1950) has analysed the locomotory movements of *Peripatus* and discussed the function and possible evolution of the haemocoel in this animal. Unlike most polychaetes, *Peripatus* crawls solely by movements of the appendages, without direct intervention of the longitudinal muscles. However, the longitudinal muscles become involved indirectly. Change of speed is effected by a change of gait rather than by a change in the duration of each pace, and when faster gaits are used, the body is extended; with slower gaits, the body is contracted. Changes in length involve the fluid skeleton represented by the haemocoel, but it is the whole body which is elongated and there are no differential changes of shape of individual segments during the locomotory cycle, such as we observe in annelids. A coelom subdivided by septa and mesenteries permits only local changes of shape, but, it is argued, the vascular system provides the origins of a fluid system that is continuous throughout the length of the body. Manton claims that by dilation of the vascular system at the expense of the coelom, an appropriate mechanism to suit the needs of a soft-bodied animal crawling solely by movements of appendages was evolved.

In a later paper, Manton (1953) pointed out that an important difference between the polychaete parapodium and the leg of *Peripatus* is that the former is supported by acicula in errant worms and in many of them, muscles attached to the base of the aciculum are responsible for the movement of the whole appendage. This is incompatible with the development of both the musculature and the type of movement that we observe in the leg of *Peripatus*. Manton suggests, therefore, that the Onychophora arose from annelids 'in which simple ventro-laterally directed limbs raised the body off the ground, as in *Peripatus* and the middle Cambrian *Aysheaia*. This line of annelids did not necessarily pass through a stage possessing the polychaete type of locomotory mechanism and presumably lacked parapodial acicula'.

These interpretations present certain difficulties: (1) They depend upon the view that the common ancestors of annelids and arthropods possessed parapodia of some type. There is little evidence that the most primitive annelids possessed parapodial extensions of the body wall; it is more probable that parapodia arose as a later specialization in worms that were already segmented and had a body-form comparable to that of the modern oligochaetes. If this is so, it is more likely that segmental appendages were independently evolved in annelids and arthropods than that the appendages were evolved in a common ancestor of arthropods and polychaetes. If the latter alternative is accepted, this would have the effect of placing a dichotomy between oligochaetes and leeches on the one hand, and polychaetes and arthropods on the other, instead of between annelids and arthropods. It would follow, then, that the Arthropoda should be contained within the phylum Annelida. Apart from a common prejudice against this conclusion, it is ruled out by the many important fundamental differences between polychaetes and arthropods. The evolution of segmental appendages appears to present no great difficulty, particularly when it is remembered that the prolegs of insect larvae, which are turgid appendages comparable to parapodia, have been independently evolved many times over (Hinton, 1955).

(2) In those polychaetes in which a quasi-skeletal parapodial musculature has been evolved, the need for pressure differentials in the coelom of neighbouring segments disappears. The septa are then reduced and cease to act as

water-tight barriers between adjacent segments. The evolution of comparable muscles in the Onychophora have, it is claimed, resulted in an expansion of the blood vascular system at the expense of the coelom, to provide for longitudinal conduction of pressure changes in the haemocoel. This argument raises the problem of why this modification should have taken place when reduction of the septa would have served the same purpose. If the parapodial musculature was so arranged that the production of local pressure differences was no longer essential for locomotion (the only condition under which a longitudinal transmission of pressure changes would be appropriate), the septa would already have lost their function and could hardly have necessitated the evolution of an alternative fluid system to bypass them.

More recently, Manton (1961b) has demonstrated additional properties of the haemocoel which may be relevant to a discussion of its origins. Geophilomorph centipedes burrow in the soil in a similar manner to earthworms (Manton, 1952b), that is by insinuating the head into a crevice which is then widened by thickening and shortening the segments posterior to the head. The radial thrust that can be generated by *Orya*, a large geophilomorph, is much greater than that which can be developed by the polychaete *Arenicola* of a similar size, and both can exert considerably greater thrusts than an earthworm. In both *Orya* and *Arenicola*, contraction of muscles in all parts of the body can contribute to the thrust, but in the earthworm only the muscles of the segments directly concerned are involved because of the isolation of different parts of the body by septa. *Arenicola* can develop its maximum thrust only once or twice and this is accompanied by leakage of coelomic fluid and bursting of the capillaries of the gills, which reduces its efficiency. *Orya*, on the other hand, can develop its maximum thrust many times in succession without injury because the haemocoel does not communicate with the exterior and there are no external capillaries. Manton suggests, therefore, that the evolution of the haemocoel may have been associated with burrowing into the surface layers of the substratum.

Both the haemocoel and the coelom serve as hydrostatic skeletons and both, if Manton's (1961b) suggestion is accepted, evolved as adaptations to burrowing. The two fluid systems might be regarded as the products of parallel evolution but for the fact that arthropods are coelomate and, if the most primitive arthropods were soft-bodied, the haemocoel must have evolved in animals that already had a coelom. It is difficult to understand why or how one fluid skeletal system (the coelom) should be replaced by another (the haemocoel) when both serve the same function. Manton's suggestion that the overriding advantage of the haemocoel is that it does not communicate with the exterior and that there are no capillaries to be ruptured by high internal hydrostatic pressures, is not entirely satisfactory.

Under experimental conditions, the repeated development of maximal radial thrusts by *Arenicola* is followed by damage and the leakage of coelomic fluid, but it must be remembered that Manton measured the radial thrust by causing the worms to crawl under a bridge and lift weights placed upon it. Those parts of the body wall not in contact with the bridge or the ground were unprotected and it might be expected that the stress placed upon them would result in injury. When the worm burrows, however, all parts of the body wall are in contact with the surrounding substratum and are supported by it. The sphincter muscle of the nephridiopore is able to withstand twice the normal working hydrostatic pressures without leakage of the coelomic fluid (Chapman and Newell, 1947), so that there is, in fact, an ample safety margin for all but exceptional activities. The type of adaptation shown by *Arenicola* is adequate to permit sedentary polychaetes to form burrows in most marine sediments and if arthropods evolved in a marine environment, as is generally supposed, we may question whether any substratum that they could have burrowed into at all would have been so difficult to penetrate as to warrant such a radical evolutionary change as the functional replacement of the coelom by the haemocoel. Rather, it is in terrestrial environments that considerable radial thrusts are necessary to enlarge fissures in dried soils. The geophilomorphs are well adapted to do this, but they are highly specialized chilopods. They illustrate what is possible for animals with an arthropodan organization; it seems unlikely that they reflect the mode of life of the earliest arthropods.

The most we are able to conclude from our present knowledge of arthropodan phylogeny is that they evolved from an early pre-annelid stock, that is, from soft-bodied worms that were segmented and possessed a spacious coelomic cavity subdivided by septa. While the annelids have exploited the potentialities of the extremely flexible hydrostatic system that they possessed, arthropods have made much greater advances by the evolution of a true skeletal musculature and a rigid exoskeleton. Correlated with this abandonment of hydraulic mechanisms for locomotion, the coelom of arthropods is diminutive and the intersegmental septa are lost. There has also been a substantial modification of segmentation in arthropods. With the loss of internal compartmentation, organs have become specialized and tend to be concentrated rather than serially repeated; there has been a reduction in the number of segments, compared with most annelids, associated with the change in locomotory habit; and a number of segments have fused pre-orally to form an enlarged head with a large brain, to which the fused segmental ganglia contribute. The changes are obvious and the advantages they confer upon arthropods are intelligible; what remains obscure is how the initial steps in this phyletic sequence were made.

CONCLUSIONS

'Anything said on these questions lies in the realm of fantasy'; thus Hyman (1959), with reference to such phylogenetic problems as the origin of the

Metazoa from the Protozoa, or the Bilateria from the Radiata. That the same stricture might be applied to the phylogeny and interrelationships of most existing phyla is evident from the preceding section of this chapter. It is certain that many of the derivations of major groups of animals that are accepted now will be re-examined, rejected and changed in the future. The construction of phylogenies is a continuous process and, because of the nature of the evidence upon which we must work, finality or universal agreement is impossible. Nevertheless, so long as these exercises in theoretical zoology continue, certain principles must be observed. The most important and least considered of these is that hypothetical constructs which represent ancestral, generalized forms of modern groups, or stem forms from which several modern phyla diverge, must be possible animals. In other words, they must be conceived as living organisms, obeying the same principles that we have discovered in existing animals, and any new structures or properties that are postulated to have arisen in these hypothetical forms must have conferred some selective advantage upon them over their predecessors.

With these principles in mind, we have considered the evolution of the main types of morphological organization that exist in modern animals. The coelom has many functions, but of these, its role as a hydrostatic skeleton appears to be of the greatest and most general evolutionary significance. In the majority of coelomates the coelom can be seen to function as a hydrostatic organ, but in some, the need for a hydrostatic organ is lost and the coelom is then correspondingly small or may even be almost completely obliterated. This strict correlation between the existence of a spacious coelom and its use as a hydrostatic organ gives overwhelming support to the view that this is the prime and original function of the secondary body cavity. This is not to say that non-coelomic body cavities may not also serve as hydrostatic skeletons, the coelenteron of actinians, the pseudocoel of nematodes and the haemocoel of some molluscs and arthropods are all examples to the contrary. It is also true that a coelomic cavity may survive although it does not have any hydrostatic function, as perhaps in the molluscan pericardium and the brachiopod mantle canals. But these instances of alternative hydrostatic organs and the acquisition of different functions by the coelom are readily intelligible and do not conflict with the primary correlation between the coelom and its mechanical function.

The first appearance of a coelom occurred in solid-bodied worms comparable to modern turbellarians and nemerteans in that they possessed circular and longitudinal body-wall muscles. The evolution of a coelomic cavity made no difference to the principle of functioning of the body-wall musculature, nor to the changes of shape that the animal could perform. It merely permitted faster and more powerful forces to be brought to bear upon the environment and, in particular, permitted the animals to burrow into the substratum and to exploit new habitats. There is no reason *a priori* why a secondary body cavity with the same morphological attributes should not have been evolved independently in several different groups of solid-bodied animals in answer to the same selective pressure. Embryological evidence suggests that this may have been so.

The period immediately following the first appearance of the coelom also appears to have been one of important radiative evolution when fundamentally new types of body organization developed. The early coelomates were necessarily rather sedentary animals because of the difficulty of producing the localized changes of shape necessary for locomotion in a body with a coelomic cavity that extended uninterrupted from one end of the animal to the other. Subdivision of the hydrostatic skeleton followed as a locomotory adaptation.

In a limited sense, the evolution of the oligomerous animals, with three divisions of the coelom, is an example of such a development. In the more primitive lophophorate and deuterostome phyla, two coelomic compartments appear to be related to locomotion, while the third is an isolated hydrostatic organ responsible for maintaining turgor pressure in the lophophore and tentacles. These were (and their more primitive living representatives still are) tubicolous worms which appeared above the surface of the substratum and fed upon the detritus in suspension in the sea-water, which they collected in their tentacular apparatus. They exploited a different habitat from the unsegmented coelomates which lived entirely beneath the surface of the substratum and fed upon the organic material in the surface and sub-surface deposits. It is uncertain if all the modern oligomerous phyla are descended from a common oligomerous ancestor, or if the tripartite structure of the body was evolved more than once in different, if closely related, unsegmented coelomate animals. It is also possible, though perhaps less likely, that animals with a tripartite coelom evolved directly from solid-bodied worms, in parallel with the evolution of the unsegmented coelomates.

Another line of advance which developed directly from the unsegmented coelomate condition was the evolution of metameric segmentation by the subdivision of the coelom with septa, and the development of a segmented body-wall musculature and nervous system. This new organization of the hydrostatic system did not enable the worms to perform movements that could not already be performed by non-segmented coelomates, but it did facilitate the production of localized changes of shape. The improved mechanical efficiency of the segmented worms freed them from a sedentary existence, though initially they remained beneath the surface of the substratum and lived as burrowers in much the same manner as modern earthworms.

Concurrently with the initial radiation of the Metazoa, there was a tendency towards the evolution of larger body size and a longer period of somatic growth before reproduction. Initially, the control of the time of onset of sexual maturity was imperfect and precocious maturation must have been a relatively common event. Sometimes this was the source of new types of structure. The pseudocoelomates, in which the secondary body cavity is not coelomic but is a persistent blastocoel, were derived, probably polyphyletically, from among the early coelomate animals in this way. The chordates were similarly evolved from an oligomerous stock by paedogenesis, though in this case from a more advanced larval stage in which the coelomic compartments were already incipient.

It is clear that metameric segmentation, as it is commonly used, is a term with little precision. The two groups of animals to which the concept might be applied are of entirely distinct derivation and the manner in which parts of the body are serially repeated and the adaptations they represent in chordates and annelids, are quite different. Animals may be more reasonably grouped together as coelomate, pseudocoelomate or oligomerous, but even so, there is a strong presumption that each group has been polyphyletically derived and that these terms therefore refer to grades of construction rather than to taxonomic entities.

In all these major divisions of the Metazoa, however derived, further important radiation has taken place, with the evolution of new types of anatomical organization and the adoption of new habitats. Some of these changes have had profound consequences for the coelom. The adoption of a completely sessile habit, as in some lophophorates and deuterostomes, removes the need for locomotory structures and the fluid skeleton associated with them. The spacious coelom may be retained as a hydrostatic system for the eversion of polypides or of a food-collecting structure, but it is no longer subdivided into separate compartments. In arthropods, the evolution of skeletal muscles and a rigid skeleton reduces the need for hydrostatic organs and again the coelom declines in importance and is small.

This analysis of metazoan evolution provides no more than a framework within which the detailed phylogenies of animals must be set, and we are left with much uncertainty about the precise derivations and the interrelationships of the metazoan phyla. So long as we confine our attention to the functional aspects of the course of evolution, several alternative derivations of a phylum can be equally well accommodated within this framework. For the elucidation of the interrelationships and the selection of the most probable line of descent, we must turn to the evidence of comparative anatomy, embryology and palaeontology, the basis of all classical systematics. But here we are embarrassed by the equivocal nature of much of the evidence. It is often difficult to assess the systematic value of characters and, particularly when considering the phylogeny of major groups of animals, it is difficult to distinguish recent adaptive features from those which are genuinely primitive.

It is an indispensable principle that structure must be considered in relation to function; in isolation it is meaningless. Reconstructing the evolutionary history of animals is one of the most fruitful unifying disciplines in zoology, but it can progress and unify only if some attempt is made to view the whole of zoological knowledge. Perhaps the volume of information at our disposal is already too great for this to be possible, but occasionally, by raising our eyes from the immediate problems that confront us, we can achieve a partial synthesis which gives fresh direction to our enquiries.

BIBLIOGRAPHY

ADAM, H. 1960 Different types of body movement in the hagfish, *Myxine glutinosa* L. *Nature, Lond.* **188**, 595–6.
AFFLECK, R. J. 1950 Some points in the function, development and evolution of the tail in fishes. *Proc. zool. Soc. Lond.* **120**, 349–68.
AGERSBORG, H. P. K. 1919 Notes on *Melibe leonina* (Gould). *Publ. Puget Sd Mar. (biol.) Sta.* **2**, 269–77.
— 1921 Contribution to the knowledge of the nudibranchiate mollusk, *Melibe leonina* (Gould). *Amer. Nat.* **55**, 222–53.
— 1923 The morphology of the nudibranch mollusk, *Melibe* (syn. *Chioraera*) *leonina* (Gould). *Quart. J. micr. Sci.* **67**, 507–92.
AHLBORN, F. 1895 Über die Bedeutung der Heterocerkie und ähnlicher unsymmetrischer Schwanzformen schwimmender Wirbelthiere für die Ortsbewegung. *Z. wiss. Zool.* **60**, 1–15.
ALLEN, E. J. 1893 Nephridia and body-cavity of some decapod Crustacea. *Quart. J. micr. Sci.* **34**, 403–26.
ALLMAN, G. J. 1843 On the muscular system of certain fresh water ascidian zoophytes. *Proc. R. Irish Acad.* **2**, 319–32.
— 1856 *A monograph of the fresh-water Polyzoa.* London: Ray Soc.
ANDERSON, D. T. 1959 The embryology of the polychaete *Scoloplos armiger*. *Quart. J. micr. Sci.* **100**, 89–166.
ANDERSSON, K. A. 1907 Die Pterobranchier der Schwedischen Südpolarexpedition, 1901–1903. Nebst Bemerkungen über *Rhabdopleura normanii* Allman. *Wiss. Ergebn. schwed. Südpolarexped.* **5** (10), 1–122.
ASHWORTH, J. H. 1912 *Catalogue of the Chaetopoda in the British Museum (Natural History).* A. Polychaeta; Part 1. Arenicolidae. London: Brit. Mus.
— 1915 On the larvae of *Lingula* and *Pelagodiscus* (*Discinisca*). *Trans. roy. Soc. Edinb.* **51**, 45–69.
ASSHETON, R. 1908 A new species of *Dolichoglossus*. *Zool. Anz.* **33**, 517–20.
ATKINS, D. 1959 The growth stages of the lophophore of the brachiopods *Platidia davidsoni* (Eudes Deslongchamps) and *P. anomioides* (Philippi), with notes on the feeding mechanism. *J. mar. biol. Ass. U.K.* **38**, 103–32.
— 1960 The ciliary feeding mechanism of the Megathyridae (Brachiopoda), and the growth stages of the lophophore. *J. mar. biol. Ass. U.K.* **39**, 459–79.
AX, P. 1956 Die Gnathostomulida, eine rätselhafte Wurmgruppe aus dem Meeressand. *Abh. Akad. Wiss. Lit., Mainz, Math-Nat. Kl.* Jhg. 1956, (8), 531–62.
BAHL, K. N. 1919 On a new type of nephridia found in Indian earthworms of the genus *Pheretima*. *Quart. J. micr. Sci.* **64**, 67–119.
— 1947 Excretion in Oligochaeta. *Biol. Rev.* **22**, 109–47.
BAINBRIDGE, R. 1960 Speed and stamina in three fish. *J. exp. Biol.* **37**, 129–53.
BALFOUR, F. M. 1878 *A monograph of the development of elasmobranch fishes.* London: Macmillan.

BIBLIOGRAPHY

— 1880 Larval forms: their nature, origin and affinities. *Quart. J. micr. Sci.* **20**, 381–407.

— 1881 *A treatise on comparative embryology*. London: Macmillan. **2**.

BARETS, A. 1961 Contribution à l'étude des systèmes moteurs 'lent' et 'rapide' du muscle latéral des Téléostéens. *Arch. Anat. micr. Morph. exp.* **50**, 91–187.

BARNES, H. 1957 Processes of restoration and synchronization in marine ecology. The spring diatom increase and the spawning of the common barnacle, *Balanus balanoides*. *Année biol.* **33**, 67–85.

BARTH, R. 1937 Muskulatur und Bewegungsart der Raupen zugleich ein Beitrag zur Spannbewegung und Schruckstellung der Spannerraupen. *Zool. Jb. (Anat.)* **62**, 507–66.

BATESON, W. 1884 The early stages in the development of *Balanoglossus* (sp. incert.). *Quart. J. micr. Sci.* **24**, 208–36.

— 1885 The later stages in the development of *Balanoglossus kowalevskii*, with a suggestion as to the affinities of the Enteropneusta. *Quart. J. micr. Sci.* **25** (suppl.), 81–122.

— 1886 The ancestry of the Chordata. *Quart. J. micr. Sci.* **26**, 535–71.

BATHAM, E. J. and PANTIN, C. F. A. 1950 Muscular and hydrostatic action in the sea-anemone *Metridium senile* (L.). *J. exp. Biol.* **27**, 264–89.

— and PANTIN, C. F. A. 1951 The organization of the muscular system of *Metridium senile*. *Quart. J. micr. Sci.* **92**, 27–54.

BATHER, F. A. 1900 Echinodermata. In: LANKESTER, E. R. (ed.), *A treatise on zoology*. London: Black. **3**.

BEAUCHAMP, P. de 1909 Recherches sur les Rotifères: les formations tégumentaires et l'appareil digestif. *Arch. Zool. exp. gén.* iv, **10**, 1–410.

— 1911 Conceptions récentes sur l'anatomie et l'embryologie comparées des Vers et des groupes voisins. Les théories du trophocoele. *Bull. Sci. Fr. Belg.* **46**, 106–48.

— 1959 Généralités. In: GRASSÉ, P. P. (ed.), *Traité de zoologie*. Paris: Masson. **5** (1), 3–11.

— 1961 Classe des Turbellariés. Turbellaria (Ehrenberg, 1831). In: GRASSÉ, P. P. (ed.), *Traité de zoologie*. Paris: Masson. **4** (1), 35–212.

BECHER, S. 1907 *Rhabdomolgus ruber* Keferstein und die Stammform der Holothurien. *Z. wiss. Zool.* **88**, 545–689.

BENENDEN, E. van 1891 Recherches sur le développement des *Aracnactis*. Contribution à la morphologie des Cérianthides. *Arch. Biol., Paris* **11**, 115–47.

BENHAM, W. B. 1899 *Balanoglossus otagoensis* n. sp. *Quart. J. micr. Sci.* **42**, 497–504.

BERGH, R. S. 1885 Die Exkretionsorgane der Würmer. *Kosmos, Lwow* **17**, 97–122.

BERRILL, N. J. 1935 Studies in tunicate development. Part III. Differential retardation and acceleration. *Philos. Trans.* B, **225**, 255–326.

— 1951 Regeneration and budding in tunicates. *Biol. Rev.* **26**, 456–75.

— 1955 *The origin of the vertebrates*. Oxford: Clarendon Press.

— 1961 *Growth, development and pattern*. San Francisco-London: Freeman.

BHATIA, M. L. 1941 *Hirudinaria* (the Indian cattle leech). *Indian zool. Mem.* **3**.

BIEDERMANN, W. 1905 Studien zur vergleichenden Physiologie der peristaltischen Bewegungen. II. Die lokomotorischen Wellen der Schneckensohle. *Pflüg. Arch. ges. Physiol.* **107**, 1–56.

BIRD, A. F. 1956 Chemical observations of the nematode cuticle. Observations of the whole cuticle. *Exp. Parasitol.* **5,** 350-8.
— 1958 Further observations on the structure of nematode cuticle. *Parasitology* **48,** 32-7.
— and DEUTSCH, K. 1957 The structure of the cuticle of *Ascaris lumbricoides* var. *suis. Parasitology* **47,** 319-28.
BLOCHMANN, F. 1892 *Untersuchungen über den Bau der Brachiopoden. I. Die Anatomie von* Crania anomala *O.F.M.* Jena: Fischer.
— 1900 *Untersuchungen über den Bau der Brachiopoden. II. Die Anatomie von* Discinisca *und* Lingula. Jena: Fischer.
BODDEKE, R., SLIJPER, E. J. and STELT, A, van der 1959 Histological characteristics of the body-musculature of fishes in connection with their mode of life. *Proc. K. ned. Akad. Wetensch.* C, **62,** 576-88.
BOETTGER, C. R. 1952 Die Stämme des Tierreichs in ihrer systematischengliederung. *Abh. braunschw. wiss. Ges.* **4,** 238-300.
BÖHMIG, L. 1895 Die Turbellaria Acoela der Plankton Expedition. *Ergebn. Plankton Exped.* **2** (H.g.), 3-48.
— 1929 Nemertini In: KÜKENTHAL, W. and KRUMBACH, T. (eds.), *Handbuch der Zoologie.* Berlin-Leipzig: De Gruyter. **2** (1: 3), 1-110.
BONE, Q. 1960 The origin of the chordates. *J. Linn. Soc. (Zool.)* **44,** 252-69.
BORELLI, G. A. 1680 *De motu animalium.* Rome.
BORG, F. 1923 On the structure of cyclostomatous Bryozoa. *Ark. Zool.* **15** (11), 1-17.
— 1926 Studies on recent cyclostomatous Bryozoa. *Zool. Bidr. Uppsala* **10,** 181-507.
BÖSIGER, E. 1950 Vergleichende Untersuchungen über die Brustmuskulatur von Huhn, Wachtel und Star. *Acta anat.* **10,** 385-429.
BRADLEY, C. E. 1959 *The movement of* Ascaris lumbricoides. Thesis, University of Bristol.
BRAEKKAN, O. R. 1956 Function of the red muscle in fish. *Nature, Lond.* **178,** 747-8.
BRAEM, F. 1895 Was ist ein Keimblatt? *Biol. Cbl.* **15,** 427-43; 466-76; 491-506.
BREDER, C. M. 1926 The locomotion of fishes. *Zoologica, N.Y.* **4,** 159-297.
BRESSLAU, E. 1928-1930 Turbellaria. In: KÜKENTHAL, W. and KRUMBACH, T. (eds.), *Handbuch der Zoologie.* Berlin-Leipzig: De Gruyter. **2** (1:1), 52-320.
BRIEN, P. 1948 Embranchement des Tuniciers. In: GRASSÉ, P. P. (ed.), *Traité de zoologie.* Paris: Masson. **11,** 553-930.
— 1959 Classe des Endoproctes ou Kamptozoaires. In: GRASSÉ, P. P. (ed.), *Traité de zoologie.* Paris: Masson. **5** (1), 927-1007.
— 1960 Classe des Bryozoaires. In: GRASSÉ, P. P. (ed.), *Traité de zoologie.* Paris: Masson. **5** (2), 1053-1335.
BRØNDSTED, H. V. 1955 Planarian regeneration. *Biol. Rev.* **30,** 65-126.
BROOKS, C. M. 1929 Notes on the statoblasts and polypids of *Pectinatella magnifica. Proc. Acad. nat. Sci. Philad.* **81,** 427-41.
BROOKS, W. K. 1893 The genus *Salpa. Mem. biol. Lab. Johns Hopk. Univ.* **2.**
BROWN, A. C. 1961 Physiological-ecological studies on two sandy-beach Gastropoda from South Africa; *Bullia digitalis* Meuschen and *Bullia laevissima* (Gmelin). *Z. Morph. Ökol. Tiere* **49,** 629-57.

— and TURNER, L. G. W. 1962 Expansion of the foot in *Bullia* (Gastropoda). *Nature, Lond.* **195**, 98–9.
BROWN, R. S. 1938 The anatomy of the polychaete *Ophelia cluthensis* McGuire 1935. *Proc. roy. Soc. Edinb.* **58**, 135–60.
BUDINGTON, R. A. 1937 The normal spontaneity of movement of the respiratory muscles of *Thyone briareus* Leseur. *Physiol. Zool.* **10**, 141–55.
BULLOCK, T. H. 1940 The functional organization of the nervous system of the Enteropneusta. *Biol. Bull. Woods Hole* **79**, 91–113.
— 1945. The anatomical organization of the nervous system of Enteropneusta. *Quart. J. micr. Sci.* **86**, 55–111.
BURDON-JONES, C. 1950 An enteropneust genus new to the British Isles. *Nature, Lond.* **165**, 327.
— 1952 Development and biology of the larva of *Saccoglossus horsti* (Enteropneusta). *Philos. Trans.* B, **236**, 553–90.
— 1956 Observations of the enteropneust, *Protoglossus koehleri* (Caullery and Mesnil). *Proc. zool. Soc. Lond.* **127**, 35–58.
— 1960 Unpublished. (Cited by MORTON, 1960).
BURFIELD, S. T. 1927 *Sagitta. L.M.B.C. Mem.* **28**.
BÜRGER, O. 1894 Neue Beiträge zur Entwicklungsgeschichte der Hirudineen. Zur Embryologie von *Hirudo medicinalis* und *Aulostomum gulo. Z. wiss. Zool.* **58**, 440–59.
— 1895 Die Nemertinen des Golfes von Neapel. *Fauna u. Flora Neapel* **22**.
— 1897–1907 Nemertini. In: BRONN, H. G., *Klassen und Ordnungen des Tierreichs*. Leipzig: Akad. Verlagsgesellschaft. **4** (2, suppl.), 1–542.
BURY, H. 1895 The metamorphosis of echinoderms. *Quart. J. micr. Sci.* **38**, 45–135.
BUSK, G. 1884 Report on the Polyzoa collected by H.M.S. Challenger during the years 1873–76. Part I. The Cheilostomata. *Challenger Rep.* Zool. **10** (5), 1–216.
CALDWELL, W. H. 1885 Blastopore, mesoderm and metameric segmentation. *Quart. J. micr. Sci.* **25**, 15–28.
CANNAN, R. K. 1927 Echinochrome. *Biochem. J.* **21**, 184–7.
CANNON, H. G. 1947 On the anatomy of the pedunculate barnacle *Lithotrya. Philos. Trans.* B, **233**, 89–136.
CARLSON, A. J. 1905 The physiology of locomotion in gasteropods. *Biol. Bull. Woods Hole* **8**, 85–92.
CARRUTHERS, R. C. 1906 The primary septal plan of the Rugosa. *Ann. Mag. nat. Hist.* vii, **18**, 356–63.
CARTER, G. S. 1957 Chordate phylogeny (review of 'The Origin of Vertebrates' by N. J. Berrill). *Syst. Zool.* **6**, 187–92.
CAULLERY, M. 1944 *Siboglinum* Caullery. Type nouveau d'Invertébrés d'affinités à préciser. *Siboga Exped.* **25b**, 1–26.
— and MESNIL, F. 1904 Contribution à l'étude des Entéropneusts. *Protobalanus* (n.g.) *koehleri. Zool. Jb. (Anat.)* **20**, 227–56.
CHAPMAN, G. 1949 The mechanism of opening and closing of *Calliactis parasitica. J. mar. biol. Ass. U.K.* **28**, 641–9.
— 1950 Of the movement of worms. *J. exp. Biol.* **27**, 29–39.
— 1953a Studies of the mesogloea of coelenterates. I. Histology and chemical properties. *Quart. J. micr. Sci.* **94**, 155–76.

— 1953b Studies on the mesogloea of coelenterates. II. Physical properties. *J. exp. Biol.*, **30**, 440–51.
— 1958 The hydrostatic skeleton in the invertebrates. *Biol. Rev.* **33**, 338–71.
— and NEWELL, G. E. 1947 The rôle of the body fluid in relation to movement in soft-bodied invertebrates. I. The burrowing of *Arenicola*. *Proc. roy. Soc.* B, **134**, 431–55.
— and NEWELL, G. E. 1956 The rôle of the body fluid in the movement of soft-bodied invertebrates. II. The extension of the siphons of *Mya arenaria* L. and *Scrobicularia plana* (da Costa). *Proc. roy. Soc.* B, **145**, 564–80.
CHILD, C. M. 1901 The habits and natural history of *Stichostemma*. *Amer. Nat.* **35**, 975–1006.
— 1904 Studies on regulation. IV. Some experimental modifications of form-regulation in *Leptoplana*. *J. exp. Zool.* **1**, 95–133.
— 1941 *Patterns and problems of development*. Chicago: University Press.
CHITWOOD, B. G. 1936 Observations on the chemical nature of the cuticle of *Ascaris lumbricoides* var. *suis*. *Proc. helm. Soc. Wash.* **3**, 39–49.
— and CHITWOOD, M. B. 1950 *An introduction to nematology*. Baltimore: Author. Revised edition.
CHUANG, S. H. 1956 The ciliary feeding mechanisms of *Lingula unguis* (L.) (Brachiopoda). *Proc. zool. Soc. Lond.* **127**, 167–89.
CLAPARÈDE, E. 1868 Les Annélides Chètopodes du golfe de Naples. *Mém. Soc. Phys., Genève* **19**, 313–584.
CLARK, M. E. and CLARK, R. B. 1962 Growth and regeneration in *Nephtys*. *Zool. Jb. (Physiol.)* **70**, 24–90.
CLARK, R. B. 1956 The blood vascular system of *Nephtys* (Annelida Polychaeta). *Quart. J. micr. Sci.* **97**, 235–49.
— 1958 The gross morphology of the anterior nervous system of *Nephtys*. *Quart. J. micr. Sci.* **99**, 205–20.
— 1961 The origin and formation of the heteronereis. *Biol. Rev.* **36**, 199–236.
— 1962a On the structure and functions of polychaete septa. *Proc. zool. Soc. Lond.* **138**, 543–78.
— 1962b The control of growth and reproduction in polychaetes, and its evolutionary implications. *Symp. Soc. Endocrinol.* **12**, 323–7.
— ALDER, J. and MCINTYRE, A. D. 1962 The distribution of *Nephtys* on the Scottish coast. *J. Anim. Ecol.* **31**, 359–72.
— and CLARK, M. E. 1960a The fine structure and histochemistry of the ligaments of *Nephtys*. *Quart. J. micr. Sci.* **101**, 133–48.
— and CLARK, M. E. 1960b The ligamentary system and the segmental musculature of *Nephtys*. *Quart. J. micr. Sci.* **101**, 149–76.
— and COWEY, J. B. 1958 Factors controlling the change of shape of certain nemertean and turbellarian worms. *J. exp. Biol.* **35**, 731–48.
— and HADERLIE, E. C. 1960 The distribution of *Nephtys cirrosa* and *N. hombergi* on the south-western coasts of England and Wales. *J. Anim. Ecol.* **29**, 117–47.
— and HADERLIE, E. C. 1962 The distribution of *Nephtys californiensis* and *N. caecoides* on the Californian coast. *J. Anim. Ecol.* **31**, 339–57.
CLARKE, A. H. and MENZIES, R. J. 1959 *Neopilina* (*Vema*) *ewingi*, a second living species of the Paleozoic class Monoplacophora. *Science* **129**, 1026–7.

COE, W. R. 1895 On the anatomy of a species of nemertean (*Cerebratulus lacteus* Verrill), with some remarks on certain other species. *Trans. Conn. Acad. Arts Sci.* **9**, 479–514.
— 1943 Biology of the nemerteans of the Atlantic coast of North America. *Trans. Conn. Acad. Arts Sci.* **35**, 129–328.

COLE, F. J. 1907 A monograph of the general morphology of the myxinoid fishes, based on a study of *Myxine*. Part III. The anatomy of the muscles. *Trans. roy. Soc. Edinb.* **45**, 683–757.

COPELAND, M. 1919 Locomotion in two species of the gastropod genus *Alectrion* with observations on the behavior of the pedal cilia. *Biol. Bull. Woods Hole* **37**, 126–38.
— 1922 Ciliary and muscular locomotion in the gastropod genus *Polinices*. *Biol. Bull. Woods Hole* **42**, 132–42.

COWEY, J. B. 1952 The structure and function of the basement membrane muscle system in *Amphiporus lactifloreus* (Nemertea). *Quart. J. micr. Sci.* **93**, 1–15.

CROFTS, D. R. 1929 *Haliotis*. *L.M.B.C. Mem.* **29**.

CROZIER, W. J. 1914 The orientation of a holothurian to light. *Amer. J. Physiol.* **36**, 8–20.
— 1915a The behavior of an enteropneust. *Science* **41**, 471–2.
— 1915b The sensory reactions of *Holothuria surinamensis* Ludwig. *Zool. Jb. (Physiol.)* **35**, 233–97.
— 1916 The rhythmic pulsations of the cloaca of holothurians. *J. exp. Zool.* **20**, 297–356.
— 1917 The photic sensitivity of *Balanoglossus*. *J. exp. Zool.* **24**, 211–7.
— 1918 On the method of progression in polyclads. *Proc. nat. Acad. Sci., Wash.* **4**, 379–81.
— 1919 On the use of the foot in some mollusks. *J. exp. Zool.* **27**, 359–66.
— and PILZ, G. F. 1924 The locomotion of *Limax*. I. Temperature coefficient of pedal activity. *J. gen. Physiol.* **6**, 711–22.

DAHLGREN, U. 1934 A species and genus of fresh-water bryozoon new to North America. *Science* **79**, 510.

DALES, R. P. 1955 Feeding and digestion in terebellid polychaetes. *J. mar. biol. Ass. U.K.* **34**, 55–79.
— 1962 The polychaete stomodeum and the inter-relationships of the families of Polychaeta. *Proc. zool. Soc. Lond.* **139**, 389–428.

DAVIDSON, P. 1920 The musculature of *Heptanchus maculatus*. *Univ. Calif. Publ. Zool.* **18**, 151–70.

DAWYDOFF, C. 1909 Beobachtungen über der Regenerationsprozess bei den Enteropneusta. *Z. wiss. Zool.* **93**, 237–305.
— 1928 *Traité d'embryologie comparée des Invertébrés*. Paris: Masson.
— 1948 Contribution à la connaissance de *Siboglinum*, Caullery. *Bull. biol.* **82**, 141–63.
— 1959 Classe des Echiuriens (Echiurida de Blainville, Gephyrea Armata de Quatrefages, 1847). In: GRASSÉ, P. P. (ed.), *Traité de zoologie*. Paris: Masson. **5** (1), 855–907.

DE BEER, G. R. 1954 The evolution of the Metazoa. In: HUXLEY, J. S., HARDY, A. C. and FORD, E. B. (eds.), *Evolution as a process*. London: Allen and Unwin. Pp. 24–33.

— 1955 The continuity between the cavities of the premandibular somites and of Rathke's pocket in *Torpedo*. *Quart. J. micr. Sci.* **96**, 279–83.

DEFRETIN, R. 1949 Recherches sur la musculature des Néréidiens au cours de l'épitoquie, les glandes parapodiales et la spermiogénèse. *Ann. Inst. Océanogr., Monaco* **24**, 117–257.

DELSMAN, H. C. 1921–1922. The ancestry of the vertebrates as a means of understanding the principal features of their structure and development. *Natuurk. Tijdschr. Ned.-Ind.* **81**, 187–286; **82**, 34–89, 107–89.

DIESING, K. M. 1865 Revision der Prothelminthen. *S.B. Akad. Wiss. Wien, Math. Naturw. Kl.* **52**, 287–401.

DIETZ, P. A. 1913 Über die Form der Myotome der Teleostier und ihre Beziehung zur ausseren Liebesgestalt. *Anat. Anz.* **44**, 56–64.

DOHRN, A. 1875 *Der Ursprung der Wirbelthiere und das Princip des Functionswechsels*. Leipzig: Engelmann.

DRACH, P. 1948 La notion de Procordé et les embranchements de Cordés. In: GRASSÉ, P. P. (ed.), *Traité de zoologie*. Paris: Masson. **11**, 545–51.

EGGERS, F. 1924 Zur Bewegungsphysiologie der Nemertinen. I. *Emplectonema. Z. vergl. Physiol.* **1**, 579–87.

— 1935 Zur Bewegungsphysiologie von *Malacobdella grossa* Müll. *Z. wiss. Zool.* **147**, 101–31.

EISIG, H. 1887 Monographie der Capitelliden des Golfes von Neapel. *Fauna u. Flora Neapel* **16**.

— 1914 Zur Systematik, Anatomie und Morphologie der Ariciiden nebst Beiträgen zur generallen Systematik. *Mitt. zool. Sta. Neapel* **21**, 153–600.

EKMAN, T. 1896 Beiträge zur Kenntnis des Stieles der Brachiopoden. *Z. wiss. Zool.* **62**, 169–249.

EMELIANOV, S. W. 1935 Die Morphologie der Fischrippen. *Zool. Jb. (Anat.)* **60**, 133–262.

FAGE, L. and LEGENDRE, R. 1927 Pêches planctoniques à la lumière effectuées à Banyuls sur Mer et à Concarneau. *Arch. Zool. exp. gén.* **67**, 23–222.

FARRE, A. 1837 Observations on the minute structure of some of the higher forms of polypi. *Philos. Trans.* **127**, 387–426.

FAURÉ-FRÉMIET, E. and GARRAULT, H. 1944 Propriétés physiques de l'ascarocollagen. *Bull. biol.* **78**, 206–14.

FAUROT, L. 1895 Études sur l'anatomie, l'histologie et le développement des Actinies. *Arch. Zool. exp. gén.* iii, **3**, 43–262.

FAUSSEK, V. 1899 Über die physiologische Bedeutung des Cöloms. *Trav. Soc. Nat. St-Pétersb.* **30**, 40–57. (In Russian, German summary pp. 83–4).

— 1911 Vergleichend-embryologische Studien. (Zur Frage über die Bedeutung der Cölom-hohlen). *Z. wiss. Zool.* **98**, 529–625.

FENN, W. O. 1923 A quantitative comparison between the energy liberated and the work performed by the isolated sartorius muscle of the frog. *J. Physiol.* **58**, 175–203.

FISHER, W. K. 1954 A swimming *Sipunculus*. *Ann. Mag. nat. Hist.* xii, **7**, 238–40.

— and MACGINITIE, G. E. 1928a A new echiuroid worm from California. *Ann. Mag. nat. Hist.* x, **1**, 199–204.

— and MACGINITIE, G. E. 1928b The natural history of an echiuroid worm. *Ann. Mag. nat. Hist.* x, **1**, 204–13.
FORDHAM, M. G. C. 1925 *Aphrodite aculeata*. *L.M.B.C. Mem.* **27**.
FOX, H. M. 1938 On the blood circulation and metabolism of sabellids. *Proc. roy. Soc. B*, **125**, 554–69.
— 1952 Anal and oral intake of water by Crustacea. *J. exp. Biol.* **29**, 583–99.
FOXON, G. E. H. 1936 Observations on the locomotion of some arthropods and annelids. *Ann. Mag. nat. Hist.* x, **18**, 403–19.
FRAENKEL, G. S. and GUNN, D. L. 1940 *The orientation of animals*. Oxford: Clarendon Press.
FRANÇOIS, P. 1891 Choses de Nouméa. Observations biologiques sur les Lingules. *Arch. Zool. exp. gén.* ii, **9**, 231–9.
FRETTER, V. and GRAHAM, A. 1962 *British prosobranch molluscs*. London: Ray Soc.
FRIEDLÄNDER, B. 1894 Beiträge zur Physiologie der Zentralnervensystems und Bewegungsmechanismus der Regenwürmer. *Pflüg. Arch. ges. Physiol.* **58**, 168–206.
FRIEDRICH, H. 1933 Vergleichende Studien zur Bewegungs- und Nervenphysiologie bei Nemertinen. *Zool. Jb. (Physiol.)* **52**, 537–60.
— and LANGELOH, H. P. 1936 Untersuchungen zur Physiologie der Bewegung und des Hautmuskelschlauches bei *Halicryptus spinulosus* und *Priapulus caudatus*. *Biol. Cbl.* **56**, 249–60.
GAMBLE, F. W. 1893 Contributions to a knowledge of British marine Turbellaria. *Quart. J. micr. Sci.* **34**, 433–528.
GANSEN-SEMAL, P. van 1960 Occurrence of a non-fibrillar elastin in the earthworm. *Nature, Lond.* **186**, 654–5.
GARSTANG, W. 1894 Preliminary note on a new theory of the phylogeny of the Chordata. *Zool. Anz.* **17**, 122–5.
— 1922 The theory of recapitulation: a critical re-statement of the biogenetic law. *J. Linn. Soc. (Zool.)* **35**, 81–101.
— 1928 The morphology of the Tunicata, and its bearings on the phylogeny of the Chordata. *Quart. J. micr. Sci.* **72**, 51–187.
GASKELL, W. H. 1890 The origin of vertebrates from a crustacean-like ancestor. *Quart. J. micr. Sci.* **31**, 379–441.
— 1908 *The origin of the vertebrates*. London: Longmans, Green.
GEIGY, R. 1931 Action de l'ultra-violet sur le pôle germinal dans l'œuf de *Drosophila melanogaster* (castration et mutabilité). *Rev. suisse Zool.* **38**, 187–288.
GEORGE, J. C. and NAIK, R. M. 1958a Relative distribution and chemical nature of the fuel store of two types of fibres in the pectoralis major muscle of the pigeon. *Nature, Lond.* **181**, 709.
— and NAIK, R. M. 1958b Relative distribution of the mitochondria in the two types of fibres in the M. pectoralis major of the pigeon. *Nature, Lond.* **181**, 783.
GEROULD, J. H. 1896 The anatomy and histology of *Caudina arenata* Gould. *Bull. Mus. comp. Zool. Harv.* **29**, 123–90. Also *Proc. Boston nat. Soc.* **27**, 7–74.
GERSCH, M. 1934 Zur experimentellen Veränderung der Richtung der Wellenbewegung auf der Kreichsohle von Schnecken und zur Ruckwärtsbewegung von Schnecken. *Biol. Cbl.* **54**, 511–8.

GILCHRIST, J. D. F. 1915 Observations on the Cape *Cephalodiscus* (*C. gilchristi*) and some of its early stages. *Ann. Mag. nat. Hist.* viii, **16**, 233–43.

GISLÉN, T. 1930 Affinities between the Echinodermata, Enteropneusta and Chordonia. *Zool. Bidr. Uppsala* **12**, 199–304.

GONTCHAROFF, M. 1961 Embranchement des Némertiens. (Nemertini G. Cuvier 1817; Rhynchocoela M. Schultze 1851). In: GRASSÉ, P. P. (ed.), *Traité de zoologie*. Paris: Masson. **4** (1), 783–886.

GOODRICH, E. S. 1897a On the nephridia of the Polychaeta. Part I. On *Hesione*, *Tyrrhenia* and *Nephthys*. *Quart. J. micr. Sci.* **40**, 185–95.

— 1897b Notes on the anatomy of *Sternaspis*. *Quart. J. micr. Sci.* **40**, 233–45.

— 1903 On the body cavities and nephridia of the actinotrocha larva. *Quart. J. micr. Sci.* **47**, 103–21.

— 1909 Notes on the nephridia of *Dinophilus* and of the larvae of *Polygordius*, *Echiurus* and *Phoronis*. *Quart. J. micr. Sci.* **54**, 111–8.

— 1917 'Proboscis pores' in craniate vertebrates, a suggestion concerning the premandibular somites and hypophysis. *Quart. J. micr. Sci.* **62**, 539–53.

— 1930 *Studies on the structure and development of vertebrates*. London: Macmillan.

— 1946 The study of nephridia and genital ducts since 1895. *Quart. J. micr. Sci.* **86**, 113–392.

GRAHAM-SMITH, W. 1936 The tail of fishes. *Proc. zool. Soc. Lond.* 1936, 595–608.

GRATIOLET, P. 1862 Recherches sur l'organisation du système vasculaire dans la sangsue médicinale et l'aulastome vorace. *Ann. Sci. nat. Zool.* **17**, 174–225.

GRAY, J. 1928 *Ciliary Movement*. Cambridge: University Press.

— 1933a Studies in animal locomotion. I. The movement of fish with special reference to the eel. *J. exp. Biol.* **10**, 88–104.

— 1933b Studies in animal locomotion. II. The relationship between waves of muscular contraction and the propulsive mechanism of the eel. *J. exp. Biol.* **10**, 386–90.

— 1933c Studies in animal locomotion. III. The propulsive mechanism of the whiting (*Gadus merlangus*). *J. exp. Biol.* **10**, 391–400.

— 1936 Studies in animal locomotion. IV. The neuromuscular mechanism of swimming in the eel. *J. exp. Biol.* **13**, 170–80.

— 1939 Studies in animal locomotion. VIII. *Nereis diversicolor*. *J. exp. Biol.* **16**, 9–17.

— 1946 The mechanism of locomotion in snakes. *J. exp. Biol.* **23**, 101–20.

— 1950 The role of peripheral sense organs during locomotion in the vertebrates. *Symp. Soc. exp. Biol.* **4**, 112–26.

— 1951 Undulatory propulsion in small organisms. *Nature, Lond.* **168**, 929–30.

— 1953a Undulatory propulsion. *Quart. J. micr. Sci.* **94**, 551–78.

— 1953b The locomotion of fishes. In: *Essays in marine biology, being the Richard Elmhirst memorial lectures*. Edinburgh: Oliver and Boyd. Pp. 1–16.

— 1955 The movement of sea-urchin spermatozoa. *J. exp. Biol.* **32**, 775–801.

— and HANCOCK, G. J. 1955 The propulsion of sea-urchin spermatozoa. *J. exp. Biol.* **32**, 802–14.

— and LISSMANN, H. W. 1938 Studies in animal locomotion. VII. The earthworm. *J. exp. Biol.* **15**, 506–17.

— and LISSMANN, H. W. 1950 The kinetics of locomotion of the grass-snake. *J. exp. Biol.* **26**, 354–67.

— LISSMANN, H. W. and PUMPHREY, R. J. 1938 The mechanism of locomotion in the leech (*Hirudo medicinalis* Ray). *J. exp. Biol.* **15**, 408–30.

— and SAND, A. 1936 The locomotory rhythm of the dogfish (*Scyllium caniculata*). *J. exp. Biol.* **13**, 200–9.

GREENE, C. W. and GREENE, C. H. 1913 The skeletal musculature of the king salmon. *Bull. U.S. Bur. Fish.* **33**, 21–59.

GROBBEN, K. 1908 Die systematische Einteilung des Tierreichs. *Verh. zool.-bot. Ges. Wien* **58** (10), 491–511.

— 1923 Theoretische Erörterungen begriffend die phylogenetische Ableitung der Echinodermen. *S.B. Akad. wiss. Wien, Math. Naturw. Kl.* **132**, 263–90.

GROVE, A. J. and NEWELL, G. E. 1936 A mechanical investigation into the effectual action of the caudal fin of some aquatic chordates. *Ann. Mag. nat. Hist.* x, **17**, 280–90.

— and NEWELL, G. E. 1939 The relation of the tail form in cyclostomes and fishes to specific gravity. *Ann. Mag. nat. Hist.* xi, **4**, 401–30.

GUBERLET, J. E. 1933 Observations on the spawning and development of some Pacific annelids. *Proc. V Pacific Sci. Congr.* **5**, 4213–20.

GULLAND, G. L. 1885 Evidence in favour of the view that the coxal gland of *Limulus* and of other Arachnida is a modified nephridium. *Quart. J. micr. Sci.* **25**, 511–20.

GUSTAVSON, K. H. 1957 Some new aspects of the stability and reactivity of collagen. In: TUNBRIDGE, R. E. (ed.), *Connective tissue, a symposium*. Oxford: Blackwell. Pp. 185–207.

HADŽI, J. 1944 Turbelarijska teorija knidarijev. (Die Turbellarien-Theorie der Cnidarier und ihre Stellung im zoologischen System). *Razpr. mat. prir. Akad. Ljubljana* **3**, 1–239.

— 1953 An attempt to reconstruct the system of animal classification. *Syst. Zool.* **2**, 145–54.

— 1958 Zur Diskussion über die Abstammung der Metazoen. *Zool. Anz.* suppl. 21, 169–79.

HAECKEL, E. 1872 *Die Kalkschwämme*. Berlin: Reimer.

— 1874a Die Gastraea-Theorie, die phylogenetische Classification des Thierreichs und die Homologie der Keimblätter. *Jena. Z. Naturw.* **8**, 1–55.

— 1874b The gastraea-theory, the phylogenetic classification of the animal kingdom and the homology of the germ-lamellae. *Quart. J. micr. Sci.* (n.s.) **14**, 142–65; 223–47. (Translation of Haeckel, 1874a).

— 1875 Die Gastrula und die Eifurchung der Thiere. *Jena. Z. Naturw.* **9**, 402–508.

HAMMOND, R. A. 1962 Personal communication.

HALLEZ, P. 1879 *Contributions à l'histoire naturelle des Turbellariés*. Lille: Danel.

HANCOCK, A. 1858 On the organisation of the Brachiopoda. *Philos. Trans.* **148**, 791–869.

HAND, C. 1959 The origin and phylogeny of the coelenterates. *Syst. Zool.* **8**, 191–202.

HANSON, E. D. 1958 On the origin of the Eumetazoa. *Syst. Zool.* **7**, 16–47.

HARMER, S. F. 1896 Polyzoa. In: HARMER, S. F. and SHIPLEY, A. E. (eds.), *Cambridge natural history*. London: Macmillan. **2**, 465–533.

— 1902 On the morphology of the Cheilostomata. *Quart. J. micr. Sci.* **46**, 263–350.
— 1905 The Pterobranchia of the Siboga expedition. *Siboga Exped.* **26b**, 1–132.
— 1930 Presidential address. Polyzoa. *Proc. Linn. Soc. Lond.* **141**, 68–118.
HARRIS, J. E. 1936 The role of the fins in the equilibrium of the swimming fish. I. Wind-tunnel tests on a model of *Mustelus canis* (Mitchell). *J. exp. Biol.* **13**, 476–93.
— 1937 The mechanical significance of the position and movements of the paired fins in the Teleostei. *Pap. Tortugas Lab.* **31**, 171–89.
— 1938 The role of the fins in the equilibrium of the swimming fish. II. The role of the pelvic fins. *J. exp. Biol.* **15**, 32–47.
— 1950 *Diademodus hydei*, a new fossil shark from the Cleveland shale. *Proc. zool. Soc. Lond.* **120**, 683–97.
— 1953 Fin patterns and mode of life in fishes. In: *Essays in marine biology, being the Richard Elmhirst memorial lectures*. Edinburgh: Oliver and Boyd. Pp. 17–28.
— 1962 Personal communication.
— and CROFTON, H. D. 1957 Structure and function in the nematodes: internal pressure and cuticular structure in *Ascaris*. *J. exp. Biol.* **34**, 116–30.
— and WHITING, H. P. 1954 Structure and function in the locomotory system of the dogfish embryo. The myogenic stage of movement. *J. exp. Biol.* **31**, 501–24.
HARTMAN, O. 1954 Pogonophora Johansson, 1938. *Syst. Zool.* **3**, 183–5.
HARTOG, M. 1896 Rotifera, Gastrotricha and Kinorhyncha. In: HARMER, S. F. and SHIPLEY, A. E. (eds.), *Cambridge natural history*. London: Macmillan. **2**, 197–238.
HATSCHEK, B. 1877 Embryonalentwicklung und Knospung der *Pedicellina echinata*. *Z. wiss. Zool.* **29**, 502–49.
— 1878 Studien über Entwicklungsgeschischte der Anneliden. Ein Beitrag zur Morphologie der Bilaterien. *Arb. zool. Inst. Wien* **1**, 277–404.
— 1888–1891 *Lehrbuch der Zoologie*. Jena: Fischer.
HEALEY, E. G. 1957 The nervous system. In: BROWN, M. E. (ed.), *The physiology of fishes*. New York: Academic Press. **2**, 1–119.
HEIDER, K. 1914 Phylogenie der Wirbellosen. In: HERTWIG, R. and WETTSTEIN, R. von (eds.), *Abstammungslehre, Systematik, Paläontologie, Biogeographie*. (*Die Kultur der Gegenwart*, Teil 3, Abt. 4, Bd. 4). Leipzig-Berlin: Teubner. Pp. 453–529.
HEMPELMANN, F. 1931 Archiannelida und Polychaeta. In: KÜKENTHAL, W. and KRUMBACH, T. (eds.), *Handbuch der Zoologie*. Berlin-Leipzig: De Gruyter. **2** (2:7), 1–212.
HERLAND-MEEWIS, H. 1958 La reproduction asexuée chez les Annélides. *Année biol.* **62**, 133–66.
HERTER, K. 1928 Bewegungsphysiologische Studien an dem Egel *Hemiclepsis marginata* O. F. Müller mit besonderer Berucksichtigung der Thermokinese. *Z. vergl. Physiol.* **7**, 571–605.
— 1929 Vergleichende bewegungsphysiologische Studien an deutschen Egeln. *Z. vergl. Physiol.* **9**, 145–77.
HERTWIG, O. and HERTWIG, R. 1882 Die Coelomtheorie. Versuch einer Erklärung des mittleren Keimblattes. *Jena. Z. Naturw.* **15**, 1–150.
HESSLE, C. 1917 Zur Kenntnis der terebellomorphen Polychaeten. *Zool. Bidr. Uppsala* **5**, 39–258.

HEYMONS, R. 1891 Die Entwicklungsgeschischte der weiblichen Geschlechtsorgane von *Phyllodromia (Blatta) germanica* L. *Z. wiss. Zool.* **53**, 434–536.
HILL, A. V. 1939 The mechanical efficiency of frog's muscle. *Proc. roy. Soc.* B, **127**, 434–51.
HILTON, A. W. 1922 The nervous system of Phoronidea. *J. comp. Neurol.* **34**, 381–9.
HINTON, H. E. 1955 On the structure, function and distribution of the prolegs of the Panorpoidea, with a criticism of the Berlese-Imms theory. *Trans. R. ent. Soc. Lond.* **106**, 455–540.
— and LINN, I. J. 1950 'Kinetics of locomotion of *Tabanus* larvae.' Unpublished.
HIS, W. 1874 *Unsere Körperform.* Leipzig: Vogel.
HOFFMANN, H. 1929–1930 Amphineura und Scaphopoda. Nachträge. In: BRONN, H. G., *Klassen und Ordnungen des Tierreichs.* Leipzig: Akad. Verlagsgesellschaft. **3** (1), 1–511.
HOLST, E. von 1935 Erregungsbildung und Erregungsleitung im Fischrückmark. *Pflüg. Arch. ges. Physiol.* **235**, 345–59.
HORNY, R. 1957 Problematic molluscs (? Amphineura) from the lower Cambrian of south and east Siberia. *Sborn. Ustredniho Ustavo geol.* **23**, 397–432.
HORST, C. J. van der 1927–1939 Hemichordata. In: BRONN H. G., *Klassen und Ordnungen des Tierreichs.* Leipzig: Akad. Verlagsgesellschaft. **4** (4:ii), 601–737.
— 1940 The Enteropneusta from Inyack Island, Delagoa Bay. *Ann. S. Afr. Mus.* **32**, 293–380.
HOWELL, A. B. 1933 The architecture of the pectoral appendage of the dogfish. *J. Morph.* **54**, 399–413.
HOYLE, G. 1957 *Comparative physiology of the nervous control of muscular contraction.* Cambridge: University Press.
HUBRECHT, A. A. W. 1883 On the ancestral form of the Chordata. *Quart. J. micr. Sci.* **23**, 349–68.
— 1887 The relation of the Nemertinea to the Vertebrata. *Quart. J. micr. Sci.* **27**, 605–44.
— 1904 Die Abstammung der Anneliden und Chordaten und die Stellung der Ctenophoren und Plathelminthen im System. *Jena. Z. Naturw.* **39**, 151–76.
HUXLEY, T. H. 1853 *Lacinularia socialis.* A contribution to the anatomy and physiology of the Rotifera. *Trans. micr. Soc.* (n.s.), **1**, 1–19.
HYMAN, L. H. 1940 *The Invertebrates: Protozoa through Ctenophora.* New York: McGraw-Hill.
— 1951a *The Invertebrates: Platyhelminthes and Rhynchocoela. The acoelomate Bilateralia.* New York: McGraw-Hill.
— 1951b *The Invertebrates: Acanthocephala, Aschelminthes, and Entoprocta.* New York: McGraw-Hill.
— 1955 *The Invertebrates: Echinodermata.* New York: McGraw-Hill.
— 1958 The occurrence of chitin in the lophophorate phyla. *Biol. Bull. Woods Hole* **114**, 106–12.
— 1959 *The Invertebrates: smaller coelomate groups.* New York: McGraw-Hill.
IJIMA, I. 1887 Über einige Tricladen Europa's. *J. Coll. Sci. Tokyo* **1**, 337–58.
IKEDA, I. 1901 Observations on the development, structure and metamorphosis of Actinotrocha. *J. Coll. Sci. Tokyo* **13**, 507–92.

INADA, C. 1950 Unpublished. (Cited by PROSSER, 1950.)
IRWIN-SMITH, V. A. 1918 On the Chaetosomatidae, with descriptions of new species, and a new genus from the coast of New South Wales. *Proc. Linn. Soc. N.S.W.* **42,** 757–814.
IVANOV, A. V. 1952 New Pogonophora from the far eastern seas. *Zool. Zh.* **31,** 372–91. (In Russian).
— 1955a On external digestion in the Pogonophora. *C.R. Acad. Sci. U.S.S.R.* **100,** 381–3. (In Russian, transl. in *Syst. Zool.* **4,** 174–6).
— 1955b On the assignment of the class Pogonophora to a separate phylum of the Deuterostomia-Brachiata A. Ivanov, phyl. nov. *C.R. Acad. Sci. U.S.S.R.* **100,** 595–6. (In Russian, transl. in *Syst. Zool.* **4,** 177–8).
— 1960 Embranchement des Pogonophores. In: GRASSÉ, P. P. (ed.), *Traité de zoologie.* Paris: Masson. **5** (2), 1521–1622.
IWANOFF, P. P. 1928 Die Entwicklung der Larvalsegmente bei den Anneliden. *Z. Morph. Ökol. Tiere* **10,** 62–161.
— 1933 Die embryonale Entwicklung von *Limulus moluccanus. Zool. Jb. (Anat.)* **56,** 163–348.
— 1944 Primary and secondary metamery of the body. *J. gen. Biol., Moscow* **5,** 61–95. (In Russian with English summary).
JÄGERSTEN, G. 1947 On the structure of the pharynx of the Archiannelida with special reference to there-occurring muscle cells of aberrant type. *Zool. Bidr. Uppsala* **25,** 551–70.
— 1955 On the early phylogeny of the Metazoa. The bilaterogastraea theory. *Zool. Bidr. Uppsala* **30,** 321–54.
— 1956 Investigation on *Siboglinum ekmani* n.sp., encountered in the Skagerak, with some general remarks on the group Pogonophora. *Zool. Bidr. Uppsala* **31,** 211–52.
— 1959 Further remarks on the early phylogeny of the Metazoa. *Zool. Bidr. Uppsala* **33,** 79–108.
JARMAN, G. M. 1961 A note on the shape of fish myotomes. *Symp. zool. Soc. Lond.* **5,** 33–5.
JENNINGS, J. B. 1957 Studies on feeding, digestion, and food storage in free-living flatworms (Platyhelminthes-Turbellaria). *Biol. Bull. Woods Hole* **112,** 63–84.
— 1959 Observations on the nutrition of the land planarian *Orthodemus terrestris* (O. F. Müller). *Biol. Bull. Woods Hole* **117,** 119–24.
JENSEN, D. D. 1960 Hoplonemertines, Myxinoids and deuterostome origins. *Nature, Lond.* **188,** 649–50.
JHERING, H. von 1877 *Vergleichende Anatomie des Nervensystems und Phylogenie der Mollusken.* Leipzig: Englemann.
JOHANSSON, K. E. 1937 Über *Lamellisabella zachsi* und ihre systematische Stellung. *Zool. Anz.* **117,** 23–6.
— 1939 *Lamellisabella zachsi* Uschakow, ein Vertreter einer neuer Tierklasse Pogonophora. *Zool. Bidr. Uppsala* **18,** 253–68.
JORDON, H. 1901 Die Physiologie der Lokomotion bei *Aplysia limacina. Z. Biol.* **41,** 196–238.
— 1905 The physiology of locomotion in Gasteropods. A reply to A. J. Carlson. *Biol. Bull. Woods Hole* **9,** 138–40.

JULLIEN, J. 1888a Sur la sortie et la rentrée du polypide dans les zooécies chez les Bryozoaires cheilostomiens monodermiés. *Bull. Soc. zool. Fr.* **13**, 67–8.
— 1888b Observations anatomiques sur les Caténicelles. *Mém. Soc. zool. Fr.* **1**, 274–80.
KAISER, F. 1954 Beiträge zur Bewegungsphysiologie der Hirudineen. *Zool. Jb. (Physiol.)* **65**, 59–90.
KARLING, T. 1940 Zur Morphologie und Systematik der Alloeocoela Annulata und Rhabdocoela Lecithophora. *Acta zool. fenn.* **26**, 1–260.
KAWAGUTI, S. 1941 Haemerythrin in *Lingula*. *Mem. Fac. Sci. Agric. Taihoku (Zool.)* **23**, 95–8.
KERKUT, G. A. 1953 The forces exerted by the tube feet of the starfish during locomotion. *J. exp. Biol.* **30**, 575–83.
KERMACK, K. A. 1943 The functional significance of the hypocercal tail in *Pteraspis rostrata*. *J. exp. Biol.* **20**, 23–7.
KILIAN, R. 1932 Zur Morphologie und Systematik der Gigantorhynchidae. *Z. wiss. Zool.* **141**, 246–345.
KISHINOUYE, K. 1923 Contributions to the comparative study of so-called scombroid fishes. *J. Coll. Agric. Tokyo* **8**, 293–475.
KLEINENBERG, N. 1886 Die Entstehung des Annelides aus der Larve von *Lopadorhynchus*. *Z. wiss. Zool.* **44**, 1–227.
KNIGHT-JONES, E. W. 1952 On the nervous system of *Saccoglossus cambrensis* (Enteropneusta). *Philos. Trans.* B, **236**, 315–54.
— 1953 Feeding in *Saccoglossus* (Enteropneusta). *Proc. zool. Soc. Lond.* **123**, 637–54.
KOMAI, T. 1949 The internal structure of *Atubaria heterolopha* Sato, with an appendix on the homology of the 'notochord'. *Proc. imp. Acad. Japan* **25** (7), 19–24.
KOZLOWSKI, R. 1947 Les affinités des Graptolithes. *Biol. Rev.* **22**, 93–108.
— 1948 Les Graptolithes du Trémadoc de la Pologne. *Palaeont. polon.* **3**, 195–203.
LAM, H. J. 1920 Ueber den Bau und die Verwandtschaft der *Protannelis meyeri* nov. gen., nov. spec., eine neue Archiannelide. *Tijdschr. ned. dierk. Ver.* ii, **18**, 44–84.
LAMEERE, A. 1932a Origine du coelome. *Arch. Zool. (ital.), Napoli* **16**, 197–206.
— 1932b *Précis de zoologie*. Liège: Desoer. **2**.
LANG, A. 1881 Der Bau von *Gunda segmentata* und die Verwandtschaft der Plathelminthen mit Coelenteraten und Hirudineen. *Mitt. zool. Sta. Neapel* **3**, 187–251.
— 1894 Die Polycladen (Seeplanarien) des Golfes von Neapel und der anfresende Meeresabschnitte. *Fauna u. Flora Neapel* **11**.
— 1903 Beiträge zu einer Trophocoltheorie. *Jena. Z. Naturw.* **38**, 1–373.
LANG, K. 1953 Die Entwicklung des Eies von *Priapulus caudatus* Lam. und die systematische Stellung der Priapuliden. *Ark. Zool.* ii, **5**, 321–48.
LANGELAAN, J. W. 1905 On the form of the trunk-myotome. *Proc. Kon. Akad. Wetensch. Amsterdam* **7**, 34–40.
LANKESTER, E. R. 1874 Observations on the development of the pond snail (*Lymnaea stagnalis*), and on the early stages of other Mollusca. *Quart. J. micr. Sci.* **14**, 365–91.
— 1875a On the invaginate planula, or diploblastic phase of *Paludina vivipara*. *Quart. J. micr. Sci.* **15**, 159–66.
— 1875b Contributions to the developmental history of the Mollusca. *Philos. Trans.* B, **165**, 1–45.

— 1876 An account of Professor Haeckel's recent additions to the gastraea theory. *Quart. J. micr. Sci.* (n.s.) **16**, 51–66. (Summary of Haeckel, 1875).
— 1877 Notes on the embryology and classification of the animal kingdom; comprising a revision of speculations relative to the origin and significance of the germ layers. *Quart. J. micr. Sci.* **17**, 399–454.
— 1884 The supposed taking-in and shedding-out of water in relation to the vascular system of molluscs. *Zool. Anz.* **7**, 343–6.
— 1893 Note on the coelom and vascular system of Mollusca and Arthropoda. *Quart. J. micr. Sci.* **34**, 427–32.
— 1900 The Enterocoela and the Coelomocoela. In: LANKESTER, E. R. (ed.), *A treatise on zoology*. London: Black. **2**, 1–37.
— 1904 The structure and classification of the Arthropoda. *Quart. J. micr. Sci.* **47**, 523–82.

LEHNERT, G. H. 1891 Beobachtungen an Landplanarien. *Arch. Naturgesch.* **57**. Jhg. 306–50.

LEMCHE, H. 1957 A new living deep-sea mollusc of the Cambro-Devonian class Monoplacophora. *Nature, Lond.* **179**, 413–6.
— 1959 Protostomian relationships in the light of *Neopilina*. *Proc. XV int. Congr. Zool.* pp. 381–9.
— 1960 A possible central place for *Stenothecoides* Resser, 1939 and *Cambridium* Horny, 1957 (Mollusca Monoplacophora) in invertebrate phylogeny. *Proc. int. pal. Un.* (*XXI int. geol. Congr.*) pt. 22, 92–101.
— and WINGSTRAND, K. G. 1959 The anatomy of *Neopilina galatheae* Lemche, 1957. *Galathea Rep.* **3**, 9–71.
— and WINGSTRAND, K. G. 1960 Classe des Monoplacophores. Monoplacophora (Odhner, 1940). In: GRASSÉ, P. P. (ed.), *Traité de zoologie*. Paris: Masson. **5** (2), 1787–1821.

LEUCKHART, R. 1848 *Ueber die Morphologie und die Verwandtschaftsverhältnisse der Wirbellosen Thiere*. Braunschweig. (Cited by Hyman, 1959, and Sarvaas, 1933.)

LISSMANN, H. W. 1945a The mechanism of locomotion in gastropod molluscs. I. Kinematics. *J. exp. Biol.* **21**, 58–69.
— 1945b The mechanism of locomotion in gastropod molluscs. II. Kinetics. *J. exp. Biol.* **22**, 37–50.
— 1946 The neurological basis of the locomotory rhythm in the spinal dogfish (*Scyllium caniculum, Acanthias vulgaris*). II. The effect of de-afferentiation. *J. exp. Biol.* **23**, 162–76.
— 1950 Rectilinear locomotion in a snake (*Boa occidentalis*). *J. exp. Biol.* **26**, 368–79.

LOTMAR, W. and PICKEN, L. E. R. 1950 A new crystallographic modification of chitin and its derivatives. *Experientia* **6**, 58–9.

MACBRIDE, E. W. 1896 The development of *Asterina gibbosa*. *Quart. J. micr. Sci.* **38**, 339–411.
— 1914 *A text-book of embryology*. London: Macmillan. **1**.

MALAQUIN, A. 1893 *Recherches sur les Syllidiens. Morphologie, anatomie, reproduction et développement*. Lille: Danel.

MANN, K. H. 1962 *Leeches (Hirudinea), their structure, physiology, ecology and embryology.* Oxford: Pergamon.

MANTON, S. M. 1949 Studies on the Onychophora. VII. The early embryonic stages of *Peripatopsis*, and some general considerations concerning the morphology and phylogeny of the Arthropoda. *Philos. Trans.* B, **233**, 483–580.

— 1950 The evolution of arthropodan locomotory mechanisms. Part I. The locomotion of *Peripatus. J. Linn. Soc. (Zool.)* **41**, 529–70.

— 1952a The evolution of arthropodan locomotory mechanisms. Part 2. General introduction to the locomotory mechanisms of the Arthropoda. *J. Linn. Soc. (Zool.)* **42**, 93–117.

— 1952b The evolution of arthropodan locomotory mechanisms. Part 3. The locomotion of the Chilopoda and Pauropoda. *J. Linn. Soc. (Zool.)* **42**, 118–66.

— 1953 Locomotory habits and the evolution of the larger arthropodan groups. *Symp. Soc. exp. Biol.* **7**, 339–76.

— 1954 The evolution of arthropodan locomotory mechanisms. Part 4. The structure, habits and evolution of the Diplopoda. *J. Linn. Soc. (Zool.)* **42**, 299–368.

— 1957 The evolution of arthropodan locomotory mechanisms. Part 5. The structure, habits and evolution of the Pselaphognatha (Diplopoda). *J. Linn. Soc. (Zool.)* **43**, 153–87.

— 1958a The evolution of arthropodan locomotory mechanisms. Part 6. Habits and evolution of the Lysiopetaloidea (Diplopoda), some principles of leg design in Diplopoda and Chilopoda, and limb structure of Diplopoda. *J. Linn. Soc. (Zool.)* **43**, 487–556.

— 1958b Habits of life and evolution of body design in Arthropoda. *J. Linn. Soc. (Zool.)* **44**, 58–72.

— 1958c Hydrostatic pressure and leg-extension in Arthropoda, with special reference to arachnids. *Ann. Mag. nat. Hist.* xiii, **1**, 161–82.

— 1958d Embryology of Pogonophora and classifications of animals. *Nature, Lond.* **181**, 748–51.

— 1961a The evolution of arthropodan locomotory mechanisms. Part 7. Functional requirements and body design in Colobognatha (Diplopoda), together with a comparative account of diplopod burrowing, trunk musculature and segmentation. *J. Linn. Soc. (Zool.)* **44**, 383–462.

— 1961b Experimental zoology and problems of arthropod evolution. In: RAMSAY, J. A. and WIGGLESWORTH, V. B. (eds.), *The cell and the organism.* Cambridge: University Press. Pp. 234–55.

MARCUS, E. 1926a Beobachtungen und Versuche an lebenden Meeresbryozoen. *Zool. Jb. (System.)* **52**, 1–102.

— 1926b Beobachtungen und Versuche an lebenden Süsswasserbryozoen. *Zool. Jb. (System.)* **52**, 279–350.

— 1939 Bryozoarios marinhos brasilieros, III. *Bol. Fac. Filos. Ciênc. S. Paulo, Zoologia* **3**, 111–353.

— 1941 Sôbre Bryozoa do Brasil. *Bol. Fac. Filos, Ciênc. S. Paulo, Zoologia* **5**, 3–208.

— 1958 On the evolution of the animal phyla. *Quart. Rev. Biol.* **33**, 24–58.

MAREY, E. J. 1894 *Le Mouvement*. (Paris). English edition 1895, *Movement* (Transl. E. Pritchard). London: Heinemann.

MARTINI, E. 1909 Über die Segmentierung des Appendicularienschwanzes. *Verh. dtsch. zool. Ges.* **19**, 300–7.

MAST, O. S. 1911 *Light and the behaviour of animals*. New York: Wiley.

MASTERMAN, A. T. 1896 Preliminary note on the structure and affinities of *Phoronis*. *Proc. roy. Soc. Edinb.* **21**, 59–71.

— 1897 On the Diplochorda. 1. The structure of *Actinotrocha*. 2. The structure of *Cephalodiscus*. *Quart. J. micr. Sci.* **40**, 281–366.

— 1898 On the theory of archimeric segmentation and its bearing upon the phyletic classification of the Coelomata. *Proc. roy. Soc. Edinb.* **22**, 270–310.

MCCONNAUGHEY, B. H. and FOX, D. L. 1949 The anatomy and biology of the marine polychaete *Thoracophelia mucronata* (Treadwell) Opheliidae. *Univ. Calif. Publ. Zool.* **47**, 319–40.

MCLENDON, J. F. 1906 On the locomotion of a sea-anemone (*Metridium marginatum*). *Biol. Bull. Woods Hole* **10**, 66–7.

MENZIES, R. J., EWING, M., WORZEL, J. L. and CLARKE, A. H. 1959 Ecology of the recent Monoplacophora. *Oikos* **10**, 168–82.

METSCHNIKOFF, E. 1883 Untersuchungen über die intracelluläre Verdauung bei wirbellosen Thieren. *Arb. zool. Inst. Wien* **5**, 141–68. (Transl. (1884) as: Researches on the intracellular digestion of invertebrates. *Quart. J. micr. Sci.* **24**, 89–111).

MEYER, E. 1890 Die Abstimmung der Anneliden. Der Ursprung der Metamerie und die Bedeutung des Mesoderms. *Biol. Cbl.* **10**, 296–308. (Transl. in *Amer. Nat.* **24**, 1143–65).

— 1901 Studien über die Körperbau der Anneliden. V. Das Mesoderm der Ringelwürmer. *Mitt. zool. Stat. Neapel* **14**, 247–585.

MICHAELSEN, W. 1914 Die Oligochaeten Colombias. *Mem. Soc. Neuchâtel. Sci. nat.* **5**, 202–52.

MINOT, C. S. 1897 Cephalic homologies. A contribution to the determination of the ancestry of vertebrates. *Amer. Nat.* **31**, 927–43.

MONTGOMERY, T. H. 1897 On the connective tissues and body cavities of the nemerteans, with notes on classification. *Zool. Jb. (Anat.)* **10**, 1–46.

MORGAN, T. H. 1894 The development of *Balanoglossus*. *J. Morph.* **9**, 1–86.

MORRIS, M. C. 1950 Dilation of the foot of *Uber* (*Polinices*) *stragei* (Mollusca, class Gastropoda). *Proc. Linn. Soc. N.S.W.* **75**, 70–80.

MORSE, E. S. 1902 Observations on living Brachiopoda. *Mem. Boston Soc. nat. Hist.* **5**, 313–86.

MORTON, J. E. 1954 The biology of *Limacina retroversa*. *J. mar. biol. Ass. U.K.* **33**, 297–312.

— 1960 The functions of the gut in ciliary feeders. *Biol. Rev.* **35**, 92–140.

— and HOLME, N. A. 1955 The occurrence at Plymouth of the opisthobranch *Akera bullata*, with notes on its habits and relationships. *J. mar. biol. Ass. U.K.* **34**, 101–12.

MOSAUER, W. 1932a Über die Ortsbewegung der Schlagen. Eine Kritik und Ergänzung der Arbeit Wiedemann's. *Zool. Jb. (Physiol.)* **52**, 191–215.

— 1932b On the locomotion of snakes. *Science* **76**, 583–5.

MOSELEY, H. N. 1874 On the anatomy and histology of the land-planarians of Ceylon, with some account of their habits, and a description of two new species, and with notes on the anatomy of some European aquatic species. *Philos. Trans.* **164**, 105–71.

— 1877a On *Stylochus pelagicus*, a new species of pelagic planarian, with notes on other pelagic species, on the larval forms of *Thysanozoon*, and of a gymnosomatous pteropod. *Quart. J. micr. Sci.* **17**, 23–34.

— 1877b Notes on the structures of several forms of land planarians, with a description of two new genera and several new species, and a list of all species at present known. *Quart. J. micr. Sci.* **17**, 273–92.

NAEF, A. 1924 Studien zur generellen Morphologie der Mollusken. 3. Teil: Die typischen Beziehungen der Weichtierklassen untereinander und das Verhältnis ihrer Urformen zu anderen Cölomaten. *Ergebn. Zool.* **6**, 27–124.

— 1931 Phylogenie der Tiere. In: BAUR, E. and HARTMANN, M. (eds.), *Handbuch der Vererbungswissenschaft*. Berlin: Borntraeger. **3**, 1–200.

NEWELL, G. E. 1948 A contribution to our knowledge of the life history of *Arenicola marina* L. *J. mar. biol. Ass. U.K.* **27**, 554–80.

— 1950 The role of the coelomic fluid in the movements of earthworms. *J. exp. Biol.* **27**, 110–21.

NICHOLS, D. 1959a The histology of the tube-feet and clavulae of *Echinocardium cordatum*. *Quart. J. micr. Sci.* **100**, 73–87.

— 1959b The histology and activities of the tube-feet of *Echinocyamus pusillus*. *Quart. J. micr. Sci.* **100**, 539–55.

— 1960 The histology and activities of the tube-feet of *Antedon bifida*. *Quart. J. micr. Sci.* **101**, 105–17.

— 1961 A comparative histological study of the tube-feet of two regular echinoids. *Quart. J. micr. Sci.* **102**, 157–80.

NICOL, E. A. T. 1931 The feeding mechanism, formation of the tube, and physiology of digestion in *Sabella pavonina*. *Trans. roy. Soc. Edinb.* **56**, 537–98.

NICOL, J. A. C. 1948 The giant axons of annelids. *Quart. Rev. Biol.* **23**, 291–323.

NITSCHE, H. 1871 Beiträge zur Kenntniss der Bryozoen. III. Ueber die Anatomie und Entwicklungsgeschischte von *Flustra membranacea*. *Z. wiss. Zool.* **21**, 1–53.

NOMURA, E. 1913 On two species of aquatic Oligochaeta, *Limnodrilus gotoi* Hatai and *Limnodrilus willeyi* n. sp. *J. Coll. Sci. Tokyo* **35** (4), 1–49.

— 1915 On the aquatic oligochaete *Monopylephorus limosus* (Hatai). *J. Coll. Sci. Tokyo* **35** (9), 1–46.

NURSALL, J. R. 1956 The lateral musculature and the swimming of fish. *Proc. zool. Soc. Lond.* **126**, 127–43.

— 1958 The caudal fin as a hydrofoil. *Evolution* **12**, 116–20.

ODELL, W. 1899 Notes on fresh-water Polyzoa. *Ottawa Nat.* **13**, 107–13.

OHUYE, T. 1936 On the coelomic corpuscles in the body fluid of some invertebrates. VI. A note on the formed elements in the coelomic fluid of a brachiopod, *Terebratalia coreanica*. *Sci. Rep. Tohoku Univ.* iv, **11**, 231–8.

— 1937 On the coelomic corpuscles in the body fluid of some invertebrates. VIII. Supplementary note on the formed elements in the coelomic fluid of some Brachiopods. *Sci. Rep. Tohoku Univ.* iv, **12**, 241–53.

OLMSTED, J. M. D. 1917a Notes on the locomotion of certain Bermudan mollusks. *J. exp. Zool.* **24**, 223–36.
— 1917b The comparative physiology of *Synaptula hydriformis* (Lessueur). *J. exp. Zool.* **17**, 333–79.
— 1922 The role of the nervous system in the locomotion of certain marine polyclads. *J. exp. Zool.* **36**, 57–66.
ORTON, J. H. 1914 On ciliary mechanisms in brachiopods and some polychaetes with a comparison of the ciliary mechanisms on the gills of molluscs, Protochordata, brachiopods and cryptocephalous polychaetes and an account of the crystalline style of *Crepidula* and its allies. *J. mar. biol. Ass. U.K.* **10**, 283–311.
PANTIN, C. F. A. 1960 Diploblastic animals. *Proc. Linn. Soc. Lond.* **171**, 1–14.
— and SAWAYA, P. 1953 Muscular action in *Holothuria grisea*. *Bol. Fac. Filos. Ciênc. S. Paulo, Zoologia* **18**, 51–9.
PARKER, G. H. 1911 The mechanism of locomotion in gastropods. *J. Morph.* **22**, 155–70.
— 1914 The locomotion of chiton. *Contr. Bermuda biol. Sta.* **2** (31), 1–2.
— 1917a Pedal locomotion in actinians. *J. exp. Zool.* **22**, 111–24.
— 1917b The pedal locomotion of the sea-hare *Aplysia californica*. *J. exp. Zool.* **24**, 139–45.
— 1921 The locomotion of the holothurian *Stichopus panamensis* Clark. *J. exp. Zool.* **33**, 205–8.
— and BURNETT, F. L. 1900 The reactions of planarians, with and without eyes, to light. *Amer. J. Physiol.* **4**, 373–85.
PARRY, D. A. and BROWN, R. H. J. 1959a The hydraulic mechanism of the spider leg. *J. exp. Biol.* **36**, 423–33.
— and BROWN, R. H. J. 1959b The jumping mechanism of salticid spiders. *J. exp. Biol.* **36**, 654–64.
PATTEN, W. 1890 On the origin of vertebrates from arachnids. *Quart. J. micr. Sci.* **31**, 317–78.
— 1912. *The evolution of the vertebrates and their kin.* Philadelphia: Blakiston.
PAX, F. 1954 Die Abstammung der Cölenteraten nach der Theorie von Jovan Hadži. *Naturw. Rdsch.* **7**, 288–90.
PEARL, R. 1903 The movements and reactions of fresh-water planarians: a study in animal behaviour. *Quart. J. micr. Sci.* **46**, 509–714.
PEARSE, A. S. 1908 Observations on the behavior of the holothurian, *Thione briareus* (Leseur). *Biol. Bull. Woods Hole* **15**, 259–88.
PEDERSEN, K. J. 1959a Some features of the fine structure and histochemistry of planarian sub-epidermal gland cells. *Z. Zellforsch.* **50**, 121–42.
— 1959b Cytological studies in the planarian neoblast *Z. Zellforsch.* **50**, 799–817.
PEEBLES, F. and FOX, D. L. 1933 The structure, functions and general reactions of the marine sipunculid worm, *Dendrostoma zostericola*. *Bull. Scripps Instn Oceanogr. tech.* **3**, 201–34.
PEJLER, B. 1957 On variation and evolution in planktonic Rotatoria. *Zool. Bidr. Uppsala* **32**, 1–66.
PELSENEER, P. 1898–1899 Recherches morphologiques et phylogénétiques sur les Mollusques archaïques. *Mém. Sav. étr. Acad. R. Belg.* **57** (3), 1–113.

— 1906 Mollusca. In: LANKESTER, E. R. (ed.), *A treatise on zoology*. London: Black. 5.

PERRIER, E. 1882 *Les colonies animals et la formation des organisms*. Paris: Masson.

PETTIGREW, J. B. 1873 *Animal locomotion*. London: King.

PICKEN, L. E. R. 1936 The mechanism of urine formation in invertebrates. I. The excretion mechanism in certain arthropods. *J. exp. Biol.* **13**, 309–28.

— 1937 A new species of rhabdocoel (*Ependytes*). *J. Linn. Soc. (Zool.)* **40**, 273–7.

— PRYOR, M. G. M. and SWANN, M. M. 1947 Orientation of fibrils in natural membranes. *Nature, Lond.* **159**, 434.

PORTMANN, A. 1926 Die Kriechbewegung von *Aiptasia carnea*. Ein Beitrag zur Kenntnis der neuromuskularen Organisation der Aktinien. *Z. vergl. Physiol.* **4**, 659–67.

— 1960 Généralitiés sur les Mollusques. In: GRASSÉ, P. P. (ed.), *Traité de zoologie*. Paris: Masson. **5** (2), 1625–54.

PRENANT, A. 1929 Recherches sur la structure des muscles des Annélides Polychètes et sur leur sarcolyse. *Arch. Zool. exp. gén.* **69**, 1–135.

PRENANT, M. 1922 Recherches sur la parenchyme des Plathelminthes. Essai d'histologie comparée. *Arch. Morph. gén. exp.* **5**, 1–175.

— 1928 Notes histologiques sur *Terebratulina caput-serpentis* L. *Bull. Soc. zool. Fr.* **53**, 113–25.

PROSSER, C. L. (ed.) 1950 *Comparative animal physiology*. Philadelphia-London: Saunders. 1st edition.

PRUVOT, G. 1885 Recherches anatomiques et morphologiques sur le système nerveux des Annélides Polychètes. *Arch. Zool. exp. gén.* ii, **3**, 211–336.

RAMSAY, J. A. 1949 The osmotic relations of the earthworm. *J. exp. Biol.* **26**, 46–56.

— 1953 Exchange of sodium and potassium in mosquito larvae. *J. exp. Biol.* **30**, 79–89.

RAO, K. P. 1953 The development of *Glandiceps* (Enteropneusta; Spengelidae). *J. Morph.* **93**, 1–17.

— 1954 Bionomics of *Ptychodera flava* Eschscholtz (Enteropneusta). *J. Madras Univ. B*, **24**, 1–5.

REED, R. and RUDALL, K. M. 1948 Electron microscope studies on the structure of earthworm cuticle. *Biochem. Biophys. Acta* **2**, 7–18.

REMANE, A. 1929 Gastrotricha. In: KÜKENTHAL, W. and KRUMBACH, T. (eds.), *Handbuch der Zoologie*. Berlin-Leipzig: De Gruyter. **2** (1:4), 121–86.

— 1929–1933 Rotatoria. In: BRONN, H. G., *Klassen und Ordnungen des Tierreichs*. Leipzig: Akad. Verlagsgesellschaft. **4** (2:i), 1–576.

— 1950 Die Entstehung der Metamerie der Wirbellosen. *Zool. Anz.* suppl. **14**, 16–23.

— 1952 *Die Grundlagen des natürlichen Systems, der vergleichenden Anatomie und der Phylogenetik*. Leipzig: Akad. Verlagsgesellschaft.

— 1954 Die Geschischte der Tiere. In: HEBERER, G. (ed.), *Die Evolution der Organismen*. Jena: Fischer. 2nd edition. **2**, 340–422.

— 1958 Zur Verwandtschaft und Ableitung der niederen Metazoen. *Zool. Anz.* suppl. **21**, 179–96.

RENSCH, B. 1959 *Evolution above the species level*. London: Methuen.

RIBAUCOURT, E. de 1902 Étude sur l'anatomie comparée des Lombricides. *Bull. Sci. Fr. Belg.* **35**, 211–312.

RIETSCH, M. 1882 Études sur *Sternaspis scutata*. *Ann. Sci. nat. Zool.* v, **13**, 1–84. (Summary in *Ann. Mag. nat. Hist.* v, **7**, 426–8; 493–5 (1883).)

RITTER, W. E. 1902 The movements of Enteropneusta and the mechanism by which they are accomplished. *Biol. Bull. Woods Hole* **3**, 255–61.

ROCKWELL, H., EVANS, F. G. and PHEASANT, H. C. 1938 The comparative morphology of the vertebrate spinal column. Its form as related to function. *J. Morph.* **63**, 87–117.

ROGICK, M. 1937 Studies on freshwater Bryozoa. VI. The finer anatomy of *Lophopodella carteri* var. *typica*. *Trans. Amer. micr. Soc.* **56**, 367–96.

ROULE, L. 1900 Étude sur le développement embryonnaire des Phoronidiens. *Ann. Sci. nat. Zool.* viii, **11**, 51–250.

RUDALL, K. M. 1955 The distribution of collagen and chitin. *Symp. Soc. exp. Biol.* **9**, 49–71.

SACHWATKIN, A. A. 1956 *Vergleichende Embryologie der niederen Wirbellosen*. Berlin: Deutsch. Verlag Wiss.

SAINT-JOSEPH, A. de 1898 Les Annélides Polychètes des côtes de France (Manche et Océan). *Ann. Sci. nat. Zool.* viii, **5**, 209–464.

SALENSKY, W. 1907 Morphogenetische Studien an Würmern. IV. Zur Theorie des Mesoderms. *Mém. Acad. Sci. St.-Pétersb.* viii, **19** (11), 265–340.

SARVAAS, A. E. DU MARCHIE 1933 *La théorie du coelome*. Thesis, University of Utrecht.

SATO, T. 1936 Vorläufige Mitteilung über *Atubaria heterolopha* gen. nov., sp. nov., einen in freien Zustand aufgefundenen Pterobranchier aus dem Stillen Ozean. *Zool. Anz.* **115**, 97–106.

SAVILLE-KENT, W. 1880–1882 *A manual of the Infusoria*. London: Bogue.

SCHAEFFER, C. 1926 Untersuchungen zur vergleichenden Anatomie und Histologie der Brachiopodengattung *Lingula*. *Acta. zool., Stockh.* **7**, 329–402.

SCHEPOTIEFF, A. 1907a Die Pterobranchier. Anatomische und histologische Untersuchungen über *Rhabdopleura normanii* Allman und *Cephalodiscus dodecalophus* M'Int. 1. Teil. *Rhabdopleura normanii* Allman. 1. Abschnitt. Die Anatomie von *Rhabdopleura*. *Zool. Jb. (Anat.)* **23**, 463–534.

— 1907b Die Pterobranchier. Anatomische und histologische Untersuchungen über *Rhabdopleura normanii* Allman und *Cephalodiscus dodecalophus* M'Int. 1. Teil. *Rhabdopleura normanii* Allman. 2. Abschnitt. Knospungsprozess und Gehäuse von *Rhabdopleura*. *Zool. Jb. (Anat.)* **24**, 193–238.

— 1907c Die Pterobranchier. Anatomische und histologische Untersuchungen über *Rhabdopleura normanii* Allman und *Cephalodiscus dodecalophus* M'Int. 2. Teil. *Cephalodiscus dodecalophus* M'Int. 1. Abschnitt. Die Anatomie von *Cephalodiscus*. *Zool. Jb. (Anat.)* **24**, 553–608.

— 1908 Die Pterobranchier. Anatomische und histologische Untersuchungen über *Rhabdopleura normanii* Allman und *Cephalodiscus dodecalophus* M'Int. 3. Teil. Vergleichendanatomische Teil. *Zool. Jb. (Anat.)* **25**, 418–94.

SCHIEMENZ, P. 1884 Über die Wasseraufnahme bei Lamellibranchiaten und Gastropoden. *Mitt. zool. Sta. Neapel* **5**, 509–43.

— 1887 Über die Wasseraufnahme bei Lamellibranchiaten und Gastropoden (einschliesslich der Pteropoden). II. *Mitt. zool. Sta. Neapel* **7**, 423–72.
SCHMIDT, O. 1849 Einige neue Beobachtung über die Infusorien. *Not. Hebeite Natur u. Heilkunde* **9**, 6–7. (Cited by Hanson, 1958).
SCHNEIDER, K. C. 1902 *Lehrbuch der vergleichenden Histologie der Tiere.* Jena: Fischer.
— 1908 *Histologisches Praktikum der Tiere.* Jena: Fischer.
SCHULTZ, E. 1903 Aus dem Gebiete der Regeneration. 4. Über Regenerationserscheinungen bei *Actinotrocha branchiata* Müller. *Z. wiss. Zool.* **75**, 473–94.
SCHWARZ, A. 1932 Die tierische Einfluss auf die Meeresedimente. (Besonders auf die Beziehungen zwischen Frachtung Ablagerung und zuzammensetzung von Wassernsedimenten). *Senckenbergiana* **14**, 118–72.
SCULLY, U. 1962 Personal communication.
SEDGWICK, A. 1884 On the nature of metameric segmentation and some other morphological questions. *Quart. J. micr. Sci.* **24**, 43–82.
— 1887 The development of *Peripatus capensis*. Part III. *Quart. J. micr. Sci.* **27**, 467–550.
— 1894 Further remarks on the cell theory, with a reply to Mr. Bourne. *Quart. J. micr. Sci.* **38**, 331–7.
SELYS-LONGCHAMPS, M. de 1903 Über *Phoronis* und Actinotroch bei Helgoland. *Wiss. Meeresuntersuch. Helgol.* **6** (15), 1–55.
— 1904 Développement postembryonnaire et affinités des *Phoronopsis*. *Mém. Acad. R. Belg. Cl. Sci.* **1**, 1–150.
— 1907 *Phoronis. Fauna u. Flora Neapel* **30**.
— 1938 Origine des premières ébauches cardiaques chez les Tuniciers. *Trav. Sta. zool. Wimereux* **13**, 629–34.
SÉMON, R. 1888 Die Entwicklung der *Synapta digitata* und die Stammesgeschischte der Echinodermen. *Jena. Z. Naturw.* **22**, 175–309.
SEMPER, C. 1876 Die Verwandtschaftbeziehungen der gegliederten Thiere. *Arb. zool. Inst. Wurzburg* **3**, 115–404.
SHIPLEY, A. E. 1887 On some points in the development of *Petromyzon fluviatilis*. *Quart. J. micr. Sci.* **27**, 325–70.
SILÉN, L. 1954 On the nervous system of *Phoronis*. *Ark. Zool.* ii, **6**, 1–40.
SIMPSON, M. 1962 Reproduction of the polychaete *Glycera dibranchiata* at Solomons, Maryland. *Biol. Bull. Woods Hole* **123**, 396–411.
SIMROTH, H. 1879 Die Bewegung unserer Landnachtschnecken hauptsächlichen erörtert an der Sohle des *Limax cinereoniges* Wolf. *Z. wiss. Zool.* **32**, 284–322.
SKAER, R. J. 1961 Some aspects of the cytology of *Polycelis nigra*. *Quart. J. micr. Sci.* **102**, 295–317.
SMITH, J. E. 1946 The mechanics and innervation of the starfish tube foot—ampulla system. *Philos. Trans.* B, **232**, 279–310.
— 1947 The activities of the tube feet of *Asterias rubens* L. I. The mechanics of movement and of posture. *Quart. J. micr. Sci.* **88**, 1–14.
— 1950 Some observations on the nervous mechanisms underlying the behaviour of starfishes. *Symp. Soc. exp. Biol.* **4**, 196–220.
— 1957 The nervous anatomy of the body segments of nereid polychaetes. *Philos. Trans.* B, **240**, 135–96.

SNODGRASS, R. E. 1938 Evolution of the Annelida, Onychophora and Arthropoda. *Smithson. misc. Coll.* **97** (6), 1–159.

SÖDERSTRÖM, A. 1924 Ueber die katastrophale Metamorphose der *Polygordius*-Endolarve nebst Bemerkungen über die Spiralfurchung. *Uppsala Univ. Arsskr.* 1924, *Mat. Naturv.* **1**, 1–78.

— 1925a Das Problem der Polygordius–*Endolarve*. Stockholm: Almqvist and Wiksells.

— 1925b *Die Verwandtschaftbeziehungen der Mollusken.* Uppsala-Leipzig: Köhler.

SPENGEL, J. W. 1881 Die Geruchsorgane und das Nervensystem der Mollusken. *Z. wiss. Zool.* **35**, 333–83.

STAUFFER, H. 1924 Die Lokomotion der Nematoden. Beiträge zur kausal Morphologie der Fädenwürmer. *Zool. Jb. (System.)* **49**, 119–30.

STEINBÖCK, O. 1937 Eine Theorie über den plasmodialen Ursprung der Vielzeller (Metazoa). (IV intern Congr. Cytology) *Arch. exp. Zellforsch* **19**, 343 (published in title only, cited by Hyman, 1959).

— 1958 Zur Phylogenie der Gastrotrichen. *Zool. Anz.* suppl. 21, 128–69.

STÉPHAN-DUBOIS, F. 1951 Migrations et potentialités histogénétiques des cellules indifférenciées chez les Hydres, les Planaires et les Oligochètes. *Année biol.* **55**, 733–53.

STEPHENSON, J. 1930 *The Oligochaeta.* Oxford: Clarendon Press.

STEPHENSON, T. A. 1935 *The British sea-anemones.* London: Ray Soc. **2**.

STRINGER, C. E. 1917 The means of locomotion in planarians. *Proc. nat. Acad. Sci., Wash.* **3**, 691–2.

STRUNK, C. 1930 Beiträge zur Exkretions-physiologie der Polychäten *Arenicola marina* und *Stylarioides plumosus. Zool. Jb. (Physiol.)* **47**, 259–90.

TARLO, L. B. 1960a Discussion of Bone, 1960. *J. Linn. Soc. (Zool.)* **44**, 269.

— 1960b The invertebrate origins of the vertebrates. *Proc. int. pal. Un.* (*XXI int. geol. Congr.*), pt. 22, 113–23.

TAYLOR, G. 1951 Analysis of the swimming of microscopic organisms. *Proc. roy. Soc.* A, **209**, 447–61.

— 1952a The action of waving cylindrical tails in propelling microscopic organisms. *Proc. roy. Soc.* A, **211**, 225–39.

— 1952b Analysis of the swimming of long and narrow animals. *Proc. roy. Soc.* A, **214**, 158–83.

TEN CATE, J. and TEN CATE-KAZEJEWA, B. 1933 La coordination des mouvements locomoteurs après la section transversale de la moelle épinière chez les requins. *Arch. néerl. Physiol.* **18**, 15–23.

TÉTRY, A. 1959 Classe des Sipunculiens (Sipunculidea de Quatrefages, 1866). In: GRASSÉ, P. P. (ed.), *Traité de zoologie.* Paris: Masson. **5** (1), 785–854.

THAMDRUP, H. M. 1935 Beiträge zur Ökologie der Wattenfauna auf experimentelle Grundlage. *Medd. Komm. Danm. Havundersog., Fisk.* **10** (2), 1–125.

THIELE, J. 1902 Zur Cölomfrage. *Zool. Anz.* **25**, 82–4.

— 1910 Ueber die Auffassung der Leibeshöhle von Mollusken und Anneliden. *Zool. Anz.* **35**, 682–95.

THOMAS, J. G. 1940 *Pomatoceros, Sabella* and *Amphitrite. L.M.B.C. Mem.* **33**.

THOMPSON, D'A. W. 1917 *On growth and form.* Cambridge: University Press.

THOMPSON, T. E. and SLINN, D. J. 1959 On the biology of the opisthobranch *Pleurobranchus membranaceus*. *J. mar. biol. Ass. U.K.* **38**, 507–24.

TIEGS, O. W. and MANTON, S. M. 1958 The evolution of the Arthropoda. *Biol. Rev.* **33**, 255–337.

TUNG, T. C., WU, S. C. and TUNG, Y. F. Y. 1958 The development of isolated blastomeres of amphioxus. *Actae biol. exp. Sinica* **6**, 57–90. (In Chinese, English summary.)

UBISCH, L. von 1929 Über die Lage, Entwicklung, Induktionwirkung und Funktion von Chorda und Hydrocoel. *Zool. Anz.* suppl. **4**, 83–5.

UDE, J. 1908 Beiträge zur Anatomie und Histologie der Süsswassertricladen. *Z. wiss. Zool.* **89**, 308–70.

UEXKÜLL, J. von 1903 Der biologische Bauplan des *Sipunculus*. *Z. Biol.* **44**, 269–344.

ULRICH, W. 1949 Über die systematische Stellung einer neuen Tierklasse (Pogonofora K. E. Johansson), den Begriff der Archicoelomaten und die Einteilung der Bilaterien. *S.B. dtsch. Akad. Wiss. Math.-Nat. Kl.* 1949, (2), 1–25.

— 1950 Vorschläge zu einer Revision der Grosseinteilung des Tierreichs. *Zool. Anz.* suppl. **15**, 244–71.

USCHAKOV, P. 1933 Eine neue Form aus der Familie Sabellidae (Polychaeta). *Zool. Anz.* **104**, 205–8.

VAN CLEAVE, H. J. 1952 Some host-parasite relationships of the Acanthocephala, with special reference to their organs of attachment. *Exp. Parasitol.* **1**, 305–30.

— and BULLOCK, W. L. 1950 Morphology of *Neoechinorhynchus emydis*, a typical representative of the Eoacanthocephala. I. The praesoma. *Trans. Amer. micr. Soc.* **69**, 288–308.

VAN DAM, L. 1940 The mechanism of ventilation in *Aphrodite aculeata*. *J. exp. Biol.* **17**, 1–7.

VEJDOVSKY, F. 1882 Untersuchungen über die Anatomie, Physiologie und Entwicklung von *Sternaspis*. *Denkschr. Akad. Wiss. Wien* **43**, 1–58.

VERWORN, M. 1888 Beiträge zur Kenntniss der Süsswasserbryozoen. *Z. wiss. Zool.* **46**, 99–130.

VLÈS, F. 1907 Sur les ondes pédieuses des Mollusques raptateurs. *C.R. Acad. Sci., Paris* **145**, 267–8.

— 1913 Observations sur la locomotion d'*Otina otis* Turt. *Bull. Soc. zool. Fr.* **38**, 242–50.

WALTER, H. E. 1906 The behaviour of the pond snail *Lymnaeus elodes* Say. *Cold Spr. Harb. Monogr.* **6**, 1–35.

WATERHOUSE, F. W. 1941 The Cambrian faunas of north-eastern Australia. Part 4: Early Cambrian echinoderms similar to the larval stages of recent forms. *Mem. Queensland Mus.* **12**, 1–28.

WATSON, M. R. 1958 The chemical composition of earthworm cuticle. *Biochem. J.* **68**, 416–20.

WELDON, W. F. R. 1890 The coelom and nephridia of *Palaemon serratus*. *J. mar. biol. Ass. U.K.* (n.s.) **1**, 162–8.

— 1891 The renal organs of certain decapod Crustacea. *Quart. J. micr. Sci.* **32**, 279–91.

WELLS, G. P. 1937 Studies on the physiology of *Arenicola marina* L. I. The pacemaker role of the oesophagus, and the action of adrenaline and acetylcholine. *J. exp. Biol.* **14**, 117–57.
— 1944 Mechanism of burrowing in *Arenicola marina* L. *Nature, Lond.* **154**, 396.
— 1945 The mode of life of *Arenicola marina* L. *J. mar. biol. Ass. U.K.* **26**, 170–207.
— 1948 Thixotropy, and the mechanism of burrowing in the lugworm (*Arenicola marina* L.). *Nature, Lond.* **162**, 652–3.
— 1949a Respiratory movements of *Arenicola marina* (L.): intermittent irrigation of the tube, and intermittent aerial respiration. *J. mar. biol. Ass. U.K.* **28**, 447–64.
— 1949b The behaviour of *Arenicola marina* (L.) in sand, and the role of spontaneous activity cycles. *J. mar. biol. Ass. U.K.* **28**, 465–78.
— 1950 Spontaneous activity cycles in polychaete worms. *Symp. Soc. exp. Biol.* **4**, 127–42.
— 1951 On the behaviour of *Sabella*. *Proc. roy. Soc.* B, **138**, 278–99.
— 1952a Respiratory significance of the crown of the polychaete worms *Sabella* and *Myxicola*. *Proc. roy. Soc.* B, **140**, 70–82.
— 1952b The proboscis apparatus of *Arenicola*. *J. mar. biol. Ass. U.K.* **31**, 1–28.
— 1953 Defaecation in relation to the spontaneous activity cycles of *Arenicola marina* (L.). *J. mar. biol. Ass. U.K.* **32**, 51–63.
— 1954 The mechanism of proboscis movement in *Arenicola*. *Quart. J. micr. Sci.* **95**, 251–70.
— 1961 How lugworms move. In: RAMSAY, J. A. and WIGGLESWORTH, V. B. (eds.), *The cell and the organism.* Cambridge: University Press. Pp. 209–33.
— 1962 The warm-water lugworms of the world (Arenicolidae, Polychaeta). *Proc. zool. Soc. Lond.* **138**, 331–53.
— and ALBRECHT, E. B. 1951 The integration of activity cycles in the behaviour of *Arenicola marina* (L.). *J. exp. Biol.* **28**, 41–50.
— and DALES, R. P. 1951 Spontaneous activity patterns in animal behaviour: the irrigation of the burrow in the polychaetes *Chaetopterus variopedatus* Renier and *Nereis diversicolor* O. F. Müller. *J. mar. biol. Ass. U.K.* **29**, 661–80.
WERNER, B. 1956 Über die Winterwanderung von *Arenicola marina* L. (Polychaeta sedentaria). *Helgol. wiss. Meeresuntersuch.* **5**, 353–78.
WESENBERG-LUND, C. 1896 Biologiske Studier over Ferskandsbryozoer. *Vidensk. Medd. dansk Naturhist. Foren.* v, **8**, 252–63.
WHITE, E. I. 1935 The ostracoderm *Pteraspis* Kner and the relationships of the agnathous vertebrates. *Philos. Trans.* B, **225**, 381–457.
WHITEAR, M. 1957 Some remarks on the ascidian affinities of vertebrates. *Ann. Mag. nat. Hist.* xii, **10**, 338–47.
WHITEHOUSE, R. H. 1918 The evolution of the caudal fin of fishes. *Rec. Indian Mus.* **15**, 135–42.
WIESER, W. 1959 The effect of grain size on the distribution of small invertebrates inhabiting the beaches of Puget Sound. *Limnol. Oceanogr.* **4**, 181–94.
WILCOX, A. W. 1906 Locomotion in young colonies of *Pectinatella magnifica*. *Biol. Bull. Woods Hole* **11**, 245–52.
WILHELMI, J. 1906 Untersuchungen über die Exkretionsorgane der Susswassertricladen. *Z. wiss. Zool.* **80**, 544–75.

WILLEY, A. 1894 *Amphioxus and the ancestry of the vertebrates*. New York: Columbia University.
WILSON, C. B. 1900 The habits and early development of *Cerebratulus lacteus* (Verrill). A contribution to physiological morphology. *Quart. J. micr. Sci.* **43**, 97–198.
YAMANOUCHI, T. 1929 Notes on the behavior of the holothurian, *Caudina chilensis* (J. Müller). *Sci. Rep. Tohoku Univ.* iv, **4**, 73–115.
YATSU, N. 1902a On the habits of Japanese *Lingula*. *Annot. zool. Japon.* **4**, 61–7.
— 1902b On the development of *Lingula anatina*. *J. Coll. Sci. Tokyo* **17** (4), 1–112.
— 1902c Notes on the histology of *Lingula anatina* Brugière. *J. Coll. Sci. Tokyo* **17** (5), 1–29.
YAZAKI, M. 1930 On the circulation of the perivisceral fluid in *Caudina chilensis* (J. Müller). *Sci. Rep. Tohoku Univ.* iv, **5**, 403–14.
YONGE, C. M. 1937 Evolution and adaptation in the digestive system of the Metazoa. *Biol. Rev.* **12**, 87–115.
— 1939 On the mantle cavity and its contained organs in the Loricata (Placophora). *Quart. J. micr. Sci.* **81**, 367–90.
— 1947 The pallial organs in the aspidobranch Gastropoda and their evolution throughout the Mollusca. *Philos. Trans.* B, **232**, 443–518.
— 1957 Reflexions on the monoplacophoran, *Neopilina galatheae* Lemche. *Nature, Lond.* **179**, 672–3.
ZENKEVICH, L. A. 1945 The evolution of animal locomotion. *J. Morph.* **77**, 1–52.
ZIEGLER, H. E. 1898 Ueber den derzeitigen Stand der Cölomfrage. *Verh. dtsch. zool. Ges.* **8**, 14–78.
— 1912 Leibeshöhle. *Handwörterbuch der Naturwiss.* **6**, 148–65.
ZITTEL, K. A. von 1915 *Grundzüge der Paläontologie (Paläozoologie)*. (Neuarb. F. von BROILI). München-Berlin: Oldenbourg. **1** (4).
ZUCKERKANDL, E. 1950a Coelomic pressures in *Sipunculus nudus*. *Biol. Bull. Woods Hole* **98**, 161–73.
— 1950b Unpublished. (Cited by PROSSER, 1950.)

AUTHOR INDEX

Adam, H. 191.
Affleck, R. J. 170, 192.
Agersborg, H. P. K. 61.
Ahlborn, F. 192.
Albrecht, E. B. 98.
Alder, J. 93.
Allen, E. J. 28.
Allman, G. J. 65, 66, 101, 102, 103, 104.
Anderson, D. T. 20.
Andersson, K. A. 155, 156, 227.
Ashworth, J. H. 146, 237
Assheton, R. 157.
Atkins, D. 236.
Ax, P. 209.

Bahl, K. N. 122, 123, 141, 231.
Bainbridge, R. 194.
Balfour, F. M. 12, 202, 220.
Barets, A. 194.
Barnes, H. 218.
Barth, R. 144.
Bateson, W. 27, 200, 244.
Batham, E. J. 73, 74, 75, 107, 108, 109, 112, 113, 114.
Bather, F. A. 240.
Beauchamp, P. de 4, 16, 28, 220, 221.
Becher, S. 98.
Benenden, E. van 10, 14.
Benham, W. B. 157.
Bergh, R. S. 3, 5, 17.
Berrill, N. J. 24, 25, 26, 165, 200, 201, 221, 245, 246, 247.
Bhatia, M. L. 52.
Biedermann, W. 64.
Bird, A. F. 76, 77.
Blochmann, F. 236, 237, 238.
Boddeke, R. 194, 195.
Boettger, C. R. 219.
Böhmig, L. 85, 208.
Bone, Q. 200, 201, 221, 247.
Borelli, G. A. 174.
Borg, F. 106, 107.
Bösiger, E. 194.
Bradley, C. E. 109, 114.
Braekkan, O. R. 194, 196.
Braem, F. 6, 16.
Breder, C. M. 166, 167, 174, 175, 180.
Bresslau, E. 48, 52.
Brien, P. 65, 104, 221, 246.
Brøndsted, H. V. 67.
Brooks, C. M. 66.

Brooks, W. K. 200.
Brown, A. C. 54, 61.
Brown, R. H. J. 254.
Brown, R. S. 149.
Budington, R. A. 101.
Bullock, T. H. 157, 159.
Bullock, W. L. 159.
Burdon-Jones, C. 157, 158, 159, 225, 238, 245.
Burfield, S. T. 219.
Bürger, O. 52, 146.
Burnett, F. L. 45.
Bury, H. 240, 241.
Busk, G. 106.
Bütschli, O. 140.

Caldwell, W. H. 24.
Cannan, R. K. 238.
Cannon, H. G. 107, 254.
Carlson, A. J. 64, 96, 97.
Carruthers, R. C. 21.
Carter, G. S. 200.
Caullery, M. 153, 157, 159.
Chapman, G. 32, 33, 47, 71, 72, 73, 74, 91, 99, 107, 109, 110, 112, 113, 146, 147, 149, 257.
Child, C. M. 23, 49, 50, 51, 57.
Chitwood, B. G. 76, 77.
Chitwood, M. B. 76.
Chuang, S. H. 236, 237.
Claparède, E. 149.
Clark, M. E. 73, 89, 129, 131, 132, 138, 139, 145, 183, 184, 185, 186.
Clark, R. B. 37, 38, 39, 40, 52, 73, 89, 93, 129, 131, 132, 134, 136, 138, 139, 140, 141, 145, 183, 184, 185, 186, 218.
Clarke, A. H. 249, 250.
Coe, W. R. 41, 68, 168.
Cole, F. J. 188.
Copeland, M. 61, 62, 63.
Cowey, J. B. 36, 37, 38, 39, 40, 52.
Crofton, H. D. 76, 77, 79, 82, 109, 112, 114.
Crofts, D. R. 56.
Crozier, W. J. 45, 49, 56, 57, 62, 63, 99, 100, 101, 157.
Cuvier, G. L. C. D. 22.

Dahlgren, U. 66.
Dales, R. P. 98, 233.
Davidson, P. 189.
Dawydoff, C. 6, 156, 226, 244.

AUTHOR INDEX

De Beer, G. R. 207, 247.
Defretin, R. 132.
Delsman, H. C. 200.
Deutsch, K. 76.
Diesing, K. M. 207.
Dietz, P. A. 188, 189.
Dohrn, A. 2, 200.
Drach, P. 246.

Eggers, F. 34, 46, 47, 51, 68.
Eisig, H. 134, 141, 200, 233.
Ekman, T. 237.
Emelianov, S. W. 190.
Evans, F. G. 187, 191.
Ewing, M. 250.

Fage, L. 149.
Farre, A. 102, 104.
Fauré-Frémiet, É. 76.
Faurot, L. 92, 94.
Faussek, V. 15, 16.
Fenn, W. O. 114.
Fisher, W. K. 89, 90, 93, 97, 99, 167.
Fordham, M. G. C. 135, 145.
Fox, D. L. 85, 149, 150.
Fox, H. M. 98, 108, 139.
Foxon, G. E. H. 124.
Fraenkel, G. S. 143.
François, P. 225, 237.
Fretter, V. 250, 252.
Friedländer, B. 196.
Friedrich, H. 51, 68, 87.

Gamble, F. W. 42, 46, 168.
Gansen-Semal, P. van 77.
Garrault, H. 76.
Garstang, W. 26, 200, 246, 247.
Gaskell, W. H. 200.
Geigy, R. 7.
Geoffroy Saint-Hilaire, É. 22.
George, J. C. 194.
Gerould, J. H. 91.
Gersch, M. 59, 60, 61.
Gilchrist, J. D. F. 155, 156, 227.
Gislén, T. 200, 201, 202.
Gontcharoff, M. 49.
Goodrich, E. S. 3, 4, 5, 15, 17, 18, 25, 28, 88, 165, 191, 203, 226, 231, 247.
Graham, A. 250, 252.
Graham-Smith, W. 192.
Gratiolet, P. 67.
Gray, J. 25, 42, 46, 78, 79, 118, 119, 124, 125, 126, 127, 128, 166, 167, 174, 175, 176, 178, 179, 180, 181, 182, 196, 197, 198.
Greene, C. H. 190.
Greene, C. W. 190.

Grobben, K. 219, 222, 242, 243.
Grove, A. J. 192.
Guberlet, J. E. 94.
Gulland, G. L. 28.
Gunn, D. L. 143.
Gustavson, K. H. 76.

Haderlie, E. C. 93.
Hadži, J. 22, 207, 208, 225.
Haeckel, E. 7, 12, 22, 27, 28, 207, 209, 211, 239.
Hallez, P. 67.
Hammond, R. A. 159, 160.
Hancock, A. 236.
Hancock, G. J. 166.
Hand, C. 211.
Hanson, E. D. 207.
Harmer, S. F. 66, 103, 104, 105, 106, 107, 156.
Harris, J. E. 76, 77, 79, 82, 109, 112, 114, 169, 170, 171, 172, 173, 174, 192, 193, 202.
Hartman, O. 153.
Hartog, M. 221.
Hatschek, B. 3, 22, 206, 220.
Healey, E. G. 196, 197.
Heider, K. 251, 254.
Hempelmann, F. 140.
Herlant-Meewis, H. 24.
Herter, K. 46, 47.
Hertwig, O. 4, 6, 8, 16, 28.
Hertwig, R. 4, 6, 8, 16, 28.
Hessle, C. 150.
Heymons, R. 6, 7.
Hill, A. V. 114.
Hilton, A. W. 151.
Hinton, H. E. 142, 143, 255.
His, W. 24.
Hoffmann, H. 6.
Holme, N. A. 168, 169.
Holst, E. von 196, 197.
Horny, R. 21.
Horst, C. J. van der 157.
Howell, A. B. 174.
Hoyle, G. 145.
Hubrecht, A. A. W. 8, 12, 14, 18, 68, 200, 201.
Huxley, T. H. 220.
Hyman, L. H. 2, 10, 14, 15, 17, 22, 23, 25, 28, 65, 66, 67, 73, 78, 82, 104, 150, 165, 203, 206, 207, 208, 211, 219, 220, 221, 222, 224, 225, 226, 227, 240, 243, 257.

Ijima, I. 46.
Ikeda, I. 226.
Inada, C. 110.
Irwin-Smith, V. A. 81.

AUTHOR INDEX

Ivanov, A. V. 151, 152, 153, 154, 226, 227.
Iwanoff, P. P. 11, 20.

Jägersten, G. 12, 13, 14, 151, 207, 209, 210, 212, 222, 223, 226, 233.
Jarman, G. M. 190.
Jennings, J. B. 211.
Jensen, D. D. 200, 201.
Jhering, H. von 207.
Johansson, K. E. 152, 226.
Jordan, H. 53, 64.
Jullien, J. 106.

Kaiser, F. 197.
Karling, T. 208.
Kawaguti, S. 238.
Kerkut, G. A. 99.
Kermack, K. A. 192.
Kilian, R. 159.
Kishinouye, K. 192.
Kleinenberg, N. 220.
Knight-Jones, E. W. 157, 159.
Komai, T. 155, 225.
Kozlowski, R. 201.

Lam, H. J. 6.
Lameere, A. 8, 9, 11, 14, 15.
Lang, A. 4, 5, 13, 14, 17, 167.
Lang, K. 219.
Langelaan, J. W. 188.
Langeloh, H. P. 87.
Lankester, E. R. 7, 8, 11, 12, 13, 15, 28, 207, 210, 220, 254.
Legendre, R. 149.
Lehnert, G. H. 45, 53.
Lemche, H. 10, 13, 20, 21, 22, 249, 250, 251, 252, 254.
Leuckhart, R. 8.
Linn, I. J. 143.
Lissmann, H. W. 25, 46, 54, 55, 56, 57, 58, 62, 63, 79, 118, 119, 181, 196, 197, 198.
Lotmar, W. 76.

MacBride, E. W. 14, 239, 240, 241, 242, 243.
MacGinitie, G. E. 89, 90, 93, 97, 99.
Malaquin, A. 140.
Mann, K. H. 47, 52, 68.
Manton, S. M. 20, 124, 142, 143, 226, 253, 254, 255, 256.
Marcus, E. 12, 14, 66, 102, 103, 104, 105, 206, 207, 209, 210, 212, 219, 220, 221, 222, 223, 225.
Marey, E. J. 174.
Martini, E. 202.
Mast, O. S. 99.
Masterman, A. T. 11, 12, 14, 226.
McConnaughey, B. H. 149, 150.

McIntyre, A. D. 93.
McLendon, J. F. 65.
Menzies, R. J. 249, 250.
Mesnil, F. 157, 159.
Metschnikoff, E. 207, 210, 211.
Meyer, E. 4, 5, 17, 150.
Michaelsen, W. 232.
Minot, C. S. 200.
Montgomery, T. H. 68.
Morgan, T. H. 244.
Morris, M. C. 54.
Morse, E. S. 225, 237, 238.
Morton, J. E. 15, 168, 169.
Mosauer, W. 79.
Moseley, H. N. 44, 48, 49, 168.

Naef, A. 14, 23, 250, 251, 254.
Naik, R. M. 194.
Newell, G. E. 71, 72, 73, 74, 91, 94, 107, 110, 112, 113, 121, 122, 140, 141, 146, 149, 192, 257.
Nichols, D. 99.
Nicol, E. A. T. 98, 130, 139.
Nicol, J. A. C. 139.
Nitsche, H. 104.
Nomura, E. 122.
Nursall, J. R. 188, 189, 190, 191, 192, 193.

Odell, W. 66.
Ohuye, T. 238.
Olmsted, J. M. D. 45, 54, 56, 59, 61, 63, 98.
Orton, J. H. 236.
Owen, R. 22.

Pantin, C. F. A. 73, 74, 75, 107, 108, 109, 111, 112, 113, 114, 208.
Parker, G. H. 45, 54, 55, 56, 62, 63, 65, 96, 109.
Parry, D. A. 254.
Patten, W. 200.
Pax, F. 207.
Pearl, R. 43, 45, 46.
Pearse, A. S. 85, 99.
Pedersen, K. J. 67.
Peebles, F. 85.
Pejler, B. 220, 221.
Pelseneer, P. 54, 251.
Perrier, E. 22.
Pettigrew, J. B. 174, 180.
Pheasant, H. C. 187, 191.
Picken, L. E. R. 43, 76, 77, 111.
Pilz, G. F. 56.
Portmann, A. 95, 252.
Prenant, A. 140.
Prenant, M. 67, 68, 236, 238.
Pruvot, G. 149.
Pryor, M. G. M. 76, 77.
Pumphrey, R. J. 25, 46, 197, 198.

Quatrefages, A. de 22.

Ramsay, J. A. 108, 231.
Rao, K. P. 157, 159, 244.
Reed, R. 77.
Remane, A. 12, 14, 19, 22, 29, 207, 209, 212, 220, 221, 222.
Rensch, B. 213.
Ribaucourt, E. de 122.
Rietsch, M. 88.
Ritter, W. E. 157, 159.
Rockwell, H. 187, 191.
Rogick, M. 102.
Roule, L. 226.
Rudall, K. M. 76, 77.

Sachwatkin, A. A. 207.
Saint-Joseph, A. de 149.
Salensky, W. 6, 16.
Sand, A. 197.
Sarvaas, A. E. du Marchie 6, 16, 22, 27, 28, 29, 250, 254.
Sato, T. 155, 225.
Saville-Kent, W. 207.
Sawaya, P. 75, 107, 108, 111, 113.
Schaeffer, C. 237.
Schepotieff, A. 155, 156, 225, 227, 244.
Schiemenz, P. 54.
Schmidt, O. 207.
Schneider, K. C. 123, 226.
Schultz, E. 226.
Schwarz, A. 114.
Scully, U. 121.
Sedgwick, A. 8, 9, 10, 11, 14, 19, 20, 28, 207.
Selys-Longchamps, M. de 150, 226, 246.
Sémon, R. 240, 241.
Semper, C. 2, 200.
Shipley, A. E. 187.
Silén, L. 151.
Simpson, M. 25, 167.
Simroth, H. 64.
Skaer, R. J. 67.
Slijper, E. J. 194, 195.
Slinn, D. J. 168, 169.
Smith, J. E. 99, 100, 137, 154, 244.
Snodgrass, R. E. 15, 16, 17, 25, 165, 203, 207.
Söderström, A. 12, 251.
Spengel, J. W. 252.
Stauffer, H. 79, 80, 81.
Steinböck, O. 22, 207, 208.
Stelt, A. van der 194, 195.
Stéphan-Dubois, F. 67.
Stephenson, J. 122, 123, 232.
Stephenson, T. A. 95.
Stringer, C. E. 45.
Strunk, C. 72.
Swann, M. M. 76, 77.

Tarlo, L. B. 201.
Taylor, G. 166, 174, 178, 179, 180, 182, 183.
Ten Cate, J. 196.
Ten Cate-Kazejewa, B. 196.
Tétry, A. 6.
Thamdrup, H. M. 114.
Thiele, J. 16.
Thomas, J. G. 142, 147, 150.
Thompson, D'A. W. 30.
Thompson, T. E. 168, 169.
Tiegs, O. W. 254.
Tung, T. C. 202.
Tung, Y. F. Y. 202.
Turner, L. G. W. 54, 61.

Ubisch, L. von 201.
Ude, J. 18.
Uexküll, J. von 85, 110.
Ulrich, W. 12, 14, 207, 209, 222, 225.
Uschakov, P. 153.

Van Cleave, H. J. 159.
Van Dam, L. 137.
Vejdovsky, F. 88.
Verworn, M. 66.
Vlès, F. 54, 55, 56.

Walter, H. E. 61.
Waterhouse, F. W. 239.
Watson, M. R. 76.
Weldon, W. F. A. 28.
Wells, G. P. 33, 91, 92, 98, 107, 108, 110, 112, 114, 130, 139, 146, 147, 148, 149.
Werner, B. 94.
Wesenberg-Lund, C. 66.
White, E. I. 192.
Whitear, M. 200, 201, 211, 247.
Whitehouse, R. H. 192.
Whiting, H. P. 202.
Wieser, W. 213.
Wilcox, A. W. 66.
Wilhelmi, J. 18.
Willey, A. 200.
Wilson, C. B. 85.
Wingstrand, K. G. 22, 250, 251, 252.
Worzel, J. L. 250.
Wu, S. C. 202.

Yamanouchi, T. 91.
Yatsu, N. 225, 237, 238.
Yazaki, M. 111, 113.
Yonge, C. M. 15, 252.

Zenkevich, L. A. 26.
Ziegler, H. E. 15, 17, 207.
Zittel, K. A. von 21.
Zuckerkandl, E. 85, 86, 110, 111, 112.

SUBJECT INDEX

Abarenicola claparedii, 147, 148.
Absorption (*see also* Digestion), 14–5, 216.
Acanthastids, 201.
Acanthobdellidae, 253.
Acanthocephala, 159–61, 219.
Acanthocephalus, 160.
Acanthocephalus ranae, 159.
Acanthodii, 169.
Acanthurus, 172.
Aciculum, 124, 133, 234, 255.
Acmaea, 63.
Acoela, 42, 207, 208, 209, 211, 213, 224, 225.
Acoeloid theory (*see also* Syncytial protistan theory), 212.
Acoelomate animals, 11, 15, 16, 19, 34, 35, 38, 54, 71, 83, 97, 115, 120, 210, 223, 249.
Acrania (*see also* Cephalochordata), 201.
Actinia bermudensis, 65.
Actinopterygii (*see also* Chondrostei, Holostei, Teleostei), 170.
Activity, 108, 159.
Activity cycles, 98.
Adductor muscle, 107, 133.
Adherent rugae, 95.
Adhesion to substratum, 43, 46, 48, 51, 53, 62–4, 95, 96, 97, 205, 249.
Adhesive glands, 82.
Adhesive mucus, 43, 48, 62, 96, 98, 205.
Adhesive organ, 155, 238.
Aerobic metabolism, 194.
Agitation of substratum, 91.
Agnatha (*see also* Anaspida, Cephalaspida, Cyclostomata, Pteraspida), 169.
Aiptasia, 95.
Aiptasia carnea, 95.
Aiptasia couchii, 95.
Air bladder, 169, 170.
Akera, 168, 169.
Albumin, 77.
Alectrion, 61.
Algae, 52, 98.
Alimentary canal (*see* Buccal bulb, Buccal mass, Digestive system, Gut, Oesophagus, Rectum).
Allolobophora, 124.
Alternate locomotory waves, 49–50, 56, 57, 58, 125, 235.
Ambulacral system, 240, 241, 242.
Ambulation, 126–8.

Ambulatory bristles, 80.
Amia calva, 193.
Ammocoete, 187.
Ampharetidae, 147, 150, 233.
Amphictenidae, 150.
Amphineura, 53.
Amphioxus, 188, 191, 208.
Amphiporus, 38, 39.
Amphiporus lactifloreus, 36, 40.
Amplitude, 129, 177, 179, 180, 182, 183, 186, 193.
Ampulla, 99, 153.
Amputation, 73, 91.
Anaerobic metabolism, 194.
Anal chaetae, 90.
Anal fin, 173.
Anal respiration, 108.
Anasca, 104, 106.
Anaspida, 192.
Anchor (*see also* Point d'appui), 85, 86, 88, 90, 91, 93, 96, 97, 98, 119, 129, 143, 146, 153, 154, 155, 161, 164, 237.
Angel-fish, 173.
Angle of attack, 175, 177, 178, 180.
Angular movement, 185.
Anisometric extension, 76.
Annelida (*see also* Archiannelida, Hirudinea, Oligochaeta, Polychaeta), 2, 6, 10, 12, 19, 20, 22, 28, 113, 200, 202, 203, 204, 223, 224, 225, 247–9, 251, 253, 255, 260.
— coelom, 3, 67–8, 239.
— growth, 1, 23–4, 145–6.
— musculature (*see also* Musculature), 118, 187, 203.
— origin, 8, 9, 229–35.
— primitive, 140.
— segments, 2, 22, 23, 25–6, 27, 118–42, 202, 203–4, 231, 232–5, 249.
— septum (*see also* Septum), 26, 118.
Annulata, 8, 9, 28.
Antagonistic contraction, 26, 31, 33, 34, 46, 69, 75, 82, 115, 131, 156, 235, 249.
Antagonistic force, 190.
Antagonistic muscles, 26, 31, 33, 34, 37, 69, 82–3, 120, 131.
Anthozoa (*see also* Octocorallia, Tetracorallia), 14, 66, 208.
— body wall, 112, 113.
— burrowing, 92, 95.
— gastric pockets, 8, 9, 19, 22.
— hydrostatic skeleton, 65, 73–4, 258.

— internal pressure, 107, 108, 109, 112, 113.
— locomotion, 64–5, 69, 70, 94–5.
— mesenteries, 10, 22.
— musculature, 73, 109, 114.
— symmetry, 9, 19, 208.
Anus, 75, 108, 222, 237.
Aphrodite, 76, 132, 133, 135, 137, 145, 163.
Aphroditidae (*see also* Polynoidae), 137, 248.
Aplysia, 54, 63, 64, 96, 168.
Aplysia californica, 56, 96.
Apodous holothurian, 100, 101, 116, 245.
Appendage, 10, 124, 133, 135, 142, 168, 174, 232, 254, 255.
Appendicularia, 202.
Arachnida, 254.
Arc, reflex, 196–8.
Archenteron, 7.
Archiannelida, 6, 233.
Archicoelomata, 222.
Archigastrula, 210, 223.
Archimetamere, 19, 20.
Archimollusc, 249–50.
Arenicola, burrowing, 73, 74, 91, 93, 113, 114, 117, 256–7.
— coelomic fluid, 73, 74, 124, 256–7.
— forces generated, 124, 144, 256.
— internal pressure, 110, 112, 113, 114, 146.
— irrigation, 98.
— locomotion, 94, 167.
— musculature, 71, 124.
— nephridium, 71.
— proboscis, 92, 147–9.
— septum, 124, 139, 144, 146–7, 163.
Arenicola marina, 91, 92, 110, 112, 146–9.
Arenicolidae, 142, 147, 149.
Arenicolides ecaudata, 147, 148.
Arhythmic pedal locomotion, 62.
Ariciidae, 233.
Armour, 169, 239, 245.
Arthropoda (*see also* Arachnida, Chilopoda, Crustacea, Insecta), 1, 26, 50, 129, 142, 200, 223, 224, 225, 247, 251, 258.
— appendage, 10, 254.
— coelom, 28, 254–7, 260.
— cuticle, 76.
— haemocoel, 253–7, 258.
— metamerism, 20, 22, 163.
— musculature, 144–5, 163.
Articular surface, 173.
Articulata, 236.
Articulation, intervertebral, 191.
Ascaris, 76, 77, 78, 79, 82, 83, 113.
Ascaris lumbricoides, 77, 109.
Aschelminthes (*see also* Acanthocephala, Gastrotricha, Kinorhyncha, Nematoda, Nematomorpha, Rotifera), 208, 209, 211, 219–22, 223, 225.

Ascidiacea, 25, 26, 200, 201, 202, 238, 245–7.
Ascophora, 104, 105, 106.
Asexual reproduction, 22, 23, 24, 244, 246.
Aspect ratio, 193.
Aspidobranchia, 61.
Asterias, 100.
Asteroidea, 99, 100, 205, 239, 243, 244, 245.
Asymmetry, 239.
Atrium, 241.
Attachment pit, 243.
Attachment to substratum, 53, 225, 239, 241.
Atubaria, 155, 225.
Auricularia larva, 239.
Autonomy of segments, 120, 124.
Axial complex, 158.
Axial skeleton, 199.
Axis, longitudinal, 22, 24, 33, 65, 191.
— of locomotion, 174, 175, 176, 177.
— of symmetry, 8, 12, 22.
Axocoel, 242, 244.
Axon, giant, 139, 153.
Aysheaia, 255.

Backthrust (*see also* Point d'appui), 62, 94, 119–20, 125, 165, 177, 180, 186, 205.
Balistes, 171, 172.
Balistidae, 173.
Banjo-fish, 170.
Basement membrane, 35, 40–1, 42, 52, 69, 71, 76, 137, 140, 141, 203.
Basking shark, 167.
Bass, 193.
Benthic animals, 212, 213, 215, 220, 221.
Berycoids, 172.
Bilateral symmetry, 12, 19, 206, 208, 209, 212, 239, 241.
Bilateria, 206–14, 215, 218, 227, 228, 258.
Bilaterogastraea theory, 12, 13, 14, 212.
Biogenetic law (*see also* Recapitulation theory), 27.
Bipalium, 48.
Bipalium diana, 44.
Bipalium kewense, 45.
Bipedal locomotion, 58.
Bipinnaria larva, 239.
Bird, 194.
Blastaea, 209.
Blastema, 19.
Blastocoel, 16, 219, 220, 221, 226, 260.
Blastopore, 222, 224.
Blastula, 211.
Blatta, 6.
Blood, 154, 238, **254.**
Blood spaces, 28, **53.**
Blood vascular system, 44, 140, 152, 154, 163, 194, 218, 230, 250, 251, 254, 255, 256.

SUBJECT INDEX

Blowfly, 143.
Boa, 79.
Body volume (*see also* Leakage of coelomic fluid), 36–8, 71–5, 77, 107–8, 250.
Body wall, 2, 4, 16, 71, 98, 135, 139.
— antagonistic muscles, 31, 33, 34, 37, 69, 83, 120, 131.
— dilation, 33, 35, 86, 93–4, 108, 116, 135, 143, 232–4.
— folding, 26, 40, 41, 73.
— resistance to deformation, 112, 113.
— rigid, 99.
Body-wall musculature (*see also* Circular muscles, Longitudinal muscles), 4, 31–3, 42, 52, 71, 83–4, 96, 115–6, 145, 148, 202–3, 205, 214, 215, 218, 219, 232, 245, 253, 258.
— effect of sectioning, 120.
— lateral, 131, 193–6, 201.
— orientation, 33.
— reduction, 131, 134, 137, 233–4.
— segmented, 8, 17, 25, 29, 140, 228–9, 259.
— synergic contraction, 86, 88.
Bone (*see also* Rib, Vertebral column), 31.
— intermuscular, 190.
— supernumerary, 189.
Bonito, 171, 173.
Botryoidal tissue, 67, 249.
Bowfin, 193.
Brachial fold, 236.
Brachial ossicle, 244.
Brachial skeleton, 236.
Brachiopoda, 7, 12, 19, 76, 151, 222, 223, 225, 235, 236–8, 258.
Brain, 196, 257.
Braking, 170, 173, 177.
Branchia (*see* Gill).
Branchial papilla, 245.
Branchial skeleton, 201.
Branchiobdellidae, 253.
Bream, 194.
Breeding season, 218.
Bristle (*see also* Chaeta), 80, 82.
Bryozoa (*see also* Ectoprocta), 220.
Buccal bulb, 233.
Buccal mass, 92, 147, 149.
Buccal papilla, 91, 92.
Budding (*see also* Asexual reproduction), 246.
Bulla occidentalis, 61.
Bullia, 54.
Buoyancy, 169.
Burbot, 194, 195.
Burrow, depth, 112, 248.
— excavation (*see also* Burrowing), 89, 91.
— form, 90, 93, 114.
— irrigation, 112, 130–1, 137, 139, 248.

— movement in, 93–4, 129, 158, 159, 248.
— permanence, 114, 248.
Burrowing, 84–93, 94, 114, 115, 117, 129, 142–4, 150–1, 165, 215, 217, 218, 225, 227–8, 232, 234, 235, 249, 256–7, 258.
— Anthozoa, 92, 95.
— Echiuroidea, 89–90, 91.
— Enteropneusta, 157, 159, 238.
— Holothuroidea, 85, 90–1, 101, 111, 114, 116.
— in soft substratum, 88, 91, 256.
— into substratum, 86, 93, 114, 115, 119, 150.
— Myriapoda, 142–4, 256–7.
— Nematoda, 79, 80.
— Nemertea, 85, 168, 216.
— Oligochaeta, 119, 230.
— Phoronidea, 150–1.
— Polychaeta, 88–9, 91–2, 93, 110, 112, 114, 131, 134, 257.
— Priapulida, 87, 116, 124.
— Sipunculoidea, 85–6, 110, 114.
— sustained, 124, 162, 228.

Caecum, septal, 149, 150.
Calcification, 104, 106, 244.
Calliactis, 73, 113.
Calliactis parasitica, 109, 113.
Cambrian, 10, 21, 239.
Cambridium cernysevae, 21.
Cambridium nikiforavae, 21.
Cannibalism, 215.
Capacity, 37, 38.
Capillary, 256.
Capitella, 134, 141.
Capitellidae, 134, 232.
Carangidae, 166, 191.
Caranx caballus, 193.
Carcinus maenas, 111.
Cardio-pericardial vesicle, 244.
Carnivorous habit, 143, 194, 215, 249.
Carp, 195–6.
Carpoidea, 201.
Cartilage, 236.
Caterpillar, 80, 144.
Caudal fin, 166, 167, 192–3.
Caudal musculature, 26, 191–3, 200.
Caudal peduncle, 191, 193.
Caudal region, 146, 189.
Caudina, 90, 113, 114.
Caudina chilensis, 111, 113.
Cavitation of gonad, 4.
Cellularization, 207, 210.
Central nervous system, 8, 9, 98, 196–8, 218.
Centre of gravity, 98, 170.
Centrum, 188, 189.
Cephalaspida, 170.

SUBJECT INDEX

Cephalic bristle, 82.
Cephalic lobe, 151, 154, 156.
Cephalochordata, 188, 245.
Cephalodiscus, 155.
Cephalopoda, 10, 250, 252.
Ceratopogonid larva, 179, 180
Cerebratulus, 39, 41, 85, 167
Cerebratulus lacteus, 39–41.
Cerianthidea, 9, 10.
Cestoda, 1, 2, 22, 28.
Cetorhinus, 167.
Chaeta, 76, 88, 90, 119, 124, 133, 162, 230, 231–2, 248, 249.
Chaetal sac, 232.
Chaetodon, 171, 172.
Chaetognatha, 219, 222, 223, 224.
Chaetosoma haswelli, 81.
Chain reflex, 196–8.
Change of length (*see also* Extensibility), 77, 82–3, 254–5.
Change of shape, 30, 31–5, 67–8, 69, 71, 115, 117, 120, 167, 187, 213, 214, 228, 233, 248, 258.
— Chilopoda, 142, 144.
— limitations, 24, 33, 35–42, 69, 71, 73, 119.
— local, 33, 124, 144, 161, 228, 259.
— Nematoda, 78, 144.
— Nemertea, 35–42, 168, 216.
— Pterobranchia, 156.
— reversible, 31–5, 68, 131, 233.
— Turbellaria, 41, 42, 72, 214.
Change of volume, 31, 73–4, 75, 99, 107, 161, 245.
Cheilostomata, 104, 105, 106, 236.
Chilopoda, 142, 144, 256.
Chitin, 76, 91, 92, 104.
Chloraemidae, 142, 147.
Choanichthyes, 192.
Chondrichthyes, 189, 192, 197.
Chondrostei, 188.
Chordata, 26, 222, 223, 224, 225, 247.
— coelom, 3, 12.
— evolution, 2, 8, 9, 25–6, 28, 200, 204, 221, 238, 245, 247, 260.
— growth, 23.
— locomotion, 165–82, 198–200, 202–5.
— musculature, 173, 187–96, 201–2, 203.
— segmentation, 1, 2, 10, 25, 26, 27–8, 187–91, 198–202
Cilia, 42, 44, 226.
Ciliary feeding, 217, 246.
Ciliary locomotion, 48, 205, 212, 213, 214, 220, 221, 239, 242.
— Enteropneusta, 159, 238.
— Mollusca, 53, 61–3, 250.
— Nemertea, 50–1, 68, 69, 70, 84.
— rate, 45, 61.
— transition to muscular locomotion, 42, 51, 52, 61, 214.
— Turbellaria, 42–5, 69, 213.
Ciliata, 42, 207, 208, 213.
Cinclis, 8, 9.
Ciona, 246.
Circular cross-section, 36, 38, 77, 168, 179.
Circular movements, 50, 59, 60.
Circular muscles, 25, 31–4, 41, 71, 98, 109, 113, 115, 134, 137–8, 160–1, 214–5, 217, 226, 258.
— absence, 77, 131, 156.
— of septum, 122.
— reduction, 131, 132, 162, 234, 248.
— role in locomotion, 46, 48, 51, 66, 83–4, 86, 87, 88, 93, 97, 115, 215, 235, 245.
— segmented, 25, 118.
Circulatory system (*see also* Blood vascular system, Water vascular system), 19, 154, 218, 237, 244, 250.
Circumferential muscles, 99.
Circum-oral nerve ring, 8.
Cirratulidae, 233.
Cirri, 107, 254.
Cirripedia, 107, 254.
Cladoselache, 193.
Cleavage, 222, 224.
Clitellum, 122.
Cloaca, 101, 111.
Closed system, 53, 73.
Cnidaria (*see also* Anthozoa, Hydrozoa, Scyphozoa), 10, 14, 19, 23, 206, 207, 208, 209, 210, 211, 223.
— gastric pocket, 8, 9, 11, 12, 13, 14, 15, 22.
— pedal disk, 10, 64–5, 92, 95.
— septum, 10.
— strobilation, 22, 23.
Coelenterata (*see also* Anthozoa, Cnidaria, Ctenophora, Hydroid, Hydrozoa, Medusoid), 8, 10, 12, 22, 117, 206, 208, 209, 211, 239.
Coelenteric pocket (*see also* Gastric pocket), 209–10.
Coelenteron, 12, 65, 73, 258.
Coelom, 1, 4, 16, 66, 99, 130, 135, 151, 163, 214–9, 220, 230, 250, 253–7.
— definition, 7, 28, 29.
— embryological origin, 6, 7, 15, 27, 28, 152–3, 209, 216, 222–4, 246.
— evolution, 3, 7, 8, 12, 214–9.
— excretory function, 15, 16, 218, 230.
— mechanical function, 16, 246.
— polyphyletic origin, 5, 6, 15, 27, 29, 216, 217.
— reduction, 22, 67–8, 153, 156, 157, 163, 164, 229, 236–7, 244, 246, 247, 249, 252–3, 257, 258.

SUBJECT INDEX

Coelomates, early, 4, 14, 221, 259.
Coelomic canal, 67–8, 70, 237.
Coelomic corpuscle, 150, 238. 244.
Coelomic epithelium, 3, 6, 219, 246.
Coelomic fluid, 84, 86, 89, 97, 135, 139, 141, 218, 230, 238, 254.
— leakage, 72, 73, 74, 121, 124, 139, 141, 236, 256, 257.
— pressure, 86, 147, 148.
Coelomic pouches, 3, 6, 8, 9, 11, 12, 14, 17, 19, 28, 212, 223.
Coelomoduct (see also Genital duct), 71, 73.
Coenecium, 155, 156.
Coenosarc, 76.
Coenosteum, 76.
Coiling, 25, 41.
Collagen, 76, 77, 137–8, 236.
Collar, 157, 159, 238.
Colonial animals, 65, 155, 207, 209.
Columbella mercatoria, 56.
Columella muscle, 60.
Commensalism, 46.
Compensation sac, 104, 106.
Complete septum, 121, 236.
Components of motion, 174, 175, 176, 177.
Composite locomotory waves, 56, 59, 60.
Compression, 40, 41, 167, 182, 199.
Concertina movements, 79.
Condylactis passiflora, 65.
Connective tissue (see also Basement membrane), 4, 36, 67, 137, 140, 141, 152, 156, 189, 237, 238, 244.
Constant volume, 31, 37, 53, 71–5.
Constrictor muscle, 71.
Contraction, antagonistic, 26, 31, 33, 34, 46, 69, 75, 82, 115, 156, 235, 249.
— local, 33, 71, 79, 93, 117, 123.
— unilateral, 33, 134–5, 186.
Conus, 54, 62.
Conus agassizi, 62.
Convergence, 242.
Convoluta, 42, 85.
Convoluta convoluta, 42.
Convoluta saliens, 42, 168.
Co-ordination, nervous, 33, 120, 196–8, 217, 229–30.
— breeding, 218.
— locomotory waves, 51, 120, 197.
— muscular activity, 33, 196–8, 217.
Copulation, 231.
Corals, 21–2.
Corm theory, 22–4.
Couple, 182, 187, 188, 199.
Coxal gland, 28.
Crania, 236.
Creeping welt, 143.
Crepidula, 63.

Crinoidea, 99, 243, 244.
Cristatella, 65, 66.
Crotaline side-winding, 79.
Crustacea, 28, 107, 108, 111, 254.
Cryptocyst, 104.
Ctenidium, 10, 252.
Ctenophora, 206–11.
Ctenostomata, 104, 106.
Cuticle, 34, 71, 74–8, 161.
— Annelida, 77.
— Arthropoda, 144, 254.
— Nematoda, 26, 75–8, 80, 82, 116, 144.
Cuticular plate, 153, 154.
Cuticular spine, 80.
Cyclomerism theory, 17, 19–22, 29.
Cyclostomata (Chordata), 188, 191, 199.
Cyclostomata (Ectoprocta), 106, 236.
Cypraea, 60.
Cypraea exanthina, 56, 59.
Cystid, 104.
Cytoplasmic agent, 201.

Dace, 194, 195.
Damping, 71, 122, 135.
Daphnia, 108.
Dead space, 108.
Deafferentiation, 197.
Deep lateral musculature, 193, 201.
Defaecation, 146, 147.
Defence reaction, 46, 52, 86, 114, 117, 131, 139, 153, 157, 194, 195.
Deformability, 31, 33, 34, 67–8.
Deformation, 62, 114, 116, 117, 121, 135, 137, 143, 144, 145, 162, 203, 233, 236.
Dendrocelum, 39, 41, 44.
Dendrocoelum lacteum, 18, 39–41, 45.
Dendrostoma zostericola, 85.
Density, 169, 170.
Dermis, 76.
Determinate cleavage, 222.
Detritus, 213, 215, 227.
Deuterostomia, 151, 155, 222, 224, 226, 227, 238–47, 259, 260.
Diagonal locomotory waves, 56, 60.
Diagonal muscles, 52–3.
Diameter, 73.
Diaphragm (see also Septum), 102, 138, 140, 229.
Digestion, 14–5, 153, 210, 211, 213.
— extracellular, 15, 216.
— intracellular, 15, 75, 210.
Digestive system (see also Gut, Oesophagus, Rectum), 156, 210, 211, 212, 216, 226, 250, 252.
Dilatancy, 93.
Dilation of body wall, 33, 35, 86, 93–4, 116, 135, 143, 232–4.

SUBJECT INDEX

— other structures, 57, 150, 156, 161, 232, 233.
Dilator muscle, 71, 104.
Diodon, 172.
Diodora, 63.
Dioecious animals, 230.
Dipleurula, 222, 224, 239–43, 244.
Dipnoi, 192.
Diptera, 142, 143, 144.
Direct locomotory waves, 54–5, 60, 65, 94, 95, 96, 125–6, 143.
Discontinuity of locomotion, 58, 88, 187, 200.
Distension, 108, 113, 154.
Distortion, 121–2.
Ditaxic locomotory waves, 49–50, 56, 57, 58, 126, 168.
Diurnal fish, 194.
Divergence, 200.
Diverticulum gut, 5, 8, 19, 168.
— septum, 149, 150.
Dogfish, 170, 197, 198, 202.
Dolabrifera virens, 56.
Dorsal coelomic sac, 251, 252.
Dorsal fin, 8, 168, 173, 197.
Dorsal flexure, 46, 47.
Dorsal longitudinal muscle, 134, 138.
Dorsal pore, 72.
Dorsal rib, 190.
Dorsal valve, 147, 148, 149.
Dorso-ventral muscles, 22, 41, 47, 53, 131, 137, 154, 167, 168, 186, 228.
Dorso-ventral undulations (*see also* Undulatory movements), 166, 167, 170.
Dory, 171.
Draconematidae, 80, 81, 82.
Ducts, excretory, 28, 252.
— genital, 3, 15, 16, 28, 230, 231, 250, 253.
Dugesia gonocephala, 18.
Duplicature bands, 102, 103, 104.

Earthworm (*see also Allolobophora*, *Lumbricus*), 135, 143, 145, 162, 231, 256, 259.
— co-ordination, 196.
— internal pressure, 72, 110.
— locomotion, 79, 97, 118–20, 143, 144, 163.
— musculature, 118–24, 134.
— septum, 118, 120–4, 140, 144–5, 162, 256.
Echinochrome, 238.
Echinodermata (*see also* Asteroidea, Carpoidea, Crinoidea, Echinoidea, Holothuroidea, Ophiuroidea), 7, 12, 15, 19, 96, 99–101, 111, 116, 151, 154, 200, 201, 222, 223, 224, 225, 238.
— coelom, 239–45.
— musculature, 96, 99–101, 205, 245.

Echinoidea, 99, 100, 205, 244, 245.
Echiuroidea (*see also Urechis*), 89–90, 97–8, 217, 224, 225.
Ectoderm, 24, 29, 211, 244.
Ectomesoblast, 4.
Ectoprocta (*see also* Gymnolaemata, Phylactolaemata), 19, 101, 116, 155, 220, 221, 222, 223, 225, 235, 236.
— coelomic cavity, 102–7, 151.
— eversion of polypide, 102–7, 151.
— locomotion, 65–6, 69, 70.
— musculature, 66, 102–7, 236.
— vestibulum, 103, 104, 106, 107.
Eel, 166, 174–80, 181, 193, 194, 197.
Effectiveness of septa, 121.
Efficiency of locomotion, 167, 187.
— muscle contraction, 74, 114, 117.
— parapodial musculature, 141.
Eggs, 20, 149, 201, 218, 230, 231.
Elasmobranchii (*see also* Dogfish, Skate), 189, 192, 197.
Elastic ligament, 139.
Elastic recovery, 26.
Elastin, 77.
Electra pilosa, 105.
Electrical stimulation, 114.
Elliptical cross-section, 37.
Elongation, 24, 33, 119.
Embryology (*see also* Blastula, Cleavage, Gastrula, Ectoderm, Endoderm, Mesoderm), 3, 4, 14, 15, 19, 20, 146, 150, 151, 152, 187, 201, 219, 222, 226, 227, 239, 240, 242, 244, 247, 249, 253, 258.
Embryological origin of coelom, 6, 27, 28, 152–3, 216, 224, 246.
Embryological theory of metamerism, 24–5.
Emplectonema, 51.
Endocrine mechanisms, 218, 221.
Endoderm, 6, 19, 29, 210, 211.
Endomesoderm, 6.
Endophragmal skeleton, 142, 163, 253.
Endoprocta, 219, 220, 221.
Endopterygota, 142.
Endosaccal coelom, 106, 107.
Endurance, 194–6.
Energy output, 179.
Enterocoel theory, 3, 7–15, 16, 19, 21, 22, 28, 29, 210, 216, 219, 222, 223, 229.
Enterocoely, 6, 7, 15, 209, 222, 223.
Enteropneusta, 157–9, 225, 238–9, 244, 245.
Epaxial muscle, 189.
Ependytes, 43.
Epibiont, 153.
Epicardium, 246, 247.
Epicaudal lobe, 192.
Epidermis, 24, 40, 41, 42, 43, 76, 106, 107, 137, 140, 141, 160, 161, 201, 203, 213, 218.

Epinephelis, 173.
Epineural rib, 190.
Epipodial lobe, 169.
Epistome, 150, 155–7, 226, 227, 239, 243.
Epithelium, ciliated, 42.
— coelomic, 219.
Epsilonematidae, 80, 82.
Equilibrium, 38, 170.
Erectile tissue, 53, 54, 61.
Escape reaction (*see also* Defence reaction), 46, 194, 195.
Eumetazoa, 206, 217, 220, 222, 223, 224.
Eunice, 132, 137.
Eunice rousseaui, 140.
Eunicidae, 135, 139, 141, 142, 233.
Euryleptotes cavicola, 49.
Euthynnus, 171, 173.
Euthynnus alleteratus, 193.
Euthynnus lineatus, 193.
Evaporation, 38.
Eversion, polypide, 102–7, 151.
— proboscis, 84, 85, 86, 88, 91, 93, 110, 112, 147–9, 160–1, 216.
Excavation, 89, 91.
Excitation, 196.
Excretion, 4, 15, 16, 218, 231, 246.
Excretory duct, 28, 252.
Excretory product, 15.
Excretory system (*see also* Kidney, Nephridium), 7, 8, 19, 29, 230, 231.
Exocoetus, 171.
Exosaccal coelom, 106, 107.
Exoskeleton, 108, 142, 163, 236, 253, 254, 257.
Extensibility, 35–42, 44, 73, 77–8, 83.
Extension, anisometric, 76.
— body, 33, 64.
— limiting factors, 41, 69, 73.
— muscle, 42, 64.
Extensor muscle, 31.
External hydrostatic pressure, 108, 112.
External restraint, 182.
Extracellular digestion, 15, 216.
Extrinsic parapodial muscle, 130, 131–3, 134, 135, 139, 162, 234, 235.
Extrinsic septal muscle, 122.
Extrovert (*see also* Proboscis), 87, 88, 161.
Extrusion of mucus, 63.

Faeces, 146, 147, 163.
Farrella repens, 103.
Fat, 169, 194.
Fatigue, 42.
Fertilization, 218, 231.
Fibre system, basement membrane, 35–42, 52.
— cuticle, 76–8.
File-fish, 172.

Filter-feeding, 99, 217, 246.
Fin, 8, 166, 167, 168–74, 192–3, 196, 197.
— insertion, 170.
— movement, 173–4.
— muscle, 173.
— ray, 8, 173.
Fin-fold, 168.
Fish (*see also* Actinopterygii, Agnatha, Dipnoi, Elasmobranchii), 166–81, 185, 187, 194, 196, 197.
Fission plane, 23.
Fissurella nodosa, 56.
Flabelligera, 147.
Flaccid body, 98.
Flagellata, 166, 208, 210, 211.
Flame-cell, 15.
Flatfish, 171.
Flattened body, 41, 43–4, 52, 78, 167–8, 177, 186, 214, 228.
Flexibility, 167, 180, 191, 192, 193.
Flexor muscle, 31, 124, 125, 126, 133, 134, 137.
Flexure, body, 33, 46, 47, 82, 86.
— myotome, 188–91, 199.
Fluid, movement of, 71.
— pressure (*see also* Hydrostatic pressure), 31, 33, 64, 71, 116, 120, 144, 147.
— skeleton (*see* Hydrostatic skeleton).
— transport, 218, 237.
Flying, 194.
Flying-fish, 171.
Folding of body wall, 26, 40, 41, 73.
Follicular cells, 4.
Food, collection, 43, 84, 85, 98–9, 115, 139, 143, 145, 153, 194, 195, 196, 213, 215, 217, 227, 233, 236, 245–6, 249, 260.
— current, 98–9.
— supply, 213, 215.
Foot, 52–64, 65, 107, 205, 249, 250.
Foramen, 121, 122, 134.
Forest litter, 38.
Fredericella sultana, 101.
Frenulum, 153, 154, 155, 228.
Fresh-water animals, 231.
Frictional force (*see also* Point *d'appui*), 80, 89, 94, 95, 97, 129, 143, 144, 154, 162, 166, 178, 230.
Frog, 114.
Frontal membrane, 104, 106.

Gadus, 188.
Gait, 254–5.
Gamete, 4, 150, 217–8, 230.
Ganglion (*see also* Ventral nerve cord), 230.
Gastraea, 12, 209.
Gastraea theory, 209–10, 212, 219, 222, 223, 229.

Gas transport (*see also* Respiration), 238.
Gastric pocket, 8, 9, 11, 12, 13, 14, 15, 22, 210, 216.
Gastropoda (*see also* Aspidobranchia, Nudibranchia, Opisthobranchia, Pteropoda, Pulmonata, Stenoglossa, Streptoneura, Taenioglossa, Tectibranchia), 65, 66, 96, 168, 205, 249, 250.
— ciliary locomotion, 53, 61–2, 63.
— foot, 53–64, 65, 96–7, 205.
— hydrostatic skeleton, 54, 64, 96–7.
— locomotion, 46, 53–65, 69, 70, 83, 168–9, 250.
— mucus secretion, 63, 96.
— rate of locomotion, 61, 62.
— visceral mass, 54, 60, 249.
Gastrotricha, 219, 221.
Gastrula, 209, 211.
Genital duct, 3, 15, 16, 28, 230, 231, 250, 253.
Genital follicle, 6.
Genital system, 122, 230–2.
Geodesic spiral, 35.
Geodesmus lineatus, 45.
Geonemertes, 38, 39.
Geonemertes dendyi, 40.
Geophilomorpha, 142, 143, 144, 256, 257.
Geophilus longicornis, 142.
Geoplana, 48, 49.
Geoplana notocelis, 48.
Geoplana traversii, 49.
Germ cells, primary and secondary, 6–7.
Germ layer (*see also* Ectoderm, Endoderm, Mesoderm), 7, 19.
Germinal epithelium, 4, 7.
Germinal primordia, 4, 6.
Giant axon, 139, 153.
Gibbula (*see also* Ctenidium), 59–60.
Gill (*see also* Ctenidium), 10, 22, 124, 137, 146, 201, 245–6, 251–2, 256.
Gill slit, 245–6.
Girdle, 153, 173, 189.
Gizzard, 77.
Gland cell (*see also* Mucus cell), 67.
Glossobalanus minutus, 244.
Glycera, 25, 110, 138, 139, 142, 167.
Glycera dibranchiata, 25, 167.
Glyceridae, 248.
Glycogen, 194.
Glycoprotein, 77.
Gnathostomata, 188, 189, 191, 192.
Gobius, 190.
Golfingia (*see also* Phascolosoma), 110.
Gonad, 4, 12, 28, 53, 156, 217–8, 230, 253.
— Annelida, 230, 249.
— association with coelomic epithelium, 3, 15, 16.
— cavitation, 4, 216.
— development, 6, 7.
— Echinoidea, 245.
— Mollusca, 251–2.
— pseudometameric, 17, 19, 28, 29.
— Pterobranchia, 156.
— regression, 4.
Goniada, 135.
Gonocoel theory, 2, 3–7, 8, 12, 15, 16, 28, 216, 217, 250, 251.
Gonoduct (*see* Genital duct).
Gradient, pressure, 122, 135, 162, 255, 256.
— septal development, 145.
Grain size, 91.
Graptolithina, 201.
Grouper, 173.
Growth, 23–4, 26, 145–6, 218, 221, 259.
Growth and Form, 30.
Gudgeon, 194.
Gular membrane, 147, 148, 149.
Gunda (*see also* Procerodes), 8.
Gunda segmentata, 5.
Gut, 2, 4, 8, 53, 73, 75, 108, 137, 138, 147, 152, 208, 212, 213, 219, 239, 245.
— contents, 146.
— diverticulum, 5, 8, 19, 168.
— pseudometameric, 5, 17, 19.
— suspensory muscle, 135, 136, 137, 139, 141, 163, 248.
Gymnolaemata, (*see also* Anasca, Ascophora, Cheilostomata, Ctenostomata, Cyclostomata), 104, 105, 106, 236.

Habitat (*see also* Freshwater animals, Interstitial fauna, Marine animals, Pelagic animals, Periphytic animals, Terrestrial animals), 30, 213.
Haemal spine, 190, 192.
Haemerythrin, 238.
Haemocoel, 1, 16, 142, 250, 253–7, 258.
Haemolymph, 107.
Haemopis, 146.
Halcampa chrysanthemum, 94.
Halcinia convexa, 56.
Halibut, 196.
Halicryptus, 93.
Haliotis, 56, 57, 58, 63.
Haminoea antillarum, 61.
Head, 142, 174, 177, 180, 229, 257.
— shaft, 110, 114.
— shield, 170.
Heart, 246, 247, 250, 251.
Helcion pellucidum, 61.
Helix, 63, 64, 96, 97.
Helix aspersa, 55.
Helix pomatia, 55.
Helobdella stagnalis, 47.

Hemichordata (*see also* Enteropneusta, Pterobranchia), 7, 12, 19, 151, 155–9, 164, 201, 222, 223, 224, 225, 226, 239, 242, 247.
Herbivorous habit, 196.
Hermaphroditism, 231.
Herring, 196.
Hesione, 136, 139.
Hesionidae, 135, 136, 141, 248.
Heterocercal tail, 169, 192.
Hippocamidae, 173.
Hippoglossus, 171.
Hirudinea, 25, 47, 52, 53, 67–8, 78, 141, 145, 146, 167, 168, 179, 186, 205, 214, 249, 253, 255.
— coelom, 67–8, 239.
— creeping, 46, 47, 67, 70, 197–8, 215.
— musculature, 47, 168.
— swimming, 25, 167, 179, 180.
Hirudo, 68, 197.
Holocentrus, 171, 172.
Holostei, 188, 192.
Holothuria captiva, 100.
Holothuria grisea, 75, 108, 111, 113.
Holothuria rathbuni, 99.
Holothuria surinamensis, 99.
Holothuroidea, 75, 96, 98, 100, 107, 112–3, 239, 244, 245.
— burrowing, 85, 90–1, 101, 111, 114, 116.
Homocercal tail, 192.
Hook, 161.
Hoplolaimidae, 79–80, 82–3.
Hoplolaimus, 79, 80, 83.
Hoplopagrus guntheri, 193.
Horizontal septum, 188, 189, 190.
Hormone, 218, 221.
Host, 161.
Hyalinoecia, 139, 142.
Hydrocoel, 201, 240, 241, 242.
Hydrodynamic analysis, 174.
Hydrofoil, 169, 170.
Hydroid (*see also* Hydrozoa), 24, 155.
Hydropore, 240, 244.
Hydrostatic independence, 120, 123, 124, 139, 146, 147, 149, 151, 154, 155, 157, 159, 162, 163, 164, 227, 248, 256, 259.
Hydrostatic organ (*see also* Hydrostatic skeleton), 16, 27, 102, 159–60, 212, 235, 244, 247.
Hydrostatic pressure, 72, 77, 84, 86, 107–15, 116, 146.
— change, 33, 71, 86, 88, 104, 116, 122, 144, 151, 234, 236.
— transmission, 31, 33, 71, 116, 120, 144.
Hydrostatic skeleton, 46, 71, 84, 116–7, 187, 202, 215, 217, 220, 227, 229, 235–6, 256, 257–8, 260.

— Annelida, 133, 135, 248, 249.
— Anthozoa, 73–5, 92, 108, 114.
— Echinodermata, 101, 245.
— Ectoprocta, 103–7, 116, 236.
— Insecta, 142, 250.
— Mollusca, 54, 64, 96–7, 250.
— Nemertea, 37, 68, 85, 115, 216.
— Onychophora, 250, 254–5.
— parenchymatous, 34, 67–8, 71.
— Pogonophora, 151–4, 227.
— properties, 31–5.
— Pterobranchia, 155–7, 227.
Hydrozoa, (*see also* Milleporina, Siphonophora), 208.
Hypaxial muscle, 189.
Hypertrophy, 49, 65, 138, 246.
Hypoblast, 8.
Hypocaudal lobe, 192.
Hypocercal tail, 192.
Hypothetical model, 31–8, 40, 52.
Hypotonic medium, 108.

Ilyanassa, 61.
Inarticulate brachiopod, 236.
Inclination, body, 170, 177.
— fibre, 36, 77.
— parapodium, 183, 184–6.
Incomplete septum, 121.
Incompressibility, 31, 33, 34.
Independence of fluid system, 149, 154, 155, 159, 162, 163, 164, 227.
Independent metamerism, 8.
Indeterminate cleavage, 222.
Inertia, 166.
Inextensible fibre, 35–6, 41, 76–7, 137–8.
Inflexible body, 167.
Infusoria, 207.
Initiation of swimming, 197.
Innervation of fin, 174.
Insecta (*see also* Diptera, Endopterygota), 108, 142, 253, 254, 255.
— creeping welt, 143.
— locomotion, 80, 143–4.
— musculature, 144–5.
Insertion, fin, 170.
— muscle, 187, 199, 203, 248.
— parapodium, 187, 199, 203, 248.
— septum, 137–8, 139, 140, 149, 199.
Intercalation of mesenteries, 10.
Intercellular space, 16.
Intermuscular bone, 190.
Internal pressure (*see also* Hydrostatic pressure), 71, 72, 86, 88, 108, 109–11, 112, 113, 114.
Intersegmental region, 131, 132, 133, 137, 138, 162, 232, 234.
Interstitial fauna, 85, 213, 221.

Interstitial fluid, 216.
Intervertebral articulation, 191.
Intestine (*see also* Gut), 4, 12, 137, 139, 141, 161.
— diverticulum, 5, 8, 19, 169.
Intracellular digestion, 75, 210.
Intrinsic musculature, of parapodium, 131, 133.
— of tail fin, 192.
Invaginated skeleton, 142.
Irradiation, 7.
Irrigation of burrow, 112, 130, 137, 139, 248.
Isolation, hydrostatic, 124, 139, 146, 147, 149, 151, 154, 155, 157, 163, 164, 248, 256, 259.
Isometric contraction, 74.
Isometric limit, 113–4, 117.
Isometric tension, 114.

Jack, 193.
Joint, 26, 142.

Keel, 193.
Keratin, 77.
Kidney, 8, 252.
Kinorhyncha, 219, 221.

Labial tentacles, 8.
Labridae, 170.
Lactophrys, 172.
Lacunar tissue, 53, 54, 97, 160.
Lamellibranchiata, 46, 107, 250.
Lamellisabella, 152, 153.
Lamellisabella zachsi, 151, 152.
Lamellisabellidae, 154.
Lamna cornubica, 193.
Larva, 207, 220, 221, 222, 226, 230, 231, 239, 243, 244, 246, 255.
— ammocoete, 187.
— anuran, 166, 167.
— ascidian, 25, 26, 200, 202, 221, 245, 247.
— auricularia, 239.
— bipinnaria, 239.
— dipleurula, 222, 224, 239–43, 244.
— insect, 108, 142, 143, 179, 253, 255.
— planktotrophic, 20.
— tornaria, 239.
— trochophore, 220, 221, 222, 224, 253.
Lateral compression, 167.
Lateral fin, 168, 170, 193.
Lateral flexure, 33, 199.
Lateral motion, 174, 177, 180, 181.
Lateral musculature, 193–6, 201.
Lateral pedal locomotory waves, 56, 59.
Lateral thrust, 124.
Leading edge, 177, 184.

Leakage of coelomic fluid, 72, 121, 124, 139, 141, 236, 256, 257.
Leanira, 137.
Leg (*see also* Appendage), 129.
Lemniscus, 160, 161.
Length, change of, 77, 82–3, 254–5.
— muscle fibre, 73, 114, 123.
Leptoplana, 49, 50, 57, 167.
Leptoplana lactoalba, 45, 49.
Leptoplana tremellaris, 50.
Lever, 124.
Leverage, 99, 117.
Lift, 169, 170, 192.
Ligament, 107, 189, 191.
Limax, 64.
Limicoline oligochaete, 231.
Limit, of change of shape, 36, 41, 69, 73.
— isometric, 113–4, 117.
Limnodrilus, 122.
Limpet, 53, 63.
Lineus gesserensis, 38–40.
Lineus longissimus, 38–41.
Lineus socialis, 41.
Lingula, 76, 236, 237.
Lipobranchius, 73.
Lip, 84.
Lithotrya, 107, 254.
Littorina, 56.
Lizard-fish, 171.
Lobe, cephalic, 151, 154, 156.
— epipodial, 169.
— epistomial, 155, 156, 226, 227.
Local change of shape, 124, 144, 161, 228, 259.
Local contraction, 33, 71, 93, 123.
Local volume change, 31.
Locking mechanism, 161.
Locomotion (*see also* Ambulation, Burrowing, Looping movement, Swimming), 16, 227, 235, 257, 259–60.
— Asteroidea, 99, 100, 205.
— ciliary, 42–3, 61–3, 238.
— Echiuroidea, 90, 97–8.
— Ectoprocta, 65–6, 69, 70.
— Enteropneusta, 157–9, 238.
— Gastropoda, 46, 53–64, 69, 70, 83, 168–9, 250.
— Hirudinea, 25, 46, 47, 67, 70, 167, 179, 180, 197–8, 215.
— Holothuroidea, 85, 90–1, 99–101, 116, 245.
— in burrow, 90, 93–4, 124, 129, 150, 158, 159, 248.
— in tube, 130, 150, 154, 155, 227, 236.
— Inseca, 80, 143–4.
— Myriapoda, 142–4, 256–7.
— Nematoda, 78–83, 93, 116.

— Nemertea, 42–53, 69, 70, 84–5, 98, 167, 168.
— Oligochaeta, 79, 97, 118–20, 143, 144, 163.
— on soft substratum, 53, 214.
— on substratum, 94–8, 100, 115, 119, 124–9, 141, 143, 157, 159, 167, 205, 213–14, 215, 228, 235, 249.
— on surface film, 43, 63.
— Onychophora, 254–5.
— pedal, 49–50, 54–66, 70, 83–4, 115, 214, 252.
— peristaltic (*see* Peristaltic locomotion).
— Phoronidea, 150–1.
— Polychaeta, 89, 93, 124–30, 131, 134, 162, 182–6, 187, 232–5, 248.
— Pterobranchia, 227, 243.
— rate, 44–5, 49, 61, 62, 65, 66, 179, 194–6, 254.
— reversed, 46, 51–2, 60–1, 97, 130, 174, 191.
— Turbellaria, 42–53, 69, 83, 168, 213–4.
Locomotory and turgor muscles, 144–5.
Locomotory efficiency, 43, 167, 187.
Locomotory theory of segmentation, 25–7, 29, 30, 202, 203–4.
Locomotory waves (*see also* Pedal locomotory waves, Swimming, Undulatory movements), 126–9, 174–86, 202.
— alternate, 49–50, 56, 57, 58, 125, 235.
— composite, 56, 59, 60.
— co-ordination, 51, 120, 197.
— diagonal, 56, 60.
— direct, 54–5, 60, 65, 94, 95, 96, 125–6, 143.
— ditaxic, 49–50, 56, 57, 58, 126, 168.
— lateral, 56, 59, 167.
— monotaxic, 56, 60, 126.
— opposite, 56, 126.
— retrograde, 54–5, 56, 59, 94, 125.
— transmission, 123, 191, 196–8.
— velocity, 65, 177, 178, 179, 180, 182.
Longitudinal axis, 22, 24, 33, 65, 191.
Longitudinal inextensible fibre, 41.
Longitudinal folding, 41.
Longitudinal muscles, 31–4, 64, 77, 83, 116, 138, 139, 153, 155, 156, 160, 214, 226, 233, 245, 258.
— antagonistic, 131, 143, 199.
— dorsal, 134, 138, 139.
— modification, 49, 131, 134, 162.
— role in locomotion, 46–9, 51, 83, 93, 96–7, 116, 125–8, 129, 130, 134, 135, 170, 186, 190–2, 199, 215, 234, 248.
— role in proboscis eversion, 86, 88, 89.
— segmented, 123, 141, 203, 248.
— unilateral contraction, 33, 234.
— ventral, 48, 49, 69, 97, 139, 252.

Longitudinal thrust, 83.
Looping movement, 46–7, 67–8, 69, 205, 214, 249.
Lophophorates (*see also* Brachiopoda, Ectoprocta, Phoronidea), 150–1, 155, 222, 224, 225, 226, 235–8, 247, 259, 260.
Lophophore, 66, 104, 150, 151, 153, 154, 155, 226, 227, 236–7, 238, 243, 245, 246, 259.
Lophopodella, 65, 102.
Lophopodella carteri, 66.
Lophopus, 65, 66.
Low pressure region, 146.
Lumbricidae, 231, 232.
Lumbricine chaetae, 231, 232.
Lumbricus, 76, 77, 118, 119, 121, 122, 123, 126, 140, 141.
Lumbricus terrestris, 72, 110.
Lumbrinereidae, 134.
Lumbrinereis, 72, 134.
Lungs, 169.
Lymnaea palustris, 61.

Mackerel, 189, 193, 196.
Macroscopic animals, 215.
Madreporic vesicle, 244.
Madreporite, 244.
Maintenance activity, 217.
Makaira albida, 193.
Makaira mitsukurii, 192.
Malacobdella, 34, 39, 41, 46, 47.
Malacobdella grossa, 39–41, 46, 47.
Mandibles, 143.
Mantle, 107, 237.
— canal, 237, 258.
— cavity, 46, 107.
Marginella arena, 61.
Marine animals, 38–41, 196, 230, 257.
Marlin, 192, 193.
Marphysa, 137, 138.
Matricin, 77.
Maturation, sexual, 4, 218, 221, 259–60.
Maximum pressure, 109–14.
Maximum tension, 114.
Maximum thrust, 144.
Maximum volume, 36, 77, 113.
Mechanical advantage, 133, 135, 199, 203.
Mechanical efficiency, 74, 114, 117.
Mechanical function of coelom, 4, 16.
Mechanical independence, 120–4.
Mechanical stimulation, 110, 111.
Mechanical stress, 24, 50, 197.
Medulla, 198.
Medusoid, 11, 12, 14, 20.
Megascolecidae, 119, 122, 231.
Melibe leonina, 61.
Membranipora, 105.

Membranous sac, 106, 107.
Menophylla, 21.
Mesenchyme, 4, 6, 8, 15, 26, 201, 214, 244.
Mesentery (*see also* Septum), 10, 109, 114, 152, 153, 203, 237.
Mesoblast, 8.
Mesocoel, reduction, 22, 157, 229, 238, 247.
Mesoderm, 4, 6, 7, 8, 15, 16, 24, 216.
Mesodermal band, 4, 19.
Mesodermal derivative, 24.
Mesodermal growth, 24, 26.
Mesodermal teloblast, 4, 6, 19.
Mesogloea, 73, 113.
Mesomere (*see also* Mesosome), 12.
Mesosome, 156, 222, 226–8, 243.
Mesostomum tetragonum, 45.
Metabolism, 194.
Metabolite, 230.
Metacoel (*see also* Somatocoel), 150, 151, 152, 154, 155, 156, 164, 223, 225–8, 244, 245, 247.
— reduction, 157, 238, 244.
Metamere (*see also* Metasome), 12.
Metamerism, 1, 17–27, 140, 150, 163, 164, 202, 247–57, 259, 260.
— coelom (*see also* Septum), 7, 8, 12, 16, 27.
— evolution, 11, 12, 17–30, 165, 168, 174, 200–2, 203–4, 228–35, 259.
— function, 2, 4, 17, 28, 161–4, 165, 168, 200, 202–4, 259.
— genital system, 17, 230, 231, 232.
— musculature, 24, 25, 26, 118, 123, 174, 187, 198–200, 203, 209, 229, 230, 232, 247–8, 252, 254.
— nervous system, 163, 174, 229, 249, 251, 259.
— polyphyletic origin, 2, 27, 28, 29, 202, 260.
Metamorphosis, 220.
Metanephridium, 231.
Metasome, 155, 156, 157, 222, 226–8, 235, 236.
Metazoa, 1, 2, 211–2, 259.
— origin, 1, 207–12, 258.
— primitive, 22.
Metridium, 73, 74, 75, 112, 113, 114.
Metridium marginatum, 65.
Metridium senile, 74, 108, 109, 113.
Micromere, 19.
Microphagy, 213, 215.
Micropora, 105.
Micropterus dolomieu, 193.
Microscopic animals, 165, 166.
Migration, 195.
Milleporina, 76.
Mitochondria, 194.

Mixonephridium, 72.
Model, hypothetical, 31–8, 40, 52.
Mola, 172.
Mollusca (*see also* Amphineura, Cephalopoda, Gastropoda, Lamellibranchiata, Monoplacophora, Polyplacophora), 10, 16, 19, 20, 22, 28, 52–64, 107, 205, 208, 223, 224, 225, 247, 249–53, 254, 258.
— coelom, 16, 28, 250–3.
— ctenidium, 10, 22, 251, 252.
— foot, 10, 52–64, 107, 205, 249, 250.
— haemocoel, 16, 28, 53, 250.
— locomotion, 46, 53–64, 69, 70, 83, 168–9, 250.
— metamerism, 11, 22, 251–2.
— musculature, 10, 21, 53–64, 69, 251.
— origin, 10, 251–2.
— pericardium, 16, 28, 250, 251, 252, 258.
— reno-pericardial canal, 28.
— shell, 21, 62, 250–1.
— visceral mass, 54, 60, 62, 249, 250–2.
Monacanthus, 172.
Monacanthus hispidus, 173.
Monophyletic origin of coelom, 6, 27.
Monoplacophora (*see also* Neopilina), 10, 11, 21, 22, 249–52.
Monopylephorus, 122.
Monotaxic locomotory waves, 56, 60, 126.
Morula, 209, 210.
Mosquito, 108.
Motor impulse, 174.
Mouth, 8, 9, 65, 73, 75, 84, 92, 93, 95, 100, 108, 149, 210, 212, 213, 222, 224, 226, 236, 240.
Movement, direction of, 12, 46, 51–2, 60–1, 65, 97, 130, 174, 191.
— in burrow, 93–4, 153, 227, 228.
Mucosa, 161.
Mucus, 93, 98, 213, 214, 250.
— adhesive, 46, 48, 62, 63, 96, 98, 205.
— cell, 63.
— net, 99, 217.
— secretion 38, 43, 48, 63, 65–6, 96.
Multicellular animals (*see also* Metazoa), 207, 210, 212.
Muscle bulb, 233.
Muscle cell, 194–6, 202.
Muscle fatigue, 42.
Muscle fibre, length of, 73, 114, 123.
— overlap, 187, 200.
Muscle scar, 10, 21.
Muscle tonus, 108, 113.
Muscle, type, 145, 194–6.
Muscles, antagonistic, 26, 31, 33, 34, 37, 69, 82–3, 120, 131.
— flexor, 31, 124, 125, 126, 133, 134, 137.
— hypertrophy of, 49, 65, 138, 246.

— locomotory and turgor, 144–5.
— optimal working length, 73–4, 114.
— restoring force, 26, 31, 69, 191.
— retractor, 66, 84, 86, 87, 88, 93, 144, 160.
— skeletal, 114, 145, 244.
— ventral longitudinal, 48, 49, 69, 96, 97, 133, 134.
Muscular co-ordination, 33, 51, 120, 196–8, 217.
Muscular locomotion, 42, 44–5, 52, 53, 54–5, 69, 205, 214.
Musculature, Anthozoa, 73, 109, 113.
— body-wall (*see* Body-wall musculature).
— Brachiopoda, 236.
— Chordata, 26, 173, 180, 187–96, 201–2, 203.
— Echinodermata, 96, 99–101, 116, 244–5.
— Ectoprocta, 65–6, 102–7.
— gonadial, 4.
— metamerism of, 24, 25, 26, 118, 123, 174, 187, 198–200, 203, 209, 229, 230, 232, 247–8, 252, 254.
— Mollusca, 10, 21, 53–64, 69, 251.
— Nematoda, 77–8, 79, 82–3, 187.
— Nemertea, 34, 35–42, 67, 186, 187.
— Oligochaeta, 118–24, 134, 140–1, 144–5.
— Polychaeta, 130–9, 141–2, 145–9, 162–3, 182, 184–7, 233, 248–9, 255–6.
— segmental, 8, 17, 22, 25, 29, 123, 131–3, 162, 187, 202, 203, 229, 230, 247, 248, 249, 254, 259.
Mustelus, 171.
Mya, 107.
Myocomma, 187, 189, 190, 191, 199, 203.
Myofibril, 202.
Myoglobin, 194.
Myotome, 174, 187, 188–91, 196, 199, 202, 203.
Myriapoda (*see also* Chilopoda), 142–4, 253, 254, 256–7.
Myxine, 188.
Myxocoel, 28.

Nassarius, 62.
Nassarius obsoleta, 61, 62.
Nassarius trivittata, 61.
Natica, 54.
Nautilus, 252.
Neck, 160.
— muscle, 160, 161.
Needle-fish, 171.
Negative fluid pressure, 31.
Nematoda (*see also* Ascaris, Draconematidae, Epsilonematidae, Hoplolaimidae), 26, 76–84, 108, 116, 144, 156, 219, 221, 258.
— cuticle, 26, 75–8, 80, 82, 116, 144.

— extensibility, 77–8, 83.
— internal hydrostatic pressure, 108, 109, 112, 113–4, 116.
— locomotion, 78–83, 93, 116.
— musculature, 77–8, 79, 82–3, 187.
Nematomorpha, 77, 219.
Nemertea, 3, 20, 24, 75, 200, 209, 216, 223, 224, 225, 249, 250, 252, 258.
— basement membrane, 35–41, 76.
— change of shape, 34, 35, 38–42, 71, 115.
— epidermis, 40–2.
— excretory organs, 231.
— extensibility, 35–42.
— habitat, 38–9, 51, 167, 231.
— habits, 41, 51–2.
— hypothetical model, 31–8, 40, 52.
— locomotion, 42, 45–53, 67, 68, 69, 70, 78, 84, 98, 167, 168, 186.
— mesenchyme, 26, 68, 216.
— musculature, 34, 35–42, 52, 67, 115, 186, 187.
— proboscis, 85, 116, 201.
— pseudometameric, 5, 17, 18, 24, 249, 252.
— relationships, 200, 201, 223, 224, 225, 250.
Neoblast, 67.
Neopilina (*see also* Monoplacophora), 10, 22, 250, 251, 252, 253.
Neoteny, 200, 220–2, 228, 247, 260.
Neothunnus macropterus, 193.
Nephridiopore, 71–3, 257.
Nephridium, 8, 15, 29, 71–3, 140, 163, 218, 249, 251, 257.
— Annelida, 71–3, 230–1, 257.
— Platyhelminthes, 8, 15.
— pseudometameric, 17, 18.
Nephrocoel theory, 3, 15.
Nephtyidae, 141, 202.
Nephtys, 73, 89, 129–30, 131, 132, 138–9, 167, 182–6, 187.
Nereidae, 141, 202, 248.
Nereis, 72, 110, 124–9, 132, 133, 135, 136, 137, 145, 167, 180, 182–4, 185, 187.
Nereis diversicolor, 183.
Nerine, 135, 137, 141.
Nerita nodosa, 56.
Nerita tessellata, 56.
Nerve, spinal, 196.
Nerve cord, rhythmical activity, 198.
— sheath, 133, 138, 139.
— transection, 196–8.
Nervous co-ordination, 33, 120, 196–8, 217, 229–30.
Nervous system, central, 8, 9, 98, 196–8, 218.
— Cnidaria, 8, 9, 208.
— metameric, 163, 174, 229, 249, 251, 259.

SUBJECT INDEX

— Pogonophora, 153.
— pseudometameric, 17, 18, 29, 252.
— Turbellaria, 18, 208, 214, 217.
— ventral, 8, 9, 131.
Nervous transmission, 196.
Neural spine, 190, 192.
Neuropodium, 232.
Neurosecretory cell, 218.
Nocturnal fish, 194.
Normal component, 174, 181.
Notochord, 26, 199, 201–2, 203.
Notomastus, 134.
Notopodium, 130, 131, 132.
Nudibranchiata, 61.
Nutrition (*see also* Food), 4, 12, 209, 213.

Oblique muscles, 63, 64, 122, 131, 137, 138, 139, 141.
Occlusion of coelom, 67–8, 153, 156, 157, 164, 236, 237, 238, 249, 253.
Octocorallia, 210.
Octocorallia, 13.
Oesophagus, 149, 236.
Oil, 169.
Oligobrachia, 152.
Oligochaeta (*see also* Earthworm, Lumbricidae, Megascolecidae), 130, 131, 142, 146, 163, 230–5, 248, 249, 253, 255.
— dorsal pore, 72.
— nephridium, 72, 230–1.
— reproduction, 24, 230.
— septum, 118, 120–4, 140, 144–5, 162, 256.
Oligomerous animals, 19, 22, 118, 150–9, 164, 222–8, 229, 235–47, 259.
Oligomerous condition, 12, 210, 223, 224.
Onychophora, 76, 111, 253, 254, 256.
Oocytes, 149.
Open system, 53, 73.
Ophelia, 147, 149, 163.
Ophiodromus, 136.
Ophiuroidea, 99, 244.
Opisthobranchia, 169.
Opposite locomotory waves, 56, 126.
Oral chaeta, 90.
Oral disk, 92.
Oral tentacle, 85, 90, 98, 101.
Ordovician, 21, 201.
Orya, 144, 256.
Oscillation, 193.
Osmotic regulation, 231.
Osmotic stress, 108.
Ossicle, 244.
Osteichthyes, 192.
Overlap of muscles, 187, 200.
Ovum (*see* Eggs).
Oweniidae, 233.

Oxygen (*see also* Respiration), 115, 218, 230, 238.
— diffusion, 44.
— transport, 218, 238.
Oxyuris equi, 77.

Pacemaker, 98.
Paedogenesis (*see also* Neoteny), 220–2, 247, 260.
Palaeontology, 30, 239.
Palaeopterygii, 192.
Palaeozoic, 201.
Papilla, 91, 153, 154.
Paraoesophageal cavity, 147, 148, 149.
Parapodium, musculature, 130–3, 135, 139, 141, 162.
— origin, 10, 232–5, 255.
— reduction, 134, 142, 162, 248.
— role in locomotion, 124–30, 135, 141, 182–6, 202.
Parasitism, 79, 159, 214, 249.
Parenchyma (*see also* Mesenchyme), 16, 66–8, 69, 70, 153, 156, 216, 236, 253.
— as hydrostatic skeleton, 34, 67–8, 70, 116.
— structure, 67–8.
Pargo, 193.
Parietal muscle, 103, 104, 106, 107, 109, 113.
Parieto-diaphragmatic muscle, 103.
Parieto-vestibular muscle, 103.
Parrot-fish, 173.
Particle size, 91.
Passive dragging, 50, 98, 129, 134, 157, 168, 238.
Patella, 63.
Peachia, 92, 94.
Pectinatella, 65, 66.
Pectoral fin, 166, 169, 170, 171, 172, 173, 192, 196.
Pectoral girdle, 173, 189.
Pedal disk, 10, 64–6, 92, 95.
Pedal gland, 63.
Pedal locomotory waves, 49–50, 54–66, 70, 83–4, 115, 214, 252.
— analysis, 54–5.
— Cnidaria, 64–5.
— Ectoprocta, 65–6.
— Gastropoda, 53–61, 64, 96, 101, 125, 250.
— Turbellaria, 46, 48–50, 69, 214.
Pedal muscle, 251.
Pedicel, 237.
Peduncle, 107, 191, 193.
Pelagic animals, 8, 12, 167, 196, 212, 213, 215, 220, 221, 228, 230, 231.
Pelmatozoa, 240.

SUBJECT INDEX

Penetration of substratum, 86, 90, 91, 92–3, 114, 115, 129.
Pentactula, 241, 244.
Pentaradiate symmetry, 239.
Perca flavescens, 193.
Perch, 193, 194.
Percoids, 171.
Perforation of septum, 121, 122, 134, 147–9.
Pericardium, 16, 28, 246, 247, 250, 251, 252, 258.
Perichaetine chaetae, 231–2.
Peridionites, 239.
Peripatopsis, 111.
Peripatus, 254, 255.
Peripheral stimulation, 197.
Periphytic animals, 220.
Peristalsis, 98–9, 115–6, 130, 139, 156, 214, 215, 217.
Peristalitc locomotion, 46, 92, 93–4, 97, 98, 115–7, 161–4, 217, 227, 228, 248, 252.
— Anthozoa, 94–5.
— Enteropneusta, 157–9.
— Holothuroidea, 85, 91, 96, 100, 101.
— Nematoda, 80.
— Nemertea, 46, 51, 67, 68–70, 168.
— Oligochaeta, 118–24, 162.
— Phoronidea, 151, 154.
— Pogonophora, 154.
— Polychaeta, 130, 134, 142, 248.
Peristome, 99.
Peritoneum, 6, 190, 219.
Perivisceral cavity, 75, 226, 227, 236, 244, 245, 247, 250.
Perivisceral sac, 246.
Petromyzon, 187.
Pharyngotremy, 245–6.
Pharynx, 89, 92, 245, 246.
Phascolosoma gouldi, 86, 110.
Pheretima, 122, 141.
Pheretima posthuma, 122, 123.
Phleboedesis, 254.
Phoronidea, 154, 164, 222, 227, 235–6.
— coelom, 19, 150–1, 153, 155, 226.
— movements, 150–1.
— relationships, 12, 210, 223, 225.
Phoronis, 150, 235.
Phragma 142.
Phylactolaemata, 65–6, 101, 102–4, 236.
Phyllodocidae, 141.
Phylloplana littoricola, 45, 49.
Pike, 194, 195.
Pitch, 168, 192.
Placodermi, 169, 192.
Planaria, 43, 46.
Planaria gonocephala, 45.
Planaria maculata, 44, 45.
Planaria polychroa, 45.

Planarian, 34, 75, 205, 215, 216, 221, 231, 249, 250, 252.
Plankton (*see also* Pelagic animals), 228.
Planktotrophic larvae, 20.
Planocera californica, 45.
Planula, 207, 211.
Planuloid theory, 210–2.
Plasma, 238.
Platyhelminthes (*see also* Cestoda, Turbellaria), 6, 23, 208, 211, 214, 223, 224, 225, 252.
— pseudometameric, 5, 8, 17, 249.
Plectognaths, 171, 172.
Plectrus palustris, 79.
Pleuracanthodii, 192.
Pleurobranchus, 168, 169.
Podium, 96, 99, 101, 116, 205, 244, 245.
Pogonophora, 19, 150, 151–4, 155, 156, 164, 210, 222, 223, 225, 226, 227, 238.
Point d'appui (*see also* Anchor), 62, 63, 94, 98, 101, 116, 119, 120, 125, 143, 181, 232.
Poling, 130.
Polinices, 54, 61, 62, 63.
Polinices draconis, 61.
Polybrachia, 153.
Polycelis, 39, 40, 44, 46.
Polycelis nigra, 39–41.
Polycelis tenuis, 45.
Polychaeta, 24, 50, 130, 146, 163, 202, 220, 226, 248–9, 253, 254, 256.
— burrowing, 88–9, 91–2, 93, 110, 112, 114, 131, 134, 257.
— gill, 124, 137, 146, 201, 256.
— hydrostatic pressure, 110, 113, 114.
— locomotion, 118, 124–30, 162, 167, 182–6, 187, 232–5, 248.
— musculature, 130–42, 144–7, 150, 162, 163, 202.
— nephridium, 71–3, 230–1.
— origin, 230–5.
— parapodium, 10, 125–33, 134, 135, 139, 141, 142, 162, 182–6, 202, 232–5, 248, 255.
— proboscis, 89, 91–3, 110, 112, 129, 147–50, 233.
— septum, 133, 134–9, 162, 226, 232, 234–5, 248, 255–6.
— stomodeum, 233.
— swimming, 25, 129, 182–6, 248.
Polycladida, 50, 215.
— locomotion, 44, 45, 48–50, 69, 168, 214.
Polymerous condition, 12.
Polynoidae, 135, 141.
Polyophthalmus, 149.
Polyphyletic origin of coelom, 5, 6, 27, 216, 217, 259, 260.

— metamerism, 2, 27, 28, 29, 260.
— oligomerous animals, 228, 260.
— pseudocoel, 220, 221, 260.
Polypide, 66, 102–7, 116, 151, 236, 260.
Polyplacophora, 53, 63, 69, 70, 250, 251, 252.
Pomatias, 56, 57, 58.
Pomatias elegans, 57.
Pomatobius fluviatilis, 111.
Pontobdella, 47.
Pontobdellidae, 167.
Porbeagle, 193.
Porcupine-fish, 172.
Porifera, 208, 209, 210, 211.
Power stroke, 124, 125, 126, 129, 130, 133, 182, 183, 184, 234.
Preadaptation, 215.
Pre-Cambrian, 21.
Predator, 139, 194, 215, 227.
Prehensile organ, 155, 215, 227.
Pre-oral lobe, 226, 239, 240.
Pressure, change, 33, 71, 116, 122, 234.
— gradient, 122, 135, 162, 255, 256.
— hydrostatic, 107–15.
— low, 146.
— resting, 108–14, 147.
— transmission, 116, 122–3, 144, 148, 162.
— working, 72, 109–14.
Priapulida, 219.
— locomotion, 87–8, 124.
— loss of coelomic fluid, 73.
Priapulus, 73, 87, 88, 93, 116, 124.
Primary germ cells, 6, 7,
Primary mesoderm (*see also* Mesenchyme), 4, 6, 8, 16.
Primary segments, 11, 20, 21, 22.
Proboscis, 92–3, 115, 147–9, 163, 215, 216, 239.
— Acanthocephala, 159–61.
— Echiuroidea, 89–90, 98.
— Enteropneusta, 157–8, 238.
— eversion, 84, 85, 86, 88, 91, 93, 110, 112, 147–9, 160–1, 216.
— Mollusca, 58, 84, 89.
— Nemertea, 85, 98, 115, 116, 168, 201, 216.
— Polychaeta, 89, 91–3, 110, 112, 129, 147–50, 233.
— Priapulida, 87–8.
— retraction, 86, 89, 98, 157, 160.
— Sipunculoidea, 85–6, 102, 110, 112.
— Turbellaria, 215.
Procerodes (*see also* Gunda), 8, 46.
Pro-ciliata, 209.
Productivity, 218.
Proglottid, 1, 22, 23.
Proleg, 143, 255.
Proliferation of segments, 20.

Proliferation zone, 20, 23, 24, 230.
Propodium, 61–2, 63.
Proprioceptive arc, 120, 196–7.
Proprioceptor, 198.
Propulsive force (*see also* Thrust), 168, 170, 182, 230.
Propulsive surface, 167.
Prostomium, 230.
Protannelis meyeri, 6.
Prothelmintha, 207.
Protista (*see also* Protozoa), 207, 209, 212.
Proto-annelid, 230, 232.
Proto-ascidian, 201.
Protobilateria, 212–4, 215, 217, 228.
Protocercal tail, 192.
Protochordate, 2, 200, 201, 221, 222, 245.
Protocnidaria, 14.
Protocoel (*see also* Hydrocoel), 19, 22, 150, 151, 152, 153, 155, 156, 164, 223, 227, 243, 247.
— reduction, 22, 226–7, 239, 244, 247.
Protocoelomate, 239.
Protoglossus, 157, 159.
Protoglossus koehleri, 158, 159.
Proto-hemichordate, 201.
Protomere (*see also* Protosome), 12.
Protonephridium, 15, 73, 226, 230, 231.
Protosome, 151, 154, 157, 222, 226–8, 236.
Protostomia, 151, 222, 224.
Protozoa (*see also* Ciliata, Flagellata), 207, 208, 210, 211, 258.
Protraction, 86, 130, 133, 162,
Pseudocoel, 1, 2, 77, 107, 114, 160, 216, 219–22, 258.
Pseudocoelomate, 83, 159–61, 219–22, 260.
Pseudometamerism, 8, 18, 19, 25, 29, 154, 252.
— gonads 5, 17.
— gut, 5, 17, 19.
— Nemertea, 5, 17, 18, 24.
— nephridia, 17, 18.
— nervous system, 18.
— Platyhelminthes, 5, 8, 17, 249.
Pseudometamerism theory, 8, 17–9, 22, 28, 165, 204.
Pteraspida, 192.
Pterobranchia, 155–7, 159, 164, 201, 210, 225, 226, 227, 228, 238, 239, 242–4, 245.
Pteropoda, 168.
Ptychodera, 157.
Puffer-fish, 172.
Pulmonata, 53, 56, 61.
Pumping, 98, 217.
Pygidium, 1, 11, 20, 145, 230.

Radial muscle, 122.
Radial symmetry, 12, 19, 241.
Radiata (see also Cnidaria, Ctenophora), 258.
Radiative evolution, 201, 214, 215, 216, 238, 259, 260.
Rapid ambulation, 126–8, 182.
Rate of locomotion, 179, 194–6, 254.
— Anthozoa, 65.
— Ectoprocta, 66.
— Gastropoda, 61, 62.
— Turbellaria, 44–5, 49.
Ray, 170.
Recapitulation, 4, 145, 200, 211.
Recapitulation theory, 27, 239–40.
Receptaculum, 160, 161.
Recovery stroke, 184.
Rectilinear crawling, 79.
Rectum, 146, 163, 237.
Reduction, of body-wall musculature, 131, 134, 137, 233–4.
— coelom, 22, 67–8, 153, 156, 157, 159, 163, 164, 229, 236, 237, 239, 244, 246, 247, 249, 252, 253, 257, 258.
— parapodia, 134, 142, 162, 248.
— segmentation, 118, 248, 249, 252, 257.
— septa, 135, 136, 139, 141–5, 151, 163, 234, 236, 248, 249, 255–7.
— tail, 170.
Reflex, arc, 196–8.
— chain, 196–8.
— contraction, 131, 139.
Regeneration, 18, 19, 50, 67, 244, 246.
Regional contraction, 79.
Regulation of internal pressure, 108.
Relaxation, 33, 83, 93, 149.
Reno-pericardial canal, 28.
Reproduction (see also Breeding season), 4, 12, 16, 214, 217–8, 221, 230, 231, 259.
— asexual, 22, 23, 24, 244, 246.
Reproductive system (see also Genital duct, Genital system, Gonad), 232.
Respiration, 99, 115, 238, 245, 246, 250.
— anal, 108.
Respiratory current, 98, 137, 217.
Respiratory pigment, 238.
Respiratory tree, 75, 101, 108.
Resting pressure, 108–14, 147.
Restoring force, 26, 31, 69, 191.
Restraint, 40, 138.
Retention of septum, 145–50.
Retraction, of lophophore, 104, 236.
— proboscis, 86, 89, 98, 157, 160.
Retractor muscle, 66, 84, 86, 87, 88, 93, 114, 160.
Retractor sheath, 147, 148.

Retrograde locomotory waves, 54–5, 56, 59, 94, 125.
Reversed locomotion, 46, 51–2, 60–1, 97, 130, 174, 191.
Reversible change of shape, 31–5, 68, 131, 233.
Reynold's number, 166.
Rhabditis marina, 79.
Rhabdocoela, 22, 44, 208, 213, 215.
— locomotion, 43, 45, 213.
Rhabdomolgus ruber, 98.
Rhabdopleura, 155, 156.
Rhinobatis, 170.
Rhynchocoel, 85, 115, 216.
Rhynchodemus, 38, 39.
Rhynchodemus bilineatus, 40.
Rhythmical activity of nerve, 198.
Rhythmical muscle contraction, 62.
Rib, 190.
Rigid skeleton, 31, 114, 117, 187, 247, 260.
Rigidity of septum, 121–2.
Rotation, 60.
Rotifera, 219, 220–1.
Roughness of body, 182, 183, 185.
Rudd, 194, 195.
Rudder, 170.
Rugae, 95.

Sabella, 72, 130, 131, 139.
Sabellidae, 142, 201.
Sabussowia, 46.
Saccoglossus, 157.
Saccoglossus cambrensis, 157.
Saccoglossus horsti, 158.
Sagartia luciae, 65.
Salmo, 171, 190.
Salmon, 195.
Salvelinus namaycush, 193.
Sarcomere, 194.
Sarcoplasm, 194.
Scapulo-coracoid, 173.
Scaridae, 170.
Scarus, 173.
Scavenging, 215.
Schizocoel theory, 3, 15–6.
Schizocoely, 6, 7, 216, 222.
Sclerites, 26, 142.
Scolex, 1.
Scomberomorus sierra, 193.
Scombridae, 189, 191, 193.
Scorpaena, 171.
Screw, 180.
Sculling, 166, 167, 179.
Scyphistoma, 22.
Scyphozoa, 22.

SUBJECT INDEX

Sea-anemone (*see* Anthozoa).
Searching movement, 53, 196.
Sea-snake, 167.
Secondary body cavity, 1, 2, 7, 26, 28, 71, 83, 93, 216, 219, 228.
Secondary germ cells, 7.
Secondary mesoderm, 6, 8, 16.
Secondary segments, 11, 20, 21, 22.
Secretion, 38, 43, 48, 63, 65–6, 96.
Sedentary habit, 93, 114, 117, 124, 139, 153, 161, 217, 218, 225, 226, 227, 228, 242, 246, 248, 257, 259.
Segmental nerve, 174, 198.
Segments, autonomous, 120, 124.
— hydrostatic isolation, 124, 139, 146, 147, 149, 151, 154, 155, 157, 163, 164, 248, 256, 259.
— primary and secondary, 11, 20, 21, 22.
— proliferation, 11, 20.
Segmentation, 1, 4, 6, 8, 20, 22, 24, 140, 230, 259.
— and swimming, 25.
— Annelida, 2, 22, 23, 25, 26, 27, 118–42, 202, 203–4, 231, 232–5, 249.
— Cestoda, 1, 2.
— Chordata, 2, 10, 25, 26, 27–8, 187–91, 198–202.
— coelomates, 23, 27, 118, 159.
— musculature, 8, 17, 22, 25, 29, 123, 131–3, 162, 187, 202, 203, 229, 230, 247, 248, 249, 254, 259.
— Mollusca, 11, 22, 251–2.
— oligomerous animals, 19, 222.
— reduction, 118, 248, 249, 252, 257.
Selachian, 170.
Selective feeding, 213.
Sensory input, 197.
Septum (*see also* Myocomma), 28, 118, 157, 159, 161, 187, 228, 259.
— Annelida, 203, 229–30, 249.
— Cnidaria, 10, 21.
— functions, 134–9, 144, 145–50, 256.
— incomplete, 121, 122, 134–9, 141, 147.
— insertion, 137–8, 139, 140, 149, 199, 203, 229.
— musculature, 121–3, 134–9, 140–1.
— Oligochaeta, 118, 120–4, 140, 144–5, 162, 256.
— Phoronidea, 150.
— Polychaeta, 133, 134–9, 162, 226, 232, 234–5, 248, 255–6.
— primitive, 140–1, 145.
— reduction, 135, 136, 139, 141–5, 151, 163, 234, 236, 248, 249, 255–7.
— rigidity, 121–2, 162.
— specialization, 122, 145–50.
Serous pocket, 4.

Serpulidae, 139, 142.
Sessile habit, 8, 23, 65, 151, 155, 200, 225, 226, 235–6, 238, 239, 240, 244, 245, 247, 260.
Seta, 143.
Sexual maturation, 4, 218, 259–60.
Shape, change of (*see* Change of shape).
— cross-sectional, 36–9, 77, 94, 115, 116, 131, 140, 167, 168, 179, 215.
— recovery, 31–3.
— stabilizer, 26.
Shark, 167, 169, 170, 171, 193.
Shearing force, 182.
Sheath, notochord, 203.
Shell, 21, 22, 62, 250, 251, 252.
Shell gland, 10.
Shoal, 196.
Siboglinum, 151, 152, 153.
Siboglinum ekmani, 151.
Side-winding, 79.
Sigalion, 137, 138.
Sinusoidal movement (*see also* Undulatory swimming), 52, 62, 78, 80, 125, 179, 202, 248.
Siphon, 107.
Siphonoglyph, 75.
Siphonophora, 76.
Sipunculoidea, 6, 20, 85–6, 102, 110, 112, 113, 114, 217.
Sipunculus, 86, 114, 167.
Sipunculus nudus, 85, 110, 112.
Size, of animals, 14, 42, 43, 179, 213, 259.
— body cavity, 71.
— relation to locomotion, 42–3, 69, 166, 174, 212, 214, 215, 220, 243.
Skate, 166, 170.
Skeletal muscle, 114, 133, 142, 145, 163, 244, 247, 248, 253, 254, 257.
Skeleton, axial (*see also* Vertebral column), 199.
— brachial, 236.
— branchial, 201.
— endophragmal, 142, 163, 253.
— rigid, 31, 114, 117, 187, 247, 260.
Skin, 189, 196, 197, 198.
Skin friction, 182.
Skipjack, 193.
Slipping, 62, 116, 119.
Smooth muscle, 145.
Snail (*see also* Helix), 55.
Snake, 79, 167, 179, 180, 181.
Soft-bodied animal, 31, 258.
Solid body, (*see also* Parenchyma), 69, 207, 211, 214, 216, 218, 258.
Somatocoel, 240, 242, 244.
Sorting mechanism, (*see also* Food collection), 213.

Spawning, 25, 167, 218.
Specialization, 228, 229.
Speed (*see* Rate of Locomotion).
Spermatozoa, 166, 231.
Sphaeroides, 172.
Sphincter muscle, 71–3, 84, 104, 107, 121, 141, 257.
Spicule, 236.
Spider, 254.
Spinal cord, 197.
Spinal column, 191.
Spinal nerve, 196.
Spine, 80, 143.
Spionidae, 141.
Spiral fibre, 35–6, 52, 76–8.
Spiralia, 10, 19.
Spiral muscle, 161, 237.
Spirobrachia, 151, 152.
Spirobrachiidae, 154.
Spontaneous activity, 197.
Squalus, 188.
Squirrel-fish, 171.
Stability, 170.
Stabilizer, 168.
Stagnant water, 98, 196.
Stalk, Brachiopoda, 237.
— Pterobranchia, 155, 156.
Standing waves, 48, 63.
Stationary point (*see also Point d'appui*), 54, 55, 62, 119, 120, 125, 143.
Steering, 170, 192.
Stenoglossa, 61.
Stenostomum, 23.
Stenothecoidea, 10, 21, 22.
Stepping movement, 49, 56, 57, 99.
Sterilization, 4.
Sternal sclerite, 142.
Sternaspis, 73, 88.
Stichopus, 96, 100.
Stichostemma, 51.
Stigmatogaster subterranea, 142.
Stilt bristle, 80–2.
Stimulation, 46, 110, 111, 114, 151, 197.
Stolon, 155.
Stoltella, 102, 104.
Stomochord, 201.
Stomodeum, 233.
Stone canal, 244.
Stream-lining, 166, 180, 193.
Strength of tissues, 108.
Streptoneura, 59.
Stress, 67, 140.
Striated muscle, 145, 163.
Stride, 47, 88, 254.
Strobila, 22, 23.
Strobilation, 22–4.

Stylarioides, 147.
Substratum (*see also* Burrow, Burrowing), 89, 101, 117, 129, 153, 227, 230, 233, 237, 259.
— adhesion to (*see* Adhesion to substratum).
— agitation, 91.
— attachment to, 53, 225, 239, 241.
— burrowing in, 86, 88, 91, 93, 114, 115, 119, 142, 150, 162, 215, 228, 230, 234, 248, 256–7, 258.
— contact with, 62, 94.
— locomotion on, 53, 94–8, 100, 115, 119, 124–9, 141, 143, 157, 159, 167, 205, 213–4, 215, 228, 235, 249.
— penetration of, 86, 90, 91, 92–3, 114, 115, 129.
— physical properties, 91, 92, 93, 232.
— soft, 53, 88, 91, 214, 256.
Sucker, 34, 46, 47, 63, 67, 99, 205, 249.
Suction, 96.
Sun-fish, 172.
Superficial lateral muscle, 194–6.
Supernumerary bone, 189.
Supporting structure, 26, 201.
Supra-oesophageal ganglion, 198.
Surface area, variation, 38–41.
Surface film, locomotion on, 43, 63.
Surgeon-fish, 172.
Suspensory muscle, 135, 136, 137, 139, 141, 163, 248.
Sustained burrowing, 124, 217.
Sustained swimming, 194, 195, 201.
Swimming (*see also* Undulatory movements), 69, 165–204, 213, 214, 215, 235, 239.
— Gastropoda, 168–9.
— Hirudinea, 25, 167, 179, 180, 249.
— Nematoda, 78–80, 93, 116.
— Polychaeta, 25, 129, 182–6, 248.
— relation to metamerism, 5, 17, 25–6, 27, 28, 165, 174, 187, 198–204.
— sustained, 194, 195, 201.
— Turbellaria, 42, 43, 69, 168.
Syllidae, 135, 140.
Symmetry, axis of, 8, 12, 22.
— bilateral, 12, 19, 206, 208, 209, 212, 239, 241.
— pentaradiate, 239.
— radial, 12, 19, 241.
— tetraradiate, 209.
Synaptula, 98, 100.
Synchronization of spawning, 218.
Syncytial protistan theory, 207–9, 212.
Syncytium, 67, 207, 208.
Synergic contraction, 86, 88.
Synodus, 171.

Tabanus, 143.
Tadpole, 166, 167.
Tadpole larva, 25, 26, 200, 201, 202, 221, 245, 247.
Taenioglossa, 61.
Tail, 166, 167, 168, 169, 170, 174, 179, 180, 192–3, 201, 202, 238.
Tangential component, 174.
Tectarius misricatus, 56.
Tectibranchiata, 61, 168.
Teleostei, (*see also* Balistidae, Carangidae, Hippocampidae, Labridae, Scaridae, Scombridae), 160, 170–4, 180, 189–98.
Teloblast, 4, 6, 19.
Tench, 194.
Tendon, 189, 191.
Tension, 114, 120, 190.
Tentacles, 157, 225, 229, 246, 259.
— Coelenterata, 8, 9, 10.
— Echinodermata, 85, 91, 98, 101, 240, 241, 242, 243.
— Mollusca, 250.
— Phoronidea, 150, 151, 226.
— Pogonophora, 151–4, 227.
— Polychaeta, 150.
— Pterobranchia, 155, 156, 243.
Tentacular sheath, 102, 104.
Tergal sclerite, 142.
Terminal membrane, 106, 107.
Terrestrial animals, 38, 44, 48, 49, 231.
Terrestrial habitat, 38, 44, 257.
Test, 245.
Tethys dactylomela, 56.
Tetrabranchiata, 252.
Tetracorallia, 21, 22.
Tetraradiate symmetry, 209.
Tetrataxic locomotory waves, 56.
Theca, 10.
Thickness of body, 43.
Thione briareus, 85, 99.
Thixotropy, 91, 92, 93.
Thoracic segment, 134, 146, 149.
Thoracophelia, 149, 150.
Thrust, 62, 83, 91, 94, 119–20, 124, 125, 143, 144, 162, 165, 166, 174, 177, 180, 182, 186, 205, 232, 256.
Thyone, 111.
Tide, 141.
Tonus, 33, 108, 113, 123, 145.
Tornaria larva, 239.
Torsion, mechanical, 189, 191, 203, 234.
— morphological, 240, 241, 243, 249.
Traction, 99.
Tractive force, 87, 89, 135.
Transection of nerve cord, 196–8.
Transmission, locomotory wave, 123, 191.
— pressure change, 116, 122–3, 144, 148, 162.
Transport system (*see also* Blood vascular system, Water vascular system), 218, 237.
Transverse folding, 41.
Transverse muscle, 64, 137, 138.
Transverse pressure, 41.
Tricladida, 5, 8, 18, 22, 215, 217.
— locomotion, 43–5, 48–9, 53, 67, 69, 70, 205, 214.
— reproduction, 22, 23.
— terrestrial, 38, 39, 44, 45, 48, 49, 51.
Trigger-fish, 171.
Trilobite, 10.
Triploblastic Metazoa, 12, 31, 24, 35, 208, 212.
Tritonidea tincta, 56.
Trivium, 96, 100.
Trochophore larva, 220, 221, 222, 224, 253.
Trochus, 56.
Tropidonotus, 181.
Trout, 193, 194, 195.
Trunk, 150, 151, 157, 159, 160, 226.
Trunk-fish, 172.
Tube, movement in, 130, 150, 154, 155, 227, 236.
Tube-foot (*see also* Podium), 96, 99, 153.
Tubicolous habit, 98, 130, 131, 139, 150–7, 217, 225, 227, 235, 246, 247, 259.
Tubifex, 121, 122.
Tubificidae, 122.
Tuna, 189, 192, 193, 196.
Tunicata (*see* Ascidiacea).
Turbellaria (*see also* Acoela, Polycladida, Tricladida, Rhynchocoela), 16, 20, 22, 42–53, 64, 65, 207, 209, 211, 250, 252, 258.
— asexual reproduction, 22, 23, 24.
— change of shape, 35, 38–42, 44, 71, 72.
— gonads, 4, 12, 208.
— locomotion, 42–53, 69, 83, 168, 213.
— musculature, 26, 52–3, 66–8, 187, 208, 258.
— nervous system, 18, 208.
— pseudometameric, 17, 18, 24.
— regeneration, 18, 50.
Turbellarian theory, 8, 14.
Turbo, 56.
Turbulence, 178, 179, 182.
Turgor, 53, 69, 92, 108, 114, 133, 135, 142, 143, 154, 156, 162, 163, 218, 227, 236, 242, 248, 255, 259.
Turgor muscles, 144–5.
Tylosaurus, 171.

Uncoordinated locomotory movements, 51, 62, 130.
Undifferentiated cells, 67.
Undulatory movements (*see also* Swimming), 52, 69, 78, 116, 125–9, 164–86.

SUBJECT INDEX

— co-ordination, 197.
— initiation, 196–8.
— production, 186.
Unsegmented coelomate, 130, 131, 142, 161, 229, 230, 245, 248, 249.
— burrowing, 85, 91, 114, 117, 217.
— change of shape, 83, 124, 162, 226, 259.
— locomotion, 99, 118, 124, 142, 161, 227.
— proboscis, 85.
— relationships, 4, 5, 6, 8, 11, 17, 20, 23, 26, 210, 223, 232, 252.
Urechis, 89, 91, 93, 97.
Urechis caupo, 90.
Urine, 71.
Urochordata, 223, 225.

Vacuolation, 201.
Valve, dorsal, 147, 148, 149.
Vascular system (*see* Blood vascular system, Water vascular system).
Velocity of locomotory wave, 53, 177, 178, 179, 180, 182.
Ventral fin, 173.
Ventral foramen, 121.
Ventral longitudinal muscle, 48, 49, 69, 96, 97, 133, 134.
Ventral nervous system, 8, 9, 130, 131, 138.
Ventral rib, 190.
Ventral valve, 149.
Vertebral column, 187, 188, 189, 190, 191–3.
Vertebral number, 193.
Vertebrate, 187, 196, 247.
Vesicle, 64, 244.
Vestibulum, 103, 104, 106, 107.
Viscera, 97, 98, 196.
Visceral mass, 54, 60, 62, 249, 250, 251, 252.
Viscosity, 33, 78, 166.
Viscous-elastic property, 73.

Volume, of animal, 36–8, 71–5, 77, 107–8, 250.
— change, 73–4, 75, 99, 107, 161, 245.
— conservation, 31, 37, 53, 71–5.
— local change, 31.
Volvox, 209.

Wandering cells, 67.
Water, 166, 167.
— conservation, 38, 231.
— content, 91.
— flow, 177, 178, 238.
— loss, 38, 44.
— pumping (*see also* Peristalsis, Respiration), 98, 217.
Water vascular system, 7, 240, 244.
Wave action, 93.
Wave-length, 175, 179, 180.
Weight, 245.
Whale, 167.
Whiting, 180.
Work, 42, 44, 71.
Working pressure, 72, 109–14.
Working tension, 114.
Worm, hypothetical, 31, 35–7, 118.
Wounding, 73.

X-ray, 7, 121.

Yaw, 168.
Yolk, 231.

Zenkevitchiana, 153.
Zeus, 171.
Zooecium, 102, 104.
Zooid, 22, 155, 156.
Zygapophysis, 192, 193.